AF581762

Eco-Friendly Wood Composites: Design, Characterization and Applications

Eco-Friendly Wood Composites: Design, Characterization and Applications

Editors

Pavlo Bekhta
Petar Antov
Yonghui Zhou
Viktor Savov

MDPI • Basel • Beijing • Wuhan • Barcelona • Belgrade • Manchester • Tokyo • Cluj • Tianjin

Editors
Pavlo Bekhta
Department of Wood-Based Composites, Cellulose and Paper
Ukrainian National Forestry University
Lviv
Ukraine

Petar Antov
Department of Mechanical Wood Technology
University of Forestry
Sofia
Bulgaria

Yonghui Zhou
Department of Civil and Environmental Engineering
Brunel University
London
United Kingdom

Viktor Savov
Department of Mechanical Wood Technology
University of Forestry
Sofia
Bulgaria

Editorial Office
MDPI
St. Alban-Anlage 66
4052 Basel, Switzerland

This is a reprint of articles from the Special Issue published online in the open access journal *Polymers* (ISSN 2073-4360) (available at: www.mdpi.com/journal/polymers/special_issues/Eco_friendly_Wood_Composites_Design_Characterization_Applications).

For citation purposes, cite each article independently as indicated on the article page online and as indicated below:

LastName, A.A.; LastName, B.B.; LastName, C.C. Article Title. *Journal Name* **Year**, *Volume Number*, Page Range.

ISBN 978-3-0365-7187-4 (Hbk)
ISBN 978-3-0365-7186-7 (PDF)

Contents

Editorial

Eco-Friendly Wood Composites: Design, Characterization and Applications

Viktor Savov [1,*], Petar Antov [1,*], Yonghui Zhou [2] and Pavlo Bekhta [3]

1 Faculty of Forest Industry, University of Forestry, 1797 Sofia, Bulgaria
2 Department of Civil and Environmental Engineering, Brunel University London, Uxbridge UB8 3PH, UK
3 Department of Wood-Based Composites, Cellulose and Paper, Ukrainian National Forestry University, 79057 Lviv, Ukraine
* Correspondence: victor_savov@ltu.bg (V.S.); p.antov@ltu.bg (P.A.)

The ongoing transition from a linear to a circular, low-carbon bioeconomy is crucial for reducing the consumption of global natural resources, minimizing waste generation, reducing carbon emissions, and creating more sustainable growth and jobs, the prerequisites necessary to achieve climate neutrality targets and stop biodiversity loss. In recent years, the wood-based panel industry has faced significantly increasing demands for its various products due to the rising worldwide population, shifts in land use, and growing economies. Using wood more efficiently, the optimization of natural raw material use, and sustainably converting waste into value-added products to meet projected demands for the development of wood-based composites all represent key circular economy principles requiring the reuse, recycling, or upcycling of materials.

Conventional wood composites are produced with synthetic, formaldehyde-based adhesives, commonly created from fossil-derived constituents, such as urea, phenol, melamine, etc. [1–10]. Along with their undisputable advantages, these adhesives are characterized by specific problems connected to the emission of hazardous volatile organic compounds (VOCs), including free formaldehyde emissions from finished wood-based composites, which have been linked to several major environmental issues, as well as negative effects on human health, including irritation to skin and eyes, respiratory problems, and cancer [11–17]. The shift towards a circular, low-carbon bioeconomy, growing environmental concerns, and stringent legislation related to the emission of harmful VOCs have resulted in novel requirements related to the development of sustainable and environmentally friendly wood-based composites [18–25]. In this respect, novel requirements concerning free formaldehyde emissions from wood composites have posed novel challenges for both researchers and industrial practices related to the development of sustainable and ecofriendly wood composites, the optimization of available lignocellulosic raw materials, and the use of alternative natural and renewable feedstocks [26–38]. The harmful formaldehyde released from wood composites can be reduced by adding formaldehyde scavengers to conventional adhesive systems, through the surface treatment of finished wood composites, or by using novel biobased wood adhesives as environmentally friendly alternatives to traditional thermosetting resins [39–43]. Another alternative to formaldehyde-based adhesives is the manufacture of binderless wood-based panels, since wood represents a natural polymeric material abundant in lignocellulosic constituents, such as cellulose, hemicelluloses, and lignin [44–53].

This Special Issue presents a collection of 10 high-quality original research and review papers providing examples of the most recent advances and technological developments in the fabrication, design, characteristics, and applications of ecofriendly wood and wood-based composites.

In their paper, Taghiyari et al. investigated the effects of nanosilver and heat treatment on the pull-off strength of sealer-clear finish in solid wood species [54]. They found a

Citation: Savov, V.; Antov, P.; Zhou, Y.; Bekhta, P. Eco-Friendly Wood Composites: Design, Characterization and Applications. *Polymers* **2023**, *15*, 892. https://doi.org/10.3390/polym15040892

Received: 2 February 2023
Accepted: 8 February 2023
Published: 10 February 2023

positive correlation between the density and pull-off adhesion strength in three common solid wood species, i.e., common beech, silver fir, and black poplar. A heat treatment at 145 °C increased the pull-off adhesion strength in all three species due to the formation of novel bonds in the cell walls of the polymers. The thermal degradation of the polymers at 185 °C weakened the formation of the novel bonds' positive effect, resulting in an unchanged pull-off strength compared with the control specimens. Markedly, the authors reported that the impregnation with a silver nanosuspension decreased the pull-off strength in beech specimens. It was concluded that the density was the decisive factor in determining the pull-off strength, having a significant positive correlation.

In the study by Ghozali et al., a novel in situ modification method for thermoplastic starch preparation (TPS) based on *Arenga pinnata* palm starch was proposed [55]. It was found that TPS properties could be improved with the starch modification, adding reinforcements and blending with other polymers. In this research, to prepare the modified TPS, the starch modification was carried out through an in situ modification. The modified TPS was obtained by adding *Arenga pinnata* palm starch (APPS), glycerol, and benzoyl peroxide simultaneously in a twin-screw extruder. The morphology analysis of the TPS revealed that the starch granules were damaged and gelatinized in the extrusion process. No phase separation was observed in the TPS, showing that starch granules with and without benzoyl peroxide were uniformly dispersed in the matrix. In addition, the thermal analysis showed that the addition of benzoyl peroxide increased the thermal stability of the TPS and extended the temperature range of the thermal degradation.

Another piece of research, conducted by Iswanto et al., studied the chemical, physical, and mechanical properties of belangke bamboo (*Gigantochloa pruriens*) and its application as a reinforcing material in particleboard manufacturing [56]. The results showed that this bamboo had average lignin, holocellulose, and alpha-cellulose contents of 29.78%, 65.13%, and 41.48%, respectively, with a crystallinity of 33.54%. The physical properties of bamboo, including its specific gravity, inner and outer diameter shrinkage, and linear shrinkage, were 0.59%, 2.18%, 2.26%, and 0.18%, respectively. Meanwhile, the bamboo' s mechanical properties, including compressive strength, shear strength, and tensile strength, were 42.19 MPa, 7.63 MPa, and 163.8 MPa, respectively. Despite the inferior dimensional stability, i.e., higher water absorption and thickness swelling, compared to the uncoated particleboards, the panels reinforced with bamboo strands exhibited acceptable mechanical strength. Markedly, the addition of belangke bamboo strands as a reinforcement (surface coating) in particleboards significantly improved the mechanical properties of the panels, increasing the modulus of elasticity (MOE) and bending strength (MOR) values of the fabricated composites 16 and 3 times, respectively.

A study by Makars et al. aimed to utilize suberinic acids containing residues as adhesives for particleboard manufacturing [57]. The authors investigated the chemical and thermal properties of four different adhesives obtained in different solvents (ethanol, methanol, isopropanol, and 1-butanol), as well as their performance in bonding particleboards. Based on the results of the mechanical characteristics, ethanol was chosen as the most suitable depolymerization medium. The following optimal hot-pressing parameters for manufacturing particleboards were reported: adhesive content 20 wt%; hot-pressing temperature 248 °C; hot-pressing time of 6.55 min.

In their paper, Savov et al. investigated the effect of the adhesive system on the properties of fiberboard panels bonded with hydrolysis lignin and phenol-formaldehyde (PF) resin [58]. The study proposed an alternative technological solution for manufacturing fiberboard panels using a modified hot-pressing regime with hydrolysis lignin as the main adhesive. Markedly, the main novelty of the research was the optimized adhesive system composed of unmodified hydrolysis lignin and a reduced PF resin content. It was concluded that the proposed technology was suitable for manufacturing fiberboard panels, fulfilling the strictest EN standards. Markedly, it was shown that to produce this type of panels, the minimum total content of binders should be 10.6%, and the PF resin content should be at least 14% of the adhesive system.

In another paper, Shahavi et al. investigated the feasibility of using wood leachate (WL) as an inexpensive filler in novel biodegradable poly (lactic acid)/WL composites [59]. In this research, the antibacterial, mechanical, morphological, and thermal properties of the composites were evaluated. The scanning electron microscopy (SEM) results indicated a proper filler dispersion in the polymer matrix, and WL powder improved the hydrophobic nature in the adjusted sample's contact angle experiment. Markedly, the results showed that adding WL filler improved the mechanical properties of the fabricated biocomposites. The PLA–WL biocomposites exhibited antibacterial activity according to the inhibition zone for *Escherichia coli* bacteria. The authors concluded that the developed poly (lactic acid)–WL biocomposites could be successfully used in a variety of value-added industrial applications, such as functional biopolymer materials.

The paper by Solihat et al. investigated the physical and chemical properties of *Acacia mangium* lignin isolated from pulp mill byproduct for potential applications in wood composites [60]. This research demonstrated that the lower insoluble acid content of lignin derived from a fractionated step (69.94%) rather than a single step (77.45%) correlated to the lignin yield, total phenolic content, solubility, thermal stability, and molecular distribution. It contradicted the syringyl/guaiacyl (S/G) units' ratio, where the ethanol fractionation slightly increased the syringyl unit content, increasing the S/G ratio. Hence, the fractionation step resulted in more ruptures and pores on the lignin morphological surface than the ethanol-fractionated step. The results obtained could increase the industrial valorization of lignin in manufacturing wood-based panels with improved properties and a lower environmental footprint.

In their review paper, Sydor et al. performed a comprehensive overview of the recent developments, possibilities, and challenges surrounding the efficient utilization of mycelium-based composites (MBCs) in art, architecture, and interior design applications [61]. It was found that MBCs attracted growing attention due to their role in the development of ecodesign methods. Following the synthesis of these sources of knowledge, it was concluded that MBCs are inexpensive in production, ecological, biodegradable, and offer a high artistic value. The main drawbacks of this natural material are related to its insufficient load capacity, unfavorable water affinity, and unknown durability.

Another interesting paper by Miran Merhar investigated the application of the maximum stress, Tsai–Hill, Tsai–Wu, Puck, Hoffman, and Hashin criteria to beech (*Fagus sylvatica*) plywood manufactured from differently oriented veneer sheets [62]. Specimens were cut from the manufactured panels at various angles and loaded by bending to failure. The mechanical properties of the beech veneer were also evaluated. The samples were modelled using the finite element method with a composite modulus and considering the different failure criteria, where the failure forces were calculated and compared with the measured values. The authors reported that the calculated forces based on all failure criteria were lower than those measured experimentally. The forces determined using the maximum stress criterion showed the best agreement between the calculated and measured forces.

Last, but not least, a comprehensive review of the latest advancements in the development of fire-resistant biocomposites, including a critical analysis of the flammability of wood and natural fibers as raw materials for the production of biocomposites, was conducted by Madyaratri et al. [63]. In addition, the authors investigated and discussed the feasibility of using lignin as an environmentally friendly and inexpensive flame retardant additive in the production of high-performance biocomposites with enhanced technological and fire properties. The authors concluded that the increased utilization of renewable natural feedstocks represented a prospective and viable approach to manufacturing novel biocomposite materials with engineered properties, improved fire resistance, and a lower environmental footprint.

Author Contributions: Conceptualization, P.B., P.A., Y.Z. and V.S.; Writing and editing P.B., P.A. and V.S. All authors have read and agreed to the published version of the manuscript.

Funding: This research received no external funding.

Acknowledgments: This work was supported by the project "Properties and application of innovative biocomposite materials in furniture manufacturing", no. НИС-Б-1215/04.2022, carried out at the University of Forestry, Sofia, Bulgaria.

Conflicts of Interest: The authors declare no conflict of interest.

References

1. Pizzi, A.; Papadopoulus, A.N.; Policardi, F. Wood Composites and Their Polymer Binders. *Polymers* **2020**, *12*, 12051115. [CrossRef] [PubMed]
2. Valyova, M.; Ivanova, Y.; Koynov, D. Investigation of free formaldehyde quantity in production of plywood with modified urea-formaldehyde resin. *Int. J. Wood Des. Technol.* **2017**, *6*, 72–76.
3. Lee, T.C.; Puad, N.A.D.; Selimin, M.A.; Manap, N.; Abdullah, H.Z.; Idris, M.I. An overview on development of environmental friendly medium density fibreboard. *Mater. Today* **2020**, *29*, 52–57. [CrossRef]
4. Savov, V.; Mihajlova, J. Influence of the Content of Lignosulfonate on Physical Properties of Medium Density Fiberboards. *ProLigno* **2017**, *13*, 247–251.
5. Valchev, I.; Savov, V.; Yordanov, I. Reduction of Phenol Formaldehyde Resin Content in Dry-Processed Fibreboards by Adding Hydrolysis Lignin. In Proceedings of the 2020 Society of Wood Science and Technology International Convention, Portorož, Slovenia, 12–15 July 2020; pp. 592–602.
6. Jivkov, V.; Elenska-Valchanova, D. Mechanical Properties of Some Thin Furniture Structural Composite Materials. In Proceedings of the 30th International Conference on Wood Science and Technology, Zagreb, Croatia, 12–13 December 2019; pp. 86–94.
7. Jivkov., V.; Petrova, B. Challenges for furniture design with thin structural materials. In Proceedings of the IFC, Trabzon, Turkey, 2–5 November 2020; pp. 113–123.
8. Valchev, I.; Yordanov, Y.; Savov, V.; Antov, P. Optimization of the Hot-Pressing Regime in the Production of Eco-Friendly Fibreboards Bonded with Hydrolysis Lignin. *Period. Polytech. Chem. Eng.* **2022**, *66*, 125–134. [CrossRef]
9. Karagiannidis, E.; Markessini, C.; Athanassiadou, E. Micro-Fibrillated Cellulose in Adhesive Systems for the Production of Wood-Based Panels. *Molecules* **2020**, *25*, 4846. [CrossRef]
10. Bertaud, F.; Tapin–Lingua, S.; Pizzi, A.; Navarrete, P.; Petit–Conil, M. Development of green adhesives for fibreboard manufacturing, using tannins and lignin from pulp mill residues. *Cellul. Chem. Technol.* **2012**, *46*, 449–455.
11. Zakaria, R.; Bawon, P.; Lee, S.H.; Salim, S.; Lum, W.C.; Al-Edrus, S.S.O.; Ibrahim, Z. Properties of Particleboard from Oil Palm Biomasses Bonded with Citric Acid and Tapioca Starch. *Polymers* **2021**, *13*, 3494. [CrossRef]
12. Ghahri, S.; Pizzi, A.; Hajihassani, R. A Study of Concept to Prepare Totally Biosourced Wood Adhesives from Only Soy Protein and Tannin. *Polymers* **2022**, *14*, 1150. [CrossRef]
13. Antov, P.; Savov, V.; Mantanis, G.I.; Neykov, N. Medium-density fibreboards bonded with phenol-formaldehyde resin and calcium lignosulfonate as an eco-friendly additive. *Wood Mater. Sci. Eng.* **2021**, *16*, 42–48. [CrossRef]
14. Kristak, L.; Antov, P.; Bekhta, P.; Lubis, M.A.R.; Iswanto, A.H.; Reh, R.; Sedliacik, J.; Savov, V.; Taghiayri, H.; Papadopoulos, A.N.; et al. Recent Progress in Ultra-Low Formaldehyde Emitting Adhesive Systems and Formaldehyde Scavengers in Wood-Based Panels: A Review. *Wood Mater. Sci. Eng.* **2022**. [CrossRef]
15. Nasir, M.; Gupta, A.; Beg, M.D.H.; Chua, G.K.; Kumar, A. Physical and Mechanical Properties of Medium-Density Fibreboards Using Soy-Lignin Adhesives. *J. Trop. Forest Sci.* **2014**, *26*, 41–49.
16. Ammar, M.; Mechi, N.; Hidouri, A.; Elaloui, E. Fiberboards based on filled lignin resin and petiole fibers. *J. Indian Acad. Wood Sci.* **2018**, *15*, 120–125. [CrossRef]
17. Hoareau, W.; Oliveira, F.B.; Grelier, S.; Siegmund, B.; Frollini, E.; Castellan, A. Fiberboards Based on Sugarcane Bagasse Lignin and Fibers. *Macromol. Mater. Eng.* **2006**, *291*, 829–839. [CrossRef]
18. Popovic, M.; Momčilović, M.D.; Gavrilović-Grmuša, I. New standards and regulations on formaldehyde emission from wood-based composite panels. *Zast. Mater.* **2020**, *61*, 152–160. [CrossRef]
19. Taghiyari, H.R.; Morrell, J.J.; Husen, A. Emerging Nanomaterials for Forestry and Associated Sectors: An Overview. In *Emerging Nanomaterials*; Taghiyari, H.R., Morrell, J.J., Husen, A., Eds.; Springer: Cham, Switzerland, 2023; pp. 1–24. [CrossRef]
20. Tudor, E.M.; Barbu, M.C.; Petutschnigg, A.; Réh, R.; Krišt'ák, L. Analysis of Larch-Bark Capacity for Formaldehyde Removal in Wood Adhesives. *Int. J. Environ. Res. Public Health* **2020**, *17*, 764. [CrossRef]
21. Hemmilä, V.; Adamopoulos, S.; Karlsson, O.; Kumar, A. Development of sustainable bio-adhesives for engineered wood panels—A Review. *RSC Adv.* **2017**, *7*, 38604–38630. [CrossRef]
22. Savov, V. Nanomaterials to Improve Properties in Wood-Based Composite Panels. In *Emerging Nanomaterials*; Taghiyari, H.R., Morrell, J.J., Husen, A., Eds.; Springer: Cham, Switzerland, 2023; pp. 135–153. [CrossRef]
23. Khalaf, Y.; El Hage, P.; Mihajlova, J.D.; Bargeret, A.; Lacroix, P.; El Hage, R. Influence of agricultural fibers size on mechanical and insulating properties of innovative chitosan-based insulators. *Constr. Build. Mater.* **2021**, *278*, 123071. [CrossRef]
24. Wang, Z.; Kang, H.; Liu, H.; Zhang, C.; Wang, Z.; Li, J. Dual-Network Nanocross-linking Strategy to Improve Bulk Mechanical and Water-Resistant Adhesion Properties of Biobased Wood Adhesives. *ACS Sustain. Chem. Eng.* **2020**, *8*, 16430–16440. [CrossRef]

25. Arias, A.; González-Rodríguez, S.; Vetroni Barros, M.; Salvador, R.; de Francisco, A.C.; Piekarski, C.M.; Moreira, M.T. Recent developments in bio-based adhesives from renewable natural resources. *J. Clean. Prod.* **2021**, *314*, 127892. [CrossRef]
26. Tisserat, B.; Eller, F.J.; Mankowski, M.E. Properties of Composite Wood Panels Fabricated from Eastern Redcedar Employing Furious Bio-based Green Adhesives. *BioResources* **2019**, *14*, 6666–6685.
27. Pizzi, A. Tannins: Prospectives and Actual Industrial Applications. *Biomolecules* **2019**, *9*, 344. [CrossRef] [PubMed]
28. Dunky, M. Wood Adhesives Based on Natural Resources: A Critical Review Part III. Tannin- and Lignin-Based Adhesives. *Rev. Adhes. Adhes.* **2020**, *8*, 379–525.
29. Antov, P.; Valchev, I.; Savov, V. Experimental and Statistical Modeling of the Exploitation Properties of Eco-Friendly MDF Trough Variation of Lignosulfonate Concentration and Hot-Pressing Temperature. In Proceedings of the 2nd International Congress of Biorefinery of Lignocellulosic Materials (IWBLCM 2019), Córdoba, Spain, 4–7 June 2019; pp. 104–109, ISBN 978-84-940063-7-1.
30. Savov, V.; Valchev, I.; Antov, P. Processing Factors for Production of Eco-Friendly Medium Density Fibreboards Based on Lignosulfonate Adhesives. In Proceedings of the 2nd International Congress of Biorefinery of Lignocellulosic Materials (IWBLCM 2019), Córdoba, Spain, 4–7 June 2019; pp. 165–169, ISBN 978-84-940063-7-1.
31. Savov, V.; Antov, P. Engineering the Properties of Eco-Friendly Medium Density Fibreboards Bonded with Lignosulfonate Adhesive. *Drv. Ind.* **2020**, *71*, 157–162. [CrossRef]
32. Antov, P.; Savov, V.; Krišťák, Ľ.; Réh, R.; Mantanis, G.I. Eco-Friendly, High-Density Fiberboards Bonded with Urea-Formaldehyde and Ammonium Lignosulfonate. *Polymers* **2021**, *13*, 220. [CrossRef]
33. Antov, P.; Krišťák, Ľ.; Réh, R.; Savov, V.; Papadopoulos, A.N. Eco-Friendly Fiberboard Panels from Recycled Fibers Bonded with Calcium Lignosulfonate. *Polymers* **2021**, *13*, 639. [CrossRef]
34. Antov, P.; Savov, V.; Trichkov, N.; Krišťák, Ľ.; Réh, R.; Papadopoulos, A.N.; Taghiyari, H.R.; Pizzi, A.; Kunecová, D.; Pachikova, M. Properties of High-Density Fiberboard Bonded with Urea–Formaldehyde Resin and Ammonium Lignosulfonate as a Bio-Based Additive. *Polymers* **2021**, *13*, 2775. [CrossRef]
35. Bekhta, P.; Noshchenko, G.; Réh, R.; Kristak, L.; Sedliačik, J.; Antov, P.; Mirski, R.; Savov, V. Properties of Eco-Friendly Particleboards Bonded with Lignosulfonate-Urea-Formaldehyde Adhesives and pMDI as a Crosslinker. *Materials* **2021**, *14*, 4875. [CrossRef]
36. Karthäuser, J.; Biziks, V.; Mai, C.; Militz, H. Lignin and Lignin-Derived Compounds for Wood Applications—A Review. *Molecules* **2021**, *26*, 2533. [CrossRef]
37. Yotov, N.; Savov, V.; Valchev, I.; Petrin, S.; Karatotev, V. Study on possibility for utilisation of technical hydrolysis lignin in composition of medium density fiberboard. *Innov. Wood. Ind. Eng. Des.* **2017**, *6*, 69–74.
38. Ferdosian, F.; Pan, Z.; Gao, G.; Zhao, B. Bio-Based Adhesives and Evaluation for Wood Composites Application. *Polymers* **2016**, *9*, 70. [CrossRef]
39. Kawalerczyk, J.; Walkiewicz, J.; Dziurka, D.; Mirski, R.; Brózdowski, J. APTES-Modified Nanocellulose as the Formaldehyde Scavenger for UF Adhesive-Bonded Particleboard and Strawboard. *Polymers* **2022**, *14*, 5037. [CrossRef]
40. Gonçalves, S.; Ferra, J.; Paiva, N.; Martins, J.; Carvalho, L.H.; Magalhães, F.D. Lignosulphonates as an Alternative to Non-Renewable Binders in Wood-Based Materials. *Polymers* **2021**, *13*, 4196. [CrossRef]
41. Lykidis, C. Formaldehyde Emissions from Wood-Based Composites: Effects of Nanomaterials. In *Emerging Nanomaterials*; Taghiyari, H.R., Morrell, J.J., Husen, A., Eds.; Springer: Cham, Switzerland, 2023; pp. 337–360. [CrossRef]
42. Antov, P.; Lee, S.; Lubis, M.A.R.; Yadav, S.M. Potential of Nanomaterials in Bio-Based Wood Adhesives: An Overview. In *Emerging Nanomaterials*; Taghiyari, H.R., Morrell, J.J., Husen, A., Eds.; Springer: Cham, Switzerland, 2023; pp. 25–63. [CrossRef]
43. Santos, J.; Pereira, J.; Escobar-Avello, D.; Ferreira, I.; Vieira, C.; Magalhães, F.D.; Martins, J.M.; Carvalho, L.H. Grape Canes (*Vitis vinifera* L.) Applications on Packaging and Particleboard Industry: New Bioadhesive Based on Grape Extracts and Citric Acid. *Polymers* **2022**, *14*, 1137. [CrossRef]
44. Mirski, R.; Banaszak, A.; Bekhta, P. Selected Properties of Formaldehyde-Free Polymer-Straw Boards Made from Different Types of Thermoplastics and Different Kinds of Straw. *Materials* **2021**, *14*, 1216. [CrossRef]
45. Kibleur, P.; Aelterman, J.; Boone, M.; Bulcke, J. Deep learning segmentation of wood fiber bundles in fiberboards. *Compos. Sci. Technol.* **2022**, *221*, 109287. [CrossRef]
46. Puspaningrum, T.; Haris, Y.H.; Sailah, I.; Yani, M.; Indrasti, N.S. Physical and mechanical properties of binderless medium density fiberboard (MDF) from coconut fiber. *IOP Conf. Ser. Earth Environ. Sci.* **2020**, *472*, 012011. [CrossRef]
47. Mancera, C.; Mansouri, N.E.; Pelach, M.A.; Francesc, F.; Salvadó, J. Feasibility of incorporating treated lignins in fiberboards made from agricultural waste. *Waste Manag.* **2012**, *32*, 1962–1967. [CrossRef]
48. Velasquez, J.A.; Ferrando, F.; Salvado, J. Effects of kraft lignin addition in the production of binderless fiberboard from steam exploded Miscanthus sinensis. *Ind. Crops Prod.* **2003**, *18*, 17–23. [CrossRef]
49. Bouajila, A.; Limare, A.; Joly, C.; Dole, P. Lignin plasticisation to improve binderless fiberboard mechanical properties. *Polym. Eng. Sci.* **2005**, *45*, 809–816. [CrossRef]
50. Okuda, N.; Hori, K.; Sato, M. Chemical changes of kenaf core binderless boards during hot pressing (I): Effects on the binderless board properties. *J. Wood Sci.* **2006**, *52*, 244–248. [CrossRef]
51. Okuda, N.; Hori, K.; Sato, M. Chemical changes of kenaf core binderless boards during hot pressing (II): Effects on the binderless board properties. *J. Wood Sci.* **2006**, *52*, 249–254. [CrossRef]

52. Suzuki, S.; Hiroyuki, S.; Park, S.-Y.; Saito, K.; Laemsak, N.; Okuma, M.; Iiyama, K. Preparation of binderless boards from steam exploded pulps of oil palm (*Elaeis guneensis* Jaxq.) fronds and structural characteristics of lignin and wall polysaccharides in steam exploded pulps to be discussed for self-bindings. *Holzforschung* **1998**, *52*, 417–426.
53. Quintana, G.; Velasquez, J.; Betancourt, S.; Ganan, P. Binderless fiberboard from steam exploded banana bunch. *Ind. Crops Prod.* **2009**, *29*, 60–66. [CrossRef]
54. Taghiyari, H.R.; Ilies, D.C.; Antov, P.; Vasile, G.; Majidinajafabadi, R.; Lee, S.H. Effects of Nanosilver and Heat Treatment on the Pull-Off Strength of Sealer-Clear Finish in Solid Wood Species. *Polymers* **2022**, *14*, 5516. [CrossRef]
55. Ghozali, M.; Meliana, Y.; Chalid, M. Novel In Situ Modification for Thermoplastic Starch Preparation based on Arenga pinnata Palm Starch. *Polymers* **2022**, *14*, 4813. [CrossRef]
56. Iswanto, A.H.; Madyaratri, E.W.; Hutabarat, N.S.; Zunaedi, E.R.; Darwis, A.; Hidayat, W.; Susilowati, A.; Adi, D.S.; Lubis, M.A.R.; Sucipto, T.; et al. Chemical, Physical, and Mechanical Properties of Belangke Bamboo (Gigantochloa pruriens) and Its Application as a Reinforcing Material in Particleboard Manufacturing. *Polymers* **2022**, *14*, 3111. [CrossRef]
57. Makars, R.; Rizikovs, J.; Godina, D.; Paze, A.; Merijs-Meri, R. Utilization of Suberinic Acids Containing Residue as an Adhesive for Particle Boards. *Polymers* **2022**, *14*, 2304. [CrossRef]
58. Savov, V.; Valchev, I.; Antov, P.; Yordanov, I.; Popski, Z. Effect of the Adhesive System on the Properties of Fiberboard Panels Bonded with Hydrolysis Lignin and Phenol-Formaldehyde Resin. *Polymers* **2022**, *14*, 1768. [CrossRef]
59. Shahavi, M.H.; Selakjani, P.P.; Abatari, M.N.; Antov, P.; Savov, V. Novel Biodegradable Poly (Lactic Acid)/Wood Leachate Composites: Investigation of Antibacterial, Mechanical, Morphological, and Thermal Properties. *Polymers* **2022**, *14*, 1227. [CrossRef]
60. Solihat, N.N.; Santoso, E.B.; Karimah, A.; Madyaratri, E.W.; Sari, F.P.; Falah, F.; Iswanto, A.H.; Ismayati, M.; Lubis, M.A.R.; Fatriasari, W.; et al. Physical and Chemical Properties of Acacia mangium Lignin Isolated from Pulp Mill Byproduct for Potential Application in Wood Composites. *Polymers* **2022**, *14*, 491. [CrossRef]
61. Sydor, M.; Bonenberg, A.; Doczekalska, B.; Cofta, G. Mycelium-Based Composites in Art, Architecture, and Interior Design: A Review. *Polymers* **2022**, *14*, 145. [CrossRef]
62. Merhar, M. Application of Failure Criteria on Plywood under Bending. *Polymers* **2021**, *13*, 4449. [CrossRef]
63. Madyaratri, E.W.; Ridho, M.R.; Aristri, M.A.; Lubis, M.A.R.; Iswanto, A.H.; Nawawi, D.S.; Antov, P.; Kristak, L.; Majlingová, A.; Fatriasari, W. Recent Advances in the Development of Fire-Resistant Biocomposites—A Review. *Polymers* **2022**, *14*, 362. [CrossRef]

 polymers

Article

Effects of Nanosilver and Heat Treatment on the Pull-Off Strength of Sealer-Clear Finish in Solid Wood Species

Hamid R. Taghiyari [1,*], Dorina Camelia Ilies [2], Petar Antov [3,*], Grama Vasile [2], Reza Majidinajafabadi [4] and Seng Hua Lee [5,6]

1 Wood Science and Technology Department, Faculty of Materials Engineering & New Technologies, Shahid Rajaee Teacher Training University, Tehran 16788-15811, Iran
2 Department of Geography, Tourism and Territorial Planning, Faculty of Geography, Tourism and Sport, University of Oradea, 410087 Oradea, Romania
3 Department of Mechanical Wood Technology, Faculty of Forest Industry, University of Forestry, 1797 Sofia, Bulgaria
4 Conservation Department, Moghadam Museum, The University of Tehran, Tehran 1137616687, Iran
5 Department of Wood Industry, Faculty of Applied Sciences, Universiti Teknologi MARA (UiTM) Cawangan Pahang, Bandar Tun Razak 26400, Malaysia
6 Laboratory of Biopolymer and Derivatives, Institute of Tropical Forestry and Forest Product, Universiti Putra Malaysia (UPM), Serdang 43400, Malaysia
* Correspondence: htaghiyari@sru.ac.ir (H.R.T.); p.antov@ltu.bg (P.A.)

Abstract: Pull-off strength is an important property of solid wood, influencing the quality of paints and finishes in the modern furniture industry, as well as in historical furniture and for preservation and restoration of heritage objects. The thermal modification and heat treatment of solid wood have been the most used commercial wood modification techniques over the past decades globally. The effects of heat treatment at two mild temperatures (145 and 185 °C) on the pull-off strength of three common solid wood species, i.e., common beech (*Fagus sylvatica* L.), black poplar (*Populus nigra* L.), and silver fir (*Abies alba* Mill.), were studied in the present research work. The specimens were coated with an unpigmented sealer–clear finish based on an organic solvent. The results demonstrated a positive correlation between the density and pull-off strength in the solid wood species. Heat treatment at 145 °C resulted in an increase in the pull-off strength in all three species, due to the formation of new bonds in the cell-wall polymers. Thermal degradation of the polymers at 185 °C weakened the positive effect of the formation of new bonds, resulting in a largely unchanged pull-off strength in comparison with the control specimens. Impregnation with a silver nano-suspension decreased the pull-off strength in beech specimens. It was concluded that density is the decisive factor in determining the pull-off strength, having a significant positive correlation (R-squared value of 0.89). Heat treatment at lower temperatures is recommended, to increase pull-off strength. Higher temperatures can have a decreasing effect on pull-off strength, due to the thermal degradation of cell-wall polymers.

Keywords: coating; heat treatment; nanotechnology; nanosilver; permeability; porous structure; solid wood; thermal modification

Citation: Taghiyari, H.R.; Ilies, D.C.; Antov, P.; Vasile, G.; Majidinajafabadi, R.; Lee, S.H. Effects of Nanosilver and Heat Treatment on the Pull-Off Strength of Sealer-Clear Finish in Solid Wood Species. *Polymers* **2022**, *14*, 5516. https://doi.org/10.3390/polym14245516

Academic Editor: Gianluca Tondi

Received: 12 October 2022
Accepted: 13 December 2022
Published: 16 December 2022

1. Introduction

From a biological point of view, wood has a continuous porous structure, because cells have interconnecting systems through which liquids can transfer from one to the other [1,2] In hardwood species, the main cells are fibers and vessel elements, while in softwoods, tracheids perform both functions [1–3]. The cell system and porous structure in turn affect many physical and mechanical properties. The size and diameter of fibers, vessel elements, and pits (that connect adjacent cells) are influential in determining the way fluids pass through wood, and also the way paints and resins can penetrate the wood texture, to strengthen or weaken the anchoring effect. The initial spacing between trees cultivated in a location, intercropping with a variety of indigenous plants, drying procedures to decrease moisture contents in wood [4], growing season, extractive content, and the moisture content and hygroscopicity of wood [5,6] are among the myriad factors that affect the size and dimension of cells in wood species, eventually resulting in a wide variance between solid wood species, and even within the same wood species grown in different localities. Moreover, weathering and environmental impacts on wood can also be of great importance, as they affect many wood properties, including the appearance, service life, performance of different coatings and finishes, and also their comparison [7–9]. As such, these external factors also greatly affect the preservation and conservation of heritage objects that should be maintained and preserved for future generations [10–12].

Owing to the aforementioned biological and environmental factors, uniformity in axial, radial, or tangential directions is uncommon. Therefore, engineers are constantly researching new methods and techniques for modifying various wood species, in order to improve their physical, mechanical, and aesthetic properties and provide a more homogeneous material for the wood-processing industry. Thermal modification and heat treatment are by far the most commercially advanced methods out of all the wood modification processes that have been studied. Thermal modification of wood has long been recognized as a potentially useful and environmentally-friendly wood protection method, to improve its dimensional stability, increase its bio-deterioration resistance, and enhance its resistance to UV irradiation [13–20].

Although this has negative effects on the strength properties of wood, there are various techniques for mitigating these effects [21–23]. For instance, thermal modification under different media (such as steam, water, and oil) has been investigated [22,24], and impregnation of specimens with a sodium borate solution (as an alkali-buttering medium) was also tested, to reduce the severity of the negative impact of thermal modification on the strength of wood specimens [24,25]. Temperatures lower than 130 °Conly result in slight changes in material properties, while temperatures higher than 260 °C result in unacceptable degradation of the substrate [13]. Degradation of hemicelluloses (as the main polymer in the wood cell wall) has a positive correlation with the increase in both the temperature of heat treatment and the duration of heat treatment. Degradation of hemicelluloses eventually reduces the swelling in wood indifferent directions [25,26]. However, structural modifications and chemical changes of lignin have also been suggested to be involved in this process [27,28]. In another theory about the reduction in hygroscopicity of wood, it was proposed that it can attributed to a complementary mechanism to mass loss [5]. The authors suggested that new hydrogen bonds are formed after water molecules are forced out of the polymers of the cell-wall microstructure. Some studies also focused on the effects of heat treatment on properties other than the hygroscopicity (dimensional stability) and mechanical properties. These less studied aspects include surface properties and wood printability [29–33].

Heat treatment at extremely high temperatures has a negative impact on the treated wood's strength properties. Treatment at a low temperature, on the other hand, may not be sufficient to produce satisfactory results, due to the low thermal conductivity of wood. In this connection, increasing the thermal conductivity of a piece of wood facilitates uniformity of heating of both the inners parts and the surface layers. In fact, a low thermal conductivity delays the temperature of the inner parts from increasing as fast as in the outer layer of a piece of wood; which causes overheating and a consequent unfavorable degradation of cell-wall polymers in the outer layers. As a result, metal nanoparticles with high thermal conductivity coefficients [33,34] were used to accelerate the heat transfer to the inner parts of wood bodies and to provide heat treatment uniformity between the surface layers and the core section [26]. Taghiyari et al. [35] reported that nanosilver impregnation allowed wood to be treated at lower temperatures. As wood impregnated with nanosilver has a better thermal conductivity, heat can move more easily throughout the wood samples. As a result, higher temperatures were not required for beneficial treatment effects, thus negating the negative effects induced by high temperatures. However, both processes (heat treatment and impregnation with various nano-suspensions) change the porous structure of wood species at a microscopic level. These changes have been shown to have a significant impact on the permeability (gas and liquid) of solid woods [36], as well as the penetration of coatings and paints into the porous structure, thereby altering their adhesion strength. Although impregnation with different metal nano-particles significantly affects these two properties in solid wood, as discussed above, namely an increased thermal conductivity (affecting heat treatment results) and an altered porous structure in solid wood (changing the permeability of the impregnated specimens), it should be kept in mind that different metal nano-particles and metal compounds may also form new bonds with the main cell-wall polymers (mainly cellulose, hemicelluloses, and lignin), thus also having a significant impact on the mechanical, physical, and even chemical properties of the wood. In this regard, high adsorption energy values between Ag and Cu particles with cellulose and hemicelluloses revealed the formation of new bonds, resulting in an improvement in the physical and mechanical properties both in solid wood species and wood-based composites [26,35–37].

Heat treatment techniques are not considered new methods to improve the dimensional stability and moisture content in solid woods, therefore their effects on the different properties of wood and wood-based composites have already been elaborated, including on the surface properties, coating hardness, artificial weathering, UV resistance, and pull-off adhesion strength of a variety of paints, finishes, and coatings [38–46]. In this connection, the adhesion of water-based coatings to different heat-treated wood substrates (maple, beech, and hemlock) was reported to decrease, compared to unheated samples [40]. The decreased strength was attributed to the reduced wettability that occurs in heat-treated wood samples [47–49]. Two main chemical alterations occur simultaneously, namely degradation of hemicelluloses and an increase in the cellulose crystallinity of samples. However, plasticization of lignin can also be influential in the reduction of the wettability of wood [50]. A combination of all the above-mentioned factors results in the hydroxyl groups in wood-cell polymers being too far out of reach for the paints and coatings to from a strong bond with the wood substrate.

In terms of the improvement in the thermal conductivity of wood, the authors came across few studies on the effects of enhanced and accelerated heat transfer as a result of impregnation with a silver nano-suspension, particularly its effects on the permeability and pull-off adhesion strength of treated wood. Moreover, as the porous structure and permeability value, and the type of cells that are involved in the fluid transfer process, are quite different in softwoods and hardwoods, wood species of both kinds of wood and highly favored in the industry were chosen. In this connection, beech was chosen as a globally popular and well-known wood species. Poplar is a fast-growing species that has been cultivated in different countries, including Iran. Fir is a softwood species that is exported to many countries with low forest resources, to supply the growing need for wood materials for use in low-cost furniture. Based on these facts and their favorability in the regional market, the above-mentioned wood species were chosen. Moreover, as the longitudinal permeability values of softwoods and hardwoods are substantially different to each other, the pull-off adhesion strength values were measured, to clearly demonstrate the relation of the heat treatment and its effects on the pull-off adhesion in cross-section and permeability.

Based on the above-mentioned short literature review and the different biological and environmental factors that affect the porosity system of wood, as well as the paint pull-off adhesion strength of different wood species as popular materials in both modern and historical objects [51–54], the present study aimed to investigate and evaluate the effects of heat treatment on the pull-off adhesion strength in three nanosilver-impregnated solid wood species. In this connection, two temperatures were chosen. The first temperature for heat treatment was a popular temperature (185 °C). As the specimens were small in size, it was assumed that temperatures higher than 160 °C might not clearly demonstrate the impact of the heat-transfer being facilitated by impregnation with nanosilver. Therefore, a lower temperature (145 °C) was also added to the experiment, to investigate the probable effects of increased heat-transfer caused by nanosilver impregnation on the overall pull-off adhesion strength and permeability.

2. Materials and Methods

2.1. Specimen Preparation

Beech (*Fagus orientalis* L.) and poplar (*Populus nigra* L.) are two domestic hardwood species with great industrial popularity. The density and mechanical properties of poplar is not as high as beech wood; however, as a fast-growing species, it is cultivated in many parts of Iran, to satisfy the growing requirements for wood materials. Silver fir (*Abies alba* Mill.) is a softwood species that is grown in many countries and exported worldwide for industrial purposes. In Iran, and neighboring countries as well, this wood species (fir) is imported from northern countries to supply the raw materials for the production of inexpensive furniture. Therefore, these wood species were selected in the present research work. In order to measure the pull-off strength in each wood species, thirty tangential specimens were cut. The dimension of the specimens were250 mm × 150 mm × 15 mm. Once the specimens were cut, they were closely inspected, to reject the ones with defects such as knots, fissures, and checks. The selected specimens were conditioned for eight weeks in room conditions, at a temperature of 25 ± 2 °C and 40 ± 3% relative humidity. In order to simulate the conditions in which painted furniture is actually used, the temperature and humidity were used in accordance to those of normal room conditions in indoor spaces in Tehran. Following the conditioning, the surface was sanded with a 100grit sandpaper, windblown to remove the wood dust, and coated with an unpigmented sealer–clear resin. The resin was an organic solvent finish, produced by Pars-Eshen Co. (Tehran, Iran). The technical specifications of the sealer–clear finish are provided in Table 1. As the binder, nitrocellulose was mixed into the resin. Specimens were coated using a spray in two runs. An interval of 12 h was set between the two runs. The final thickness of the coating was measured to be 110 µm using an ultrasonic coating thickness gauge (DeFlesko PosiTector 200, New York, NY 13669-2205, USA). Five dollies (20 mm in diameter) were stuck onto the painted surface of each specimen using epoxy resin. The moisture content of wood specimens was 8 ± 0.5% at the time of the pull-off tests, in order to avoid thermo-hygromechanical behavior of wood that would have affected the mechanical properties of the specimens (ASTM D4541-02) [55].

Table 1. Specifications of the sealer–clear finish used in this research work.

Coating Parts	Solids (%)	Viscosity (25 °C) cP	Density (g/cm^3)	Appearance of the Finish in Liquid Form
Sealer	38 ± 1.5	120 ± 15	0.98	Clear
Clear finish (un-pigmented coating)	39 ± 1.5	80 ± 15	0.98	Clear

2.2. Nanosilver Impregnation

A 400 ppm aqueous dispersion of silver nanoparticles was used in the present study. It was made using an electrochemical technique for impregnating specimens [56]. The size range of the silver nanoparticles was 30–80 nm. The formation and size of the silver nano-dispersion was monitored by transmission electron microscopy, for which nanosilver samples were prepared to be checked by drop-coating onto carbon-coated copper grids. The pH of the suspension was measured as 6–7; two kinds of surfactants (anionic and cationic) were used in the suspension as stabilizer; the concentration of the surfactants was two-times that of the nano-silver particles. For impregnating the specimens, the empty cell process (Rueping method) was used in a pressure vessel under 250 kPa of pressure. The vessel was manufactured by Mehrabadi Machinery Mfg. Co. (Tehran, Iran). The pressure was set at 250 kPa for 30 min. Before and after the impregnation process, al specimens were weighed with a digital scale with 0.01 g precision, and their dimensions were measured using a digital caliper with 0.01 mm precision, to measure the density and the amount of nano-suspension absorption. After the impregnation process, all specimens were collectively conditioned for twelve weeks (temperature of 25 ± 2 °C, and relative humidity of 40 ± 3%). The nanosilver-impregnated specimens were dried to a moisture content of 9 ± 0.5% before being heat-treated.

2.3. Heat Treatment Process

Specimens for heat treatment (both at 145 °C and 185 °C) were randomly arranged in a laboratory oven. Heat-treated specimens were marked with HT, and specimens impregnated with nanosilver were coded as NS. Specimens were placed on 3 mm thick wood strips, to prevent direct contact with the metal tray in the laboratory oven and consequent overheating. The heating schedule was nearly the same as in previous studies [26]. They were first heat-treated at 145 °C for twelve hours, all in a single run. Then, all HT-145 and NS-HT-145 specimens were taken out and placed under room conditions (25 ± 2 °C, and 40 ± 3% relative humidity). HT-185 and NS-HT-185 specimens continued to be heat-treated at 185 °C for four extra hours. The coding system used in the present project is summarized in Table 2.

Table 2. Description of the coding system of different treatments performed in this research work.

Coding	Description of the Treatment
Control	Specimens with no impregnation ormodification
Control-NSI	Specimens impregnated with silver nano-suspension
HT145	Heat-treated specimens at 145 °C
NSI-HT145	Nanosilver-impregnated specimens, heat-treated at 145 °C
HT185	Heat-treated specimens at 185 °C
NSI-HT185	Nanosilver-impregnated specimens, heat-treated at 185 °C

2.4. Pull-Off Adhesion Strength Testing

The pull-off strength values in wood species are measured with a test that measured the maximum force that is required to pull a dolly with a specified diameter off from a substrate. In the present research work, pull-off strength was measured in accordance with the ASTM D 4541-02 standard specifications [55]. An automatic PosiTest® pull-off adhesion testing device (Defelsko, Ogdensburg, NY, USA) was used. It contained a self-aligning spherical dolly (adhered to the wood specimens), Type V. The effective surface of the dolly was 20 mm (Figure 1). Based on Equation (1), the maximum tensile pull-off strength (X) that any specific paint, coating, or finish can adhere to the substrate could be calculated in mega Pascal (MPa) values. The breaking points were checked, to find out if the failure occurred in the substrate or in the adhesive layer between the substrate and the dolly. In case the failure occurred in the adhesive, the result of the test was eliminated from the authentic results.

$$X = \frac{4\ F}{\pi \cdot d^2} (\text{MPa}) \tag{1}$$

where F is the maximum force at failure point (kg.m.s^{-2}) and d is the diameter of the dolly (mm) (ASTM D4541-02) [55].

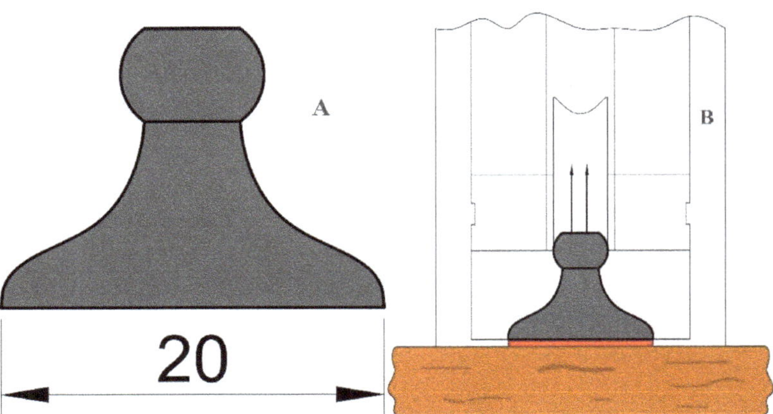

Figure 1. Schematic design of the apparatus to measure pull-off adhesion strength; (**A**) dolly with an effective diameter of 20 mm, (**B**) dolly assembled with adhesive on the substrate.

The moisture content of the specimens at the time of the pull-off adhesion tests was 9 ± 0.5%, and the temperature was 25 ± 3 °C.

2.5. Gas Permeability Measurement

The history of permeability measurement goes back to the early 18th century when Nollet (physicist, 1700–1770) sealed containers with animal bladders. It was further investigated by Richard Barrer (1910–1996), who developed the Barrer measurement technique, a non-SI unit of gas permeability [49]. This is considered one of the first scientific methods for the measurement of permeation values. Afterwards, many other methods, techniques, and devices were invented and used to measure permeability and diffusion processes in porous media, including solid wood species and wood-composite materials [57–60]. In the present study, the longitudinal gas permeability of specimens was measured using an apparatus equipped with a seven-level electronic device; the time measurement was carried out with millisecond precision [60]. Falling water was applied to measure and calculate the specific longitudinal gas permeability values. Twenty cylindrical specimens were cut in a longitudinal direction from each wood species. Specimens were 18 mm in diameter and 30 mm in length. All specimens were inspected as free from any check, knot, split, or fungal defects. The side perimeter of each specimen was covered by silicon adhesive, to allow air flow only in the longitudinal direction of the specimens. The cross-sectional ends of all specimens were trimmed using a cutter blade, to ensure that the openings of vessel elements and cells were fully open. For each specimen, the gas permeability value was separately measured. The measurement was carried out at seven different vacuum pressures and in a single run. In every run, the seven-time measurements were registered, to finally calculate the specific permeability values based on seven water columns. The water in the glass tube of the apparatus was at least 15cm above the starting point of the initial (the first step) time-measuring device (Figure 2). The necessary precautions were made to prevent any leakage from the glass tube or connections.

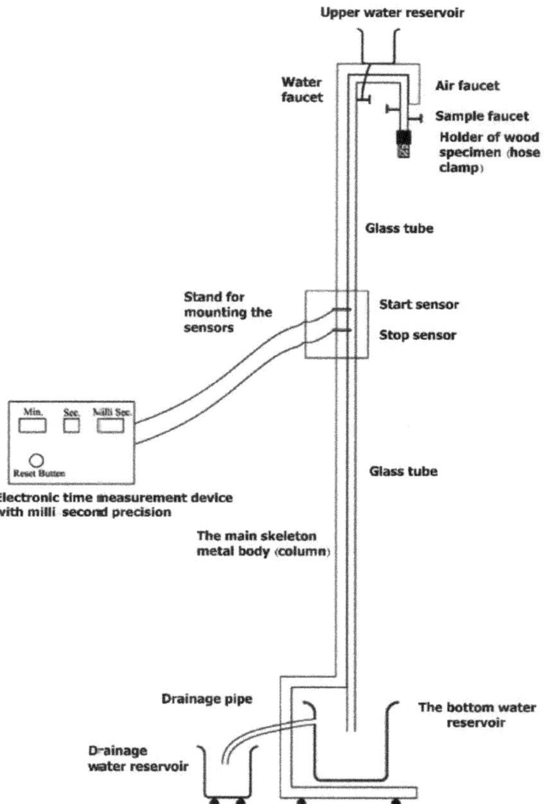

Figure 2. Schematic of the specific gas permeability measurement apparatus, equipped with a single-phase electronic time measurement device with millisecond precision (confirmed by the official certificate No. 47022; issued by The Iranian Research Organization for Scientific and Technology) (USPTO 8079249 B2) [60,61].

Each specimen was tested three times, to finally calculate the mean permeability value. Then, the mean superficial permeability coefficient was calculated using equations that were first presented by Siau [1,2]. A correction factor was used (based on the viscosity of air) to calculate the specific gas permeability [1–3].

2.6. Scanning Electron Microscopy (SEM Imaging)

Scanning electron microscope (SEM) imaging of the wood specimens was carried out using a TESCAN-VEGA II LSH apparatus at the thin-film laboratory, FE-SEM lab (Field Emission), School of Electrical & Computer Engineering (The University of Tehran, Tehran, Iran). The apparatus was produced in the Czech Republic (located in 62300 Brno). A field-emission cathode in the electron gun of a scanning electron microscope provided narrower probing beams at both low and high electron energy levels, resulting in an improved spatial resolution, and also minimized sample charging and damage.

2.7. Statistical Analysis

One-way analysis of variance (ANOVA) was conducted for statistical analysis. A significant difference at 95% level of confidence was applied, using the SAS software program (version 9.2) (2010). To distinguish the statistical difference between the different treatments, a Duncan multiple range test was carried out. Hierarchical cluster analysis was performed. This analysis clearly demonstrates potential similarities, and/or dissimilarities, between a variety of treatments [62]. The analysis in the present study was performed based on two property values simultaneously, namely unpainted and painted pull-off strength values. SPSS/20 (2010) software was utilized to perform the analyses.

3. Results

The results of density measurement indicated that beech had the highest density (0.57 g/cm^3), followed by fir (0.42 g/cm^3) and poplar (0.35 g/cm^3). Weight measurements taken immediately after the impregnation in the pressure vessel demonstrated that the NS-uptake did not correspond to the density of the three wood species. The NS-uptake was measured as 0.38, 0.27, and 0.09 (g/cm^3) for beech, poplar, and fir wood, respectively. Weight measurements of the specimens before and after the heat treatment demonstrated that the maximum and minimum mass losses were found in the HT185 beech and HT145 poplar specimens, respectively (Table 3). Beech specimens generally showed the highest mass losses in each treatment compared to their poplar and fir counterparts.

Table 3. Mass losses in the three wood species caused by heat treatments at 145 °C and 185 °C.

Heat Treatment [1]		Beech	Poplar	Fir
HT at 145 °C	HT	10.5 (±0.9) [2]	6.2 (±0.8)	7.8 (±0.9)
	NSI-HT	10.6 (±0.9)	6.3 (±0.6)	8.1 (±0.6)
HT at 185 °C	HT	11.7 (±0.7)	7.8 (±0.7)	9.2 (±0.7)
	NSI-HT	11.4 (±0.6)	8.3 (±0.5)	9.4 (±0.6)

[1] NSI = nanosilver-impregnated; HT = heat-treated at the specific temperature of either 145 °C or 185 °C. [2] Figures in parenthesis represent standard deviations.

The gas permeability measurement in the untreated wood species revealed that the highest values were for the poplar sapwood specimens (3.1 × 10^{-13} m^3.m^{-1}). The mean gas permeability in beech specimens was measured 0.74 (×10^{-13} m^3.m^{-1}), indicating that the gas permeability in poplar was four-times higher compared to that of the beech specimens. The fir wood specimens had the lowest gas permeability, which was measured at 0.003 (×10^{-13} m^3.m^{-1}).

The maximum and minimum pull-off adhesion strength values were found in the unpainted control beech specimens (9.7 MPa) and the painted nanosilver-impregnated poplar specimens heat-treated at 185 °C (2.46 MPa), respectively (Figure 3). In comparison to the other two wood species, the beech specimens had significantly higher pull-off values. It can be seen that heat treatment at 145 °C improved the pull-off strength of beech wood. However, nanosilver impregnation reduced the pull-off strength of beech wood, both painted and unpainted. The pull-off adhesion strength values of the poplar and fir specimens were generally similar, though there were fluctuations in each treatment combination. In every treatment and wood species, the values in the unpainted specimens were higher than those of the painted specimens.

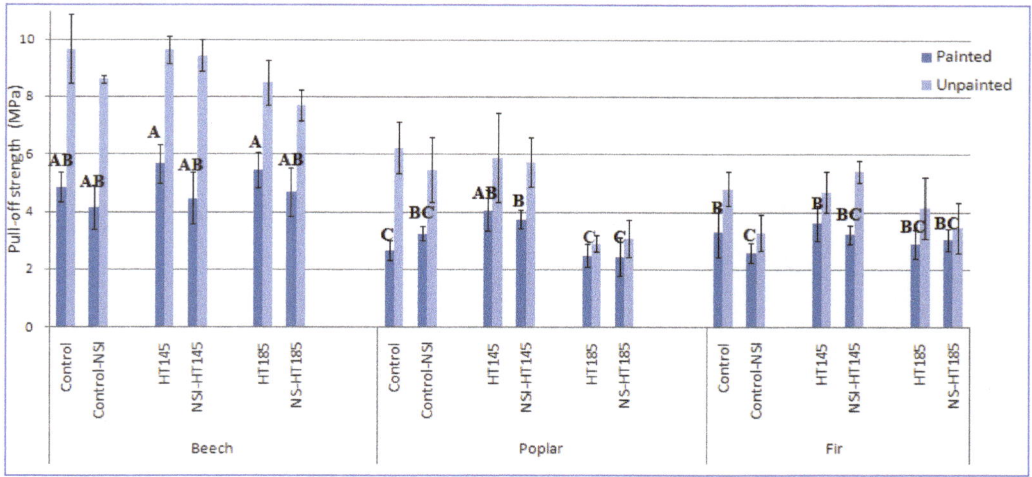

Figure 3. Pull-off adhesion strength (MPa) of the three species (beech, poplar, and fir) painted with sealer–clear finish and unpainted, letters in each column represent Duncan's multiple range test, at 95% level of confidence ($\alpha = 0.05$) based on the painted treatments (NSI = nanosilver-impregnated; HT = heat-treated at the specific temperature of either 145 or 185 °C).

4. Discussion

The poplar specimens had the highest gas permeability values, according to the measurements. The high permeability values of poplar wood were attributed to the fact that this wood species has simple perforation plates. Therefore, fluids (in this case, air) could pass through the open simple perforation plates with no physical obstacles. Moreover, the extractives in poplar sapwood were reported to be quite low, resulting in an easier fluid transfer [63]. Unlike poplar, beech wood has both simple and scalariform type perforation plates. The scalariform perforation plates are considered a physical obstacle to fluid transfer [64]. In addition to this, vessel elements in some wood species, such as beech wood, are sometimes partially or wholly blocked by different extractives, settlements, and tyloses [1–3,36]. Ultimately, fluids cannot be transferred as easily through beech wood as they can through poplar sapwood, resulting in a lower gas permeability in comparison to the permeability values of poplar. In fir wood, as a softwood species, the whole process of the transfer of fluid through the wood texture is different. Softwood species lack vessel elements that allow the transfer of liquids from leaves to the body and roots of trees, and vice versa. The pits that exist in softwoods' cell walls act as a bottleneck for the transfer of liquids, resulting in a very low permeability in softwood species [1–3]. The dimensions of pits was significantly smaller in comparison to the open poplar vessel's inner diameter (Figure 4A,B). It was reported that the permeability in hardwood species can be more than 1000-times higher than in softwood species [1,2]. Apparently, the very low NS-uptake in fir wood (0.09 g/cm^3) can be attributed to its very low permeability.

The pull-off mode of failure in all specimens showed that the failure occurred in the woody substance of the specimens. This demonstrated that the process of sticking the specimens to the dollies was correctly carried out. This also indicated that the measurement of pull-off strength values was significantly influenced by the mechanical strength of the substrate. As such, the pull-off strength values in the beech wood were all significantly higher than those of the poplar and fir wood (Figure 3). This was attributed to the higher density of the beech wood in comparison to the other two wood species. Ahigher woody mass resulted in higher mechanical properties in beech wood [26], including the pull-off strength. In fact, a greater force was needed to pull the dollies off the surface of beech specimens in comparison to the other two species (poplar and fir wood).

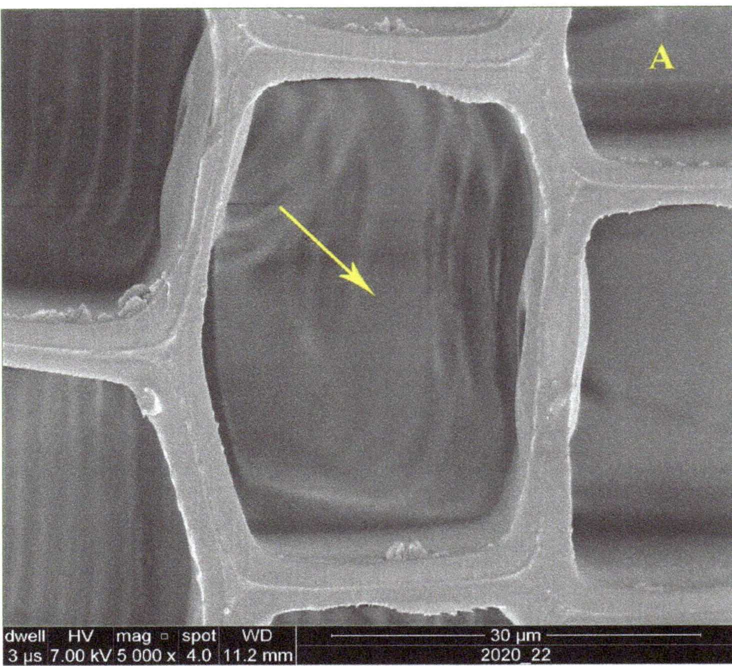

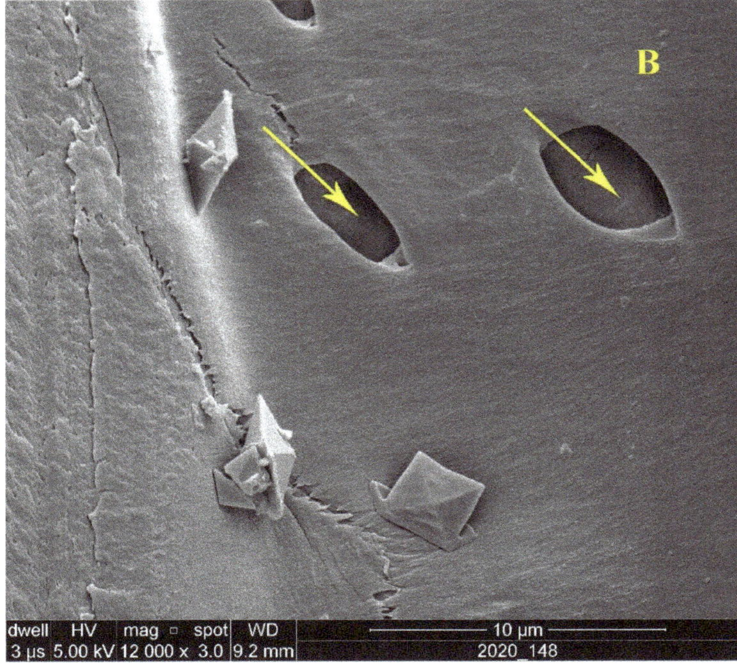

Figure 4. SEM image showing tracheid lumen in the cross-section of fir wood (**A**) and pits in the cell wall in the lateral section (**B**) (↑).

In the fir and poplar wood specimens, the pull-off strength values were rather alike. The overall similarity of these two species was attributed to their close densities. In the control (unpainted) and NS-impregnated treatments of fir and poplar, the pull-off strength values tended to be higher in the poplar specimens in comparison to the fir specimens (Figure 3). These higher values were attributed to the higher permeability in poplar, resulting in a better anchoring of the resin between the specimen and the dolly [65–69]. However, in the painted specimens, the similarity between the pull-off values in poplar and fir were more alike. The similarity in the painted treatments was attributed to the fact that the paint used in the present study was a sealer–clear finish. The sealer part blocked the openings of vessels, fibers, and tracheids, ultimately resulting in an impermeable surface to any resin and adhesive. Therefore, the pull-off strength values were fully dependent on the mechanical strength of the substrate, rather than the anchoring and penetration of the resin into the texture of the wood. This resulted in relatively similar pull-off strength values in the different treatments of the poplar and fir specimens. Thus, it is noted that the relationship between the permeability and pull-off strength in solid wood is a new topic, and therefore, further studies must be carried out to clearly evaluate possible correlations between these two properties.

Impregnation of specimens with nanosilver suspension significantly decreased the pull-off strength in all beech treatments, both painted and unpainted (Figure 3). This decrease was attributed to two main factors. The first factor was the increased permeability caused by dissolving some of the extractives, which allowed the fluid (air) to pass more easily through the vessel elements. A similar increase in permeability, as a result of impregnation with liquids and nano-suspensions, was reported previously [36,60]. The increased permeability and increased surface roughness (caused by the impregnation with the aqueous nano-suspension) allowed the adhesive (between the specimen and dolly) to penetrate deeper into the wood texture, making it unable to actively participate in the process of adhering the dolly to the specimen. The second factor that contributed to the decrease in pull-off strength in the NS-impregnated specimens was the microscopic checks that occur in the cell wall during impregnation in a pressure vessel. The formation of these micro cracks and checks was previously reported to cause a reduction in the mechanical properties of some solid wood species [36,69]. The pull-off strength in NS-impregnated specimens was eventually reduced as a result of these two factors, or better put, mechanisms that occurred concurrently. A general trend of a decrease in pull-off strength as a result of the impregnation with nanosilver was also observed in the other two wood species (poplar and fir), although some treatments showed a slight inconsistency in this regard This inconsistency from the generally decreasing trend was attributed to the great variability in strength in different specimens, which is an inherent property in solid wood species [26,68].

Heat treatment at 145 °C increased the pull-off strength in all three wood species, both in the unimpregnated and nanosilver-impregnated specimens (Figure 3). This increase was attributed to the formation of new bonds in the cell wall polymers as a result of irreversible hydrogen bonding in the course of water movement within the polymers in the cell wall [5,60]. It was reported that the formation of these new bonds resulted in an increase in the other mechanical properties, such as the bending strength and screw-withdrawal strength [5,26]. In terms of the heat treatment at 185 °C, the results showed that the pull-off strength values were not significantly different from the control values in any of the three species and treatments. Moreover, no general increasing or decreasing trend was observed. Previous studies reported a decrease in the pull-off strength as a result of heat treatment in some solid wood species [32,53], though the cited authors further added that the impact of different finishes was decisive in their final conclusions [53], as well as the thickness of the coatings [30]. Moreover, it was reported that heat treatment resulted in a decrease in coating hardness and scratch resistance in some wood species, including limba, iroka, ash, and chestnut [31]. In the present study, heat treatment at 185 °C was influenced by two simultaneous phenomena. The first phenomenon was the formation of new bonds, as discussed earlier in this section (the irreversible hydrogen bonding), having an increasing effect on the pull-off strength. However, there was a second phenomenon involved in the process, namely the condensation of lignin and thermal degradation of different cell wall polymers at 185 °C. The condensation of lignin and degradation of cell wall polymers ultimately resulted in decreased mechanical properties [13,24,68,70,71]. Microscopic imaging illustrated that the cell walls became thinner and cracked due to thermal degradation at 185 °C [13,24,60] (Figure 5). These two processes were simultaneously active in the present research study, resulting in a largely unchanged pull-off strength in the specimens heat-treated at 185 °C.

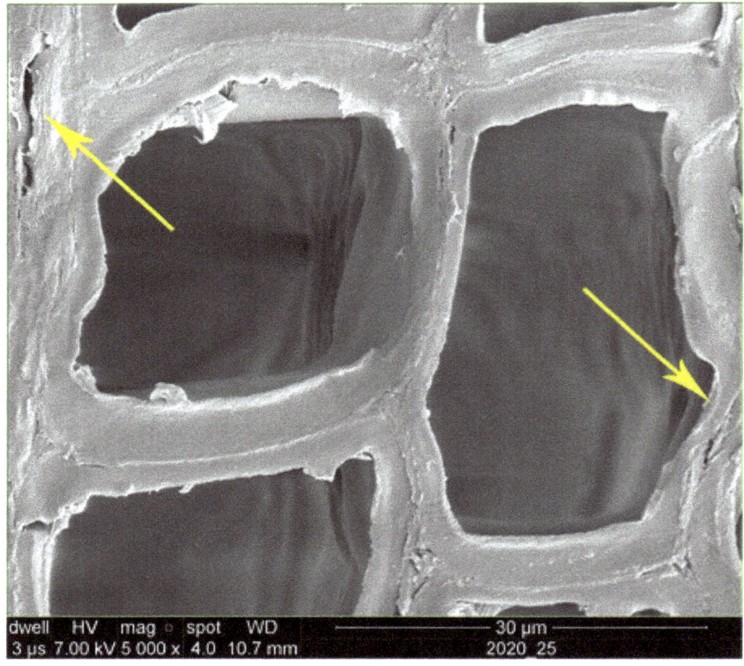

Figure 5. SEM image showing thinned walls and cracks in the cell wall of fir wood (*Abies alba*) heat-treated at 185 °C (↑).

Collective cluster analysis of the three species was carried out for the six treatments in each wood species, including 18 treatments altogether. The six treatments in each wood species comprised of the control, NS-impregnated, HT-145, NSI-HT-145, HT-185, and NSI-HT-185. The analysis was conducted based on the pull-off strength values of both painted and unpainted specimens. The results of the cluster analysis demonstrated that all beech treatments were closely clustered together, while they were remotely clustered from the other two wood species (Figure 6). This clearly demonstrated that the significantly higher density, as well as its higher mechanical properties, had a significant influence on the overall pull-off strength. Therefore, if a particular industrial application necessitates a high pull-off strength, density should be considered in the decision-making process. The similarity of cluster-grouping for the other twelve treatments of the other two wood species (poplar and fir wood) implied that density can be more influential in the eventual pull-off strength than the wood species. That is, poplar belongs to the hardwood species, while fir belongs to the softwood species, with quite different biological and chemical structures. However, not much difference was observed among the twelve treatments of these two wood species, considering their rather similar pull-off strength values. Nevertheless, it must be noted that this conclusion is applicable when the paint consists of a sealer part. This sealer part blocks the vessels, pits, and openings, making the woody substrate nearly impermeable to the adhesive that sticks the dolly to the substrate. Further studies should be carried out on a wider range of wood species and wood-finish types, to obtain a conclusive outlook on the way different paints react on substrates made of different wood species. Regression analysis of the three wood species and their treatments demonstrated high and statistically significant correlation in pull-off strength values between the painted and unpainted specimens (R-squared value of 0.89). The high and significant coefficient of determination in the three solid wood species was considered to be corroborating evidence that density was the decisive and most influential factor in the pull-off adhesion strength. Based on the cluster analysis, regression analysis, and the pull-off strength values, it can be concluded that density is the decisive factor in the ultimate pull-off strength value of wood intended for particular applications if the paint and finish consist of a sealer part. Although wood species can ultimately influence the pull-off strength results, the great effect of density puts the effect of species into context.

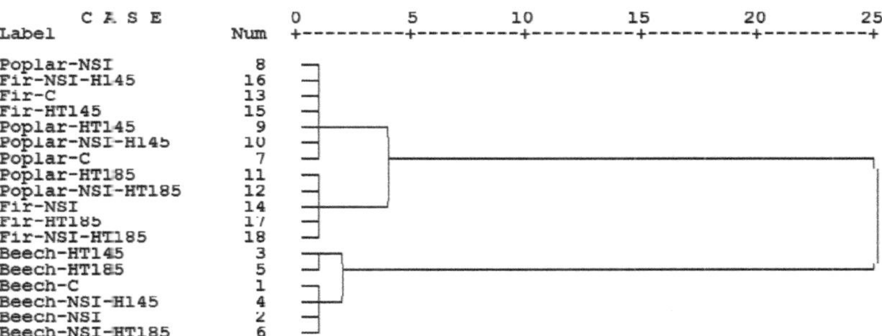

Figure 6. Cluster analysis of the three species (beech, poplar, and fir) based on the pull-off adhesion strengths of painted and unpainted specimens (C = control; HT = heat-treated at the specified temperature; NSI = nanosilver-impregnated specimens).

The present study utilized an industrial approach, and therefore an organic solvent was used, to comply with the general demands of the market. However, further studies on finishes with inorganic solvents and lower toxicity to humans should be carried out, to satisfy environmentally-friendly requirements.

5. Conclusions

The effects of nanosilver impregnation combined with heat treatment on beech, black poplar, and silver fir wood were investigated in this study. In general, heat treatment at lower temperature (145 °C) resulted in higher pull-off strength in all three wood species, most likely due to the formation of new bonds between the cell wall polymers. In comparison to untreated wood, wood treated at 185 °C showed no significant difference in pull-off strength. The degradation of cell wall polymers at high temperatures may offset the beneficial effects of new cell-wall polymer bonds. It should be noted that nanosilver impregnation did not lead to beneficial results for the pull-off strength of the treated samples. Overall, density is a decisive factor that influenced the pull-off strength of the samples, and this must therefore be taken into account if an industrial application requires a high pull-off strength. Future studies could be dedicated to the investigation of nanosilver and nanowollastonite and evaluation of their potential for improving the pull-off strength of wood used in historical and heritage objects, aiming at the better preservation and conservation of these priceless objects for future generations.

Author Contributions: Methodology, H.R.T., D.C.I., and P.A.; validation, H.R.T., G.V., and S.H.L.; investigation, H.R.T., and R.M.; writing—original draft preparation, H.R.T., D.C.I., P.A., S.H.L. and R.M.; writing—review and editing, H.R.T., D.C.I., P.A. and S.H.L.; visualization, H.R.T.; supervision, H.R.T. and P.A. All authors have read and agreed to the published version of the manuscript.

Funding: The research was partially supported by the University of Oradea, within the Grants Competition "Scientific Research of Excellence Related to Priority Areas with Capitalization through Technology Transfer: INO—TRANSFER—UO", Project No. 236/2022.

Institutional Review Board Statement: Not applicable.

Informed Consent Statement: Not applicable.

Data Availability Statement: The data presented in this study are available on request from the corresponding author.

Acknowledgments: We appreciate the kind efforts of Ali Azim-Bagirad and Mohsen Lotfalizadeh Mehrabadi for their technical electronic consultation in the design and building of the gas permeability apparatus.

Conflicts of Interest: The authors declare no conflict of interest.

References

1. Siau, J.F. *Flow in Wood;* Syracuse University Press: Syracuse, NY, USA, 1971; p. 131.
2. Siau, J.F. *Transport Processes in Wood;* Springer: Berlin/Heidelberg, Germany; GmbH & Co. KG: Berlin, Germany, 2011; p. 248.
3. Skaar, C. *Wood-Water Relations;* Springer: Berlin, Germany, 1988; p. 283.
4. Oltean, L.; Teischinger, A.; Hansmann, C. Influence of Temperature on Cracking and Mechanical Properties of Wood During Wood Drying—A Review. *BioResources* **2007**, *2*, 789–811.
5. Borrega, M.; Karenlampi, P.P. Hygroscopicity of Heat-Treated Norway Spruce (*Picea abies*) wood. *Eur. J. Wood Prod.* **2010**, *68*, 233–235. [CrossRef]
6. Hering, S.; Keunecke, D.; Niemz, P. Moisture-dependent orthotropic elasticity of beech wood. *Wood Sci. Technol.* **2012**, *46*, 927–938. [CrossRef]
7. Jirouš-Rajkovic, V.; Miklecic, J. Enhancing Weathering Resistance of Wood—A Review. *Polymers* **2021**, *13*, 1980. [CrossRef] [PubMed]
8. Van Blokland, J.; Nasir, V.; Cool, J.; Avramidis, S.; Adamopoulos, S. Machine learning-based prediction of internal checks in weathered thermally modified timber. *Constr. Build. Mater.* **2021**, *281*, 122193. [CrossRef]
9. Jones, D.; Kržišnik, D.; Hočevar, M.; Zagar, A.; Humar, M.; Popescu, C.-M.; Popescu, M.-C.; Brischke, C.; Nunes, L.; Curling, S.F.; et al. Evaluation of the Effect of a Combined Chemical and Thermal Modification of Wood though the Use of Bicine and Tricine. *Forests* **2022**, *13*, 834. [CrossRef]
10. Ilies, A.; Vasile, G. The external western Balkan border of the European Union and its borderland: Premises for building functional transborder territorial systems, in Annales. *Ann. Istrian Mediter. Stud. Ser. Hist. Sociol.* **2010**, *20*, 457–469.
11. Alexandru, I.; Olivier, D.; Ilieş, D.C. The cross-border territorial system in Romanian-Ukrainian Carpathian Area. Elements, mechanisms and structures generating premises for an integrated cross-border territorial system with tourist function. *Carp. J. Environ. Sci.* **2012**, *7*, 27–38.
12. Ilies, A.; Hurley, P.D.; Ilies, D.C.; Baias, S. Tourist animation—A chance adding value to traditional heritage: Case study's in the Land of Maramures (Romania). *Rev. Etnogr. Şi Folc. J. Ethnogr. Folk.* **2017**, *1–2*, 131–151.
13. Hill, C.A.S. *Wood Modification—Chemical, Thermal and Other Processes;* John Wiley and Sons Ltd.: West Sussex, UK, 2006.
14. Sandberg, D.; Kutnar, A.; Mantanis, G. Wood modification technologies—A review. *iForest* **2017**, *10*, 895–908. [CrossRef]
15. Kamperidou, V. The Biological Durability of Thermally- and Chemically-Modified Black Pine and Poplar Wood against Basidiomycetes and Mold Action. *Forests* **2019**, *10*, 1111. [CrossRef]
16. Hill, C.; Altgen, M.; Rautkari, L. Thermal modification of wood—A review: Chemical changes and hygroscopicity. *J. Mater. Sci.* **2021**, *56*, 6581–6614. [CrossRef]
17. Čabalová, I.; Výbohová, E.; Igaz, R.; Kristak, L.; Kačík, F.; Antov, P.; Papadopoulos, A.N. Effect of oxidizing thermal modification on the chemical properties and thermal conductivity of Norway spruce (*Picea abies* L.) wood. *Wood Mater. Sci. Eng.* **2021**, *17*, 366–375. [CrossRef]
18. Ali, M.; Abdullah, U.H.; Ashaari, Z.; Hamid, N.H.; Hua, L.S. Hydrothermal Modification of Wood: A Review. *Polymers* **2021**, *13*, 2612. [CrossRef] [PubMed]
19. Todaro, L.; Liuzzi, S.; Pantaleo, A.M.; Lo Giudice, V.; Moretti, N.; Stefanizzi, P. Thermo-modified native black poplar (*Populus nigra* L.) wood as an insulation material. *iForests* **2021**, *14*, 268–273. [CrossRef]
20. Guo, H.; Fuchs, P.; Cabane, E.; Michen, B.; Hagendorfer, H.; Romanyuk, Y.E.; Burgert, I. UV-protection of wood surfaces by controlled morphology fine-tuning of ZnO nanostructures. *Holzforschung* **2016**, *70*, 699–708. [CrossRef]

21. Awoyemi, L. Determination of Optimum Borate Concentration for Alleviating Strength Loss During Heat Treatment of Wood. *Wood Sci. Technol.* **2007**, *42*, 39–45. [CrossRef]
22. Ibrahim, U.; Ashaari, Z.; Hua, L.S.; Halis, R. Oil-heat treatment of rubberwood for optimum changes in chemical constituents and decay resistance. *J. Trop. For. Sci.* **2016**, *28*, 88–96.
23. Nasir, V.; Nourian, S.; Avramidis, S.; Cool, J. Prediction of physical and mechanical properties of thermally modified wood based oncolor change evaluated by means of "group method of data handling" (GMDH) neutral network. *Holzforschung* **2018**, *73*, 381–392. [CrossRef]
24. Awoyemi, L.; Westermark, U. Effects of borate impregnation on the response of wood strength to heat treatment. *Wood Sci. Technol.* **2005**, *39*, 484–491. [CrossRef]
25. Tjeerdsma, B.F.; Militz, H. Chemical changes in hydrothermal treated wood: FTIR analysis of combined hydrothermal and dry heat-treated wood. *Holz Roh Werkst.* **2005**, *63*, 102–111. [CrossRef]
26. Taghiyari, H.R.; Enayati, A.; Gholamiyan, H. Effects of nano-silver impregnation on brittleness, physical and mechanical properties of heat-treated hardwoods. *Wood Sci. Technol.* **2013**, *47*, 467–480. [CrossRef]
27. Repellin, V.; Guyonnet, R. Evaluation of heat treated wood swelling by differential scanning calorimetry in relation with chemical composition. *Holzforschung* **2007**, *59*, 28–34. [CrossRef]
28. Brebu, M.; Vasile, C. Thermal degradation of lignin—A review. *Cellul. Chem. Technol.* **2010**, *44*, 353–363.
29. Yildiz, S.; Gumuskaya, E. The effects of thermal modification on crystalline structure of cellulose in soft and hardwood. *Build. Environ.* **2007**, *42*, 62–67. [CrossRef]
30. Keskin, H.; Tekin, A. Abrasion resistance of cellulosic, synthetic, polyurethane, waterborne and acidhardening varnishes used woods. *Constr. Build. Mater.* **2011**, *25*, 638–643. [CrossRef]
31. Cakicier, N.; Korkut, S.; Korkut, D. Varnish layer harrdness, scratch resistance, and glossiness of various wood species as affected by heat treatment. *BioResources* **2011**, *6*, 1548–1658.
32. Atar, M.; Cinar, H.; Dongel, N.; Yalinkilic, A.C. The effect of heat treatment on the pull-off strength of optionally varnished surfaces of five wood materials. *BioResources* **2015**, *10*, 7151–7164. [CrossRef]
33. Pati, R. Molecule for electronics: A myriad of opportunities comes with daunting challenges. *J. Nanomater. Mol. Nanotechnol.* **2012**, *1*. [CrossRef]
34. Saber, R.; Shakoori, Z.; Sarkar, S.; Tavoosidana, G.H.; Kharrazi, S.H.; Gill, P. Spectroscopic and microscopic analyses of rod-shaped gold nanoparticles interacting with single-stranded DNA oligonucleotides. *IET Nanobiotechnol.* **2013**, *7*, 42–49. [CrossRef]
35. Taghiyari, H.R.; Bayani, S.; Militz, H.; Papadopoulos, A.N. Heat treatment of pine wood: Possible effect of impregnation with silver nanosuspension. *Forests* **2020**, *11*, 466. [CrossRef]
36. Taghiyari, H.R. Effects of heat-treatment on permeability of untreated and nanosilver-impregnated native hardwoods. *Maderas-Cienc. Tecnol.* **2013**, *15*, 183–194. [CrossRef]
37. Taghiyari, H.R.; Esmailpour, A.; Majidi, R.; Hassani, V.; Abdolah Mirzaei, R.; Farajpour Bibalan, O.; Papadopoulos, A.N. The effect of silver and copper nanoparticles as resin fillers on less-studied properties of UF-based particleboards. *Wood Mater. Sci. Eng.* **2020**, *17*, 317–327. [CrossRef]
38. Meijer, M. A review of interfacial aspects in wood coatings: Wetting, surface energy, substrate penetration and adhesion. European Seminar on High Performance Wood Coatings. In Proceedings of the Exterior and Interior Performance, Paris, France, 26–27 April 2004.
39. De Moura, L.F.; Brito, J.O.; Nolasco, A.M.; Uliana, L.R. Evaluation of coating performance and color stability on thermally rectified *Eucalyptus grandis* and *Pinus caribaea* Var. *Hondurensis woods. Wood Res.* **2013**, *58*, 231–242.
40. Nejad, M.; Shafaghi, R.; Ali, H.; Cooper, P. Coating performance on oil-heat treated wood for flooring. *BioResources* **2013**, *8*, 1881–1892. [CrossRef]
41. Sönmez, A.; Budakci, M.; Huseyin, P. The effect of the moisture content of wood on the layer performance of water-borne varnishes. *BioResources* **2011**, *6*, 3166–3178.
42. Gaylarde, C.C.; Morton, L.H.G.; Loh, K.; Shirakawa, M. Biodeterioration of external architectural paint films—A review. *Int. Biodeterior. Biodegrad.* **2011**, *65*, 1189–1198. [CrossRef]
43. Hazir, E.; Koc, K.H. Evaluation of wood surface coating performance using water based, solvent based and powder coating. *Maderas Cienc. Y Tecnol.* **2019**, *21*, 467–480. [CrossRef]
44. Erdinler, E.S.; Koc, K.H.; Dilik, T.; Hazir, E. Layer thickness performances of coatings on MDF: Polyurethane and cellulosic paints. *Maderas Cienc. Y Tecnol.* **2019**, *21*, 317–326. [CrossRef]
45. Kristýna, Š.; Štěpán, H.; Eliška, O.; Miloš, P.; Hakan, F. Effect of artificial weathering and temperature cycling on the adhesion strength of waterborne acrylate coating systems used for wooden windows. *J. Green Build.* **2020**, *15*, 1–14. [CrossRef]
46. Bansal, R.; Nair, S.; Pandey, K.K. UV resistant wood coating based on zinc oxide and cerium oxide dispersed linseed oil nano-emulsion. *Mater. Today Commun.* **2022**, *30*, 103177. [CrossRef]
47. Boonstra, M.J.; Tjeerdsma, B. Chemical analysis of heat treated softwoods. *Holz Als Roh Werkst.* **2006**, *64*, 204–211. [CrossRef]
48. Sivonene, H.; Maunu, S.L.; Sundholm, F.; Jämsä, S.; Viitaniemi, P. Magnetic resonance studies of thermally modified wood. *Holzforschung* **2002**, *56*, 648–654. [CrossRef]
49. Naegel, A.; Heisig, M.; Wittum, G.; Turksen, K. *Permeability Barrier: Methods and Protocols;* Humana Press: Totowa, NJ, USA, 2011; p. 467. ISBN 9781617791901.

50. Park, C.W.; Youe, W.J.; Kim, S.J.; Han, S.Y.; Park, J.S.; Lee, E.A.; Kwon, G.J.; Kim, Y.S.; Kim, N.H.; Lee, S.H. Effect of lignin plasticization on physico-mechanical properties of lignin/ply(lactic acid) composites. *Polymers* **2019**, *11*, 2089. [CrossRef] [PubMed]
51. Schultz, J.; Nardin, M. Theories and mechanism of adhesion. In *Handbook of Adhesive Technology*; Marcel Dekker: New York, NY, USA, 1994; pp. 19–35.
52. Cheng, E.; Sun, X. Effects of wood-surface roughness adhesive viscosity and processing pressure on adhesion strength of protein adhesive. *J. Adhes. Sci. Technol.* **2006**, *20*, 997–1017. [CrossRef]
53. Ozdemir, T.; Hiziroglu, S.; Kocapinar, M. Adhesion strength of cellulosic varnish coated wood species as function of their surface roughness. *Adv. Mater. Sci. Eng.* **2015**, *2015*, 525496. [CrossRef]
54. Hakkou, M.; Petrissans, M.; Zoulalian, A.; Gerardin, P. Investigation of wood wettability changes during heat treatment on the basis of chemical analysis. *Polym. Degrad.Stabil.* **2005**, *89*, 1–5. [CrossRef]
55. *ASTM D 4541-02*; Standard Test Method for Pull-Off Strength of Coatings Using Portable Adhesion Testers. Tremco Incorporated: Cuyahoga, OH, USA, 2006.
56. Khaydarov, R.A.; Khaydarov, R.R.; Gapurova, O.; Estrin, Y.; Scheper, T. Electrochemical method for the synthesis of silver nanoparticles. *J. Nanopart. Res.* **2009**, *11*, 1193–1200. [CrossRef]
57. Shi, S.H.Q. Diffusion model based on Fick's second law for the moisture absorption process in wood fiber-based composites: Is it suitable or not? *Wood Sci. Technol.* **2007**, *41*, 645–658. [CrossRef]
58. Hernandez, V.; Avramidis, S.; Navarrete, J. Albino strains of *Ophiostoma spp.* Fungi effect on radiate pine permeability. *Eur. J. Wood Wood Prod.* **2012**, *70*, 551–556. [CrossRef]
59. Choo, A.C.Y.; Tahir, M.P.; Karimi, A.; Bakar, E.S.; Abdan, K.; Ibrahim, A.; Balkis, F.A.B. Study on the longitudinal permeability of oil palm wood. *Ind. Eng. Chem. Res.* **2013**, *52*, 9405–9410. [CrossRef]
60. Taghiyari, H.R.; Moradi Malek, B. Effect of heat treatment on longitudinal gas and liquid permeability of circular and square-shaped native hardwood specimens. *Heat Mass Transf.* **2014**, *50*, 1125–1136. [CrossRef]
61. USPTO. Gas Permeability Measurement Apparatus. Patent Number 8079249 B2, 20 December 2011.
62. Ada, R. Cluster analysis and adaptation study for safflower genotypes. *Bulg. J. Agricult. Sci.* **2013**, *19*, 103–109.
63. Taghiyari, H.R.; Karimi, A.N.; Parsapajouh, D.; Pourtahmasi, K. Study on the longitudinal gas permeability of juvenile wood and mature wood. *Spec. Top. Rev. Porous Media* **2010**, *1*, 31–38. [CrossRef]
64. Lens, F.; Vos, R.A.; Charrier, G.; van der Niet, T.; Merckx, V.; Baas, P.; Aguirre Gutierrez, J.; Jacobs, B.; Chacon Dória, L.; Smets, E.; et al. Scalariform-to-simple transition in vessel perforation plates triggered by differences in climate during the evolution of *Adoxaceae*. *Ann. Bot.* **2016**, *118*, 1043–1056. [CrossRef]
65. Ekstedt, J. Studies on the Barrier Properties of Exterior Wood Coatings. Doctoral Thesis, Department of Civil and Architectural Engineering, Division of Building Materials, KTH-Royal Institute of Technology, Stockholm, Sweden, 2002; p. 75.
66. Taghiyari, H.R.; Samandarpour, A. Effects of nanosilver-impregnation and heat treatment on coating pull-off adhesion strength on solid wood. *Drvna Ind.* **2015**, *66*, 321–327. [CrossRef]
67. Taghiyari, H.R.; Esmailpour, A.; Papadopoulos, A. Paint pull-off strength and permeability in nanosilver-impregnated and heat-treated beech wood. *Coatings* **2019**, *9*, 723. [CrossRef]
68. Taghiyari, H.R.; Zolfaghari, H.; Sadeghi, M.E.; Esmailpour, A.; Jaffari, A. Correlation between specific gas permeability and sound absorption coefficient in solid wood. *J. Trop. For. Sci.* **2014**, *26*, 92–100.
69. Sandberg, D.; Haller, P.; Navi, P. Thermo-hydro and thermo-hydro-mechanical wood processing: An opportunity for future environmentally friendly wood products. *Wood Mater. Sci. Eng.* **2013**, *8*, 64–88. [CrossRef]
70. Čabalová, I.; Kačík, F.; Lagaňa, R.; Výbohová, E.; Bubeníková, T.; Čaňová, I.; Ďurkovič, J. Effect of Thermal Treatment on the Chemical, Physical, and Mechanical Properties of Pedunculate Oak (*Quercus robur* L.) Wood. *BioResources* **2018**, *13*, 157–170 [CrossRef]
71. Funaoka, M.; Kako, T.; Abe, I. Condensation of lignin during heating of wood. *Wood Sci. Technol.* **1990**, *24*, 277–288. [CrossRef]

MDPI

Article

Physical and Chemical Properties of *Acacia mangium* Lignin Isolated from Pulp Mill Byproduct for Potential Application in Wood Composites

Nissa Nurfajrin Solihat [1,*], Eko Budi Santoso [2], Azizatul Karimah [1,2], Elvara Windra Madyaratri [2], Fahriya Puspita Sari [1], Faizatul Falah [1], Apri Heri Iswanto [3,4], Maya Ismayati [1], Muhammad Adly Rahandi Lubis [1], Widya Fatriasari [1,*], Petar Antov [5], Viktor Savov [5], Milada Gajtanska [6,*] and Wasrin Syafii [2]

1 Research Center for Biomaterials, Research and Innovation Agency (BRIN), Jl. Raya Bogor KM 46, Cibinong 16911, Indonesia; karimahazizatul@gmail.com (A.K.); 103779316@student.swin.edu.au (F.P.S.); fayzaa_falah@yahoo.com (F.F.); maya_ismayati@brin.go.id (M.I.); marl@biomaterial.lipi.go.id (M.A.R.L.)
2 Department of Forest Products, Faculty of Forestry and Environment, IPB University, Bogor 16680, Indonesia; ekobudisantoso122@gmail.com (E.B.S.); elvarawindra@yahoo.com (E.W.M.); wasrinsy@indo.net.id (W.S.)
3 Department of Forest Product, Faculty of Forestry, Universitas Sumatera Utara, Medan 20155, Indonesia; apri@usu.ac.id
4 JATI-Sumatran Forestry Analysis Study Center, Jl. Tridharma Ujung No. 1, Kampus USU, Medan 20155, Indonesia
5 Faculty of Forest Industry, University of Forestry, 1797 Sofia, Bulgaria; p.antov@ltu.bg (P.A.); victor_savov@ltu.bg (V.S.)
6 Faculty of Wood Sciences and Technology, Technical University in Zvolen, 96001 Zvolen, Slovakia
* Correspondence: nissa.nurfajrin.solihat@brin.go.id (N.N.S.); widya.fatriasari@brin.go.id or widy003@brin.go.id (W.F.); gajtanska@tuzvo.sk (M.G.)

Abstract: The efficient isolation process and understanding of lignin properties are essential to determine key features and insights for more effective lignin valorization as a renewable feedstock for the production of bio-based chemicals including wood adhesives. This study successfully used dilute acid precipitation to recover lignin from black liquor (BL) through a single-step and ethanol-fractionated-step, with a lignin recovery of ~35% and ~16%, respectively. The physical characteristics of lignin, i.e., its morphological structure, were evaluated by scanning electron microscopy (SEM). The chemical properties of the isolated lignin were characterized using comprehensive analytical techniques such as chemical composition, solubility test, morphological structure, Fourier-transform infrared spectroscopy (FTIR), ^{1}H and ^{13}C Nuclear Magnetic Resonance (NMR), elucidation structure by pyrolysis-gas chromatography-mass spectroscopy (Py-GCMS), and gel permeation chromatography (GPC). The fingerprint analysis by FTIR detected the unique peaks corresponding to lignin, such as C=C and C-O in aromatic rings, but no significant differences in the fingerprint result between both lignin. The ^{1}H and ^{13}C NMR showed unique signals related to functional groups in lignin molecules such as methoxy, aromatic protons, aldehyde, and carboxylic acid. The lower insoluble acid content of lignin derived from fractionated-step (69.94%) than single-step (77.45%) correlated to lignin yield, total phenolic content, solubility, thermal stability, and molecular distribution. It contradicted the syringyl/guaiacyl (S/G) units' ratio where ethanol fractionation slightly increased syringyl unit content, increasing the S/G ratio. Hence, the fractionation step affected more rupture and pores on the lignin morphological surface than the ethanol-fractionated step. The interrelationships between these chemical and physicochemical as well as different isolation methods were investigated. The results obtained could enhance the wider industrial application of lignin in manufacturing wood-based composites with improved properties and lower environmental impact.

Keywords: acid precipitation; single and fractionation step; kraft lignin; physical and chemical properties; *A. mangium* black liquor

Citation: Solihat, N.N.; Santoso, E.B.; Karimah, A.; Madyaratri, E.W.; Sari, F.P.; Falah, F.; Iswanto, A.H.; Ismayati, M.; Lubis, M.A.R.; Fatriasari, W.; et al. Physical and Chemical Properties of *Acacia mangium* Lignin Isolated from Pulp Mill Byproduct for Potential Application in Wood Composites. *Polymers* **2022**, *14*, 491. https://doi.org/10.3390/polym14030491

Academic Editor: Nathanael Guigo

Received: 22 December 2021
Accepted: 24 January 2022
Published: 26 January 2022

1. Introduction

In 2019, Indonesia was ranked among the top 10 countries concerning pulp and paper production. In 2018, Indonesia produced 16 million tons of paper and 11 million tons of pulp [1]. For every 1 ton of pulp produced, about 7 tons of black liquor (BL) were generated as a residue at 15% solids by weight, with two-thirds of the solids consisting of organic chemicals, and the remains were inorganic chemicals [2]. In recent decades, lignin derived from BL has been considered a natural biopolymer, a viable alternative to the fossil-based chemicals due to its abundance in BL, reaching 45% dry weight [3].

Most BL is incinerated for boiler heating sources and energy, and only 5% of the BL is used for value-added applications [4]. The economical consideration of lignin isolation from BL includes recovery yield, purification, non-uniform structure, and unique reactivity [5]. The three main phenolic hydroxyl precursors in lignin are coniferyl alcohol (G), p-coumaryl alcohol (H), and sinapyl alcohol (S), which are linked to each other mostly by aryl ether linkage (β-O-4′) [6,7]. The actual properties of lignin, such as thermal stability, reactivity, molecular distribution, and solubility, depend on the ratio of these aromatic units. It varies depending on the technique of extraction and the plant source. For instance, softwoods contain mostly G units; hardwoods include both S and G, while non-wood plants have all three units [8].

Precipitation by dilute acid such as sulphuric acid is a common and feasible technique to isolate lignin from BL [9]. However, using sulphuric acid can increase lignin's ash and sulfur content. Therefore, lignin for sulfur-sensitive utilizations should be restricted [5]. Haz et al. evaluated the effect of four different dilute acids (chloric, sulphuric, acetic, and nitric) on lignin properties. Lignin precipitated by nitric and chloric acid obtained high phenolic hydroxyl both in non-conjugated and conjugated forms (>2 mmol/g), suitable for phenolic polycondensates production and rubber stabilizer [10]. Handika et al. [11] reported that high free-phenolic hydroxyl in lignin increased its reactivity to produce the high-thermal stability of polyurethane resin for textile application. As a polyphenol molecule, lignin contains high free-phenolic hydroxyl groups that are favorable for modifications, such as phenolation and methylolation [12], tailored for increasing its chemical reactivity to formaldehyde in formaldehyde-based resins used in the production of wood composites such as particleboards [13], oriented strand boards [14], and flame-retardant composites [15]. Besides, lignin modification either with poly(butylene succinate) or polypropylene biocomposites increased the thermal stability of kenaf core fiber [16,17]. According to Tejado et al., kraft lignin is suitable for phenol-formaldehyde resin due to its higher amount of free phenolic content, molecular weight, and thermal properties [18]. Therefore, it is necessary to understand the specific chemical structure of lignin to achieve optimal utilization. However, the lack of understanding of the lignin structure-property–application relationship (SPARs) is a major roadblock to further development of lignin [19]. Of these, one characterization technique is insufficient to produce coherent data to identify the feature of lignin because of its complex structure and variation. Therefore, a comprehensive analytical technique is pivotal to understanding the properties of isolated lignin, allowing its large-scale utilization.

This study emphasizes the efficient isolation method of lignin from industrial residues of the pulp and paper industry. Our fundamental comprehensive analytical technique provides knowledge for industries to produce superior lignin-based, value-added products, especially for wood composites. Lignin was isolated by two different methods of dilute chloric acid precipitation. The physical characteristics of lignin, such as morphological structure, were conducted by scanning electron microscopy (SEM). Chemical features, including its total hydroxyl phenolic content and solubility in the organic and base solvent, were determined by ultraviolet-visible spectrophotometer (UV-Vis). The functional group was identified by attenuated total reflection Fourier-transform infrared (ATR-FTIR). ^{1}H and ^{13}C nuclear magnetic resonance (NMR) were employed to predict structural properties of lignin corresponding to its fingerprint signal. The elucidation aromatic precursors unit in lignin structure was analyzed by pyrolysis-gas chromatography-mass spectrometry

(PyGC/MS). Meanwhile, the thermal features of lignin were characterized by thermogravimetric analysis (TGA) and differential scanning calorimetry (DSC). Gel permeation chromatography (GPC) was used to measure the molecular distribution of lignin.

2. Materials and Methods

2.1. Material

A derived BL from *Acacia mangium* was collected from a pulp and paper mill factory in Sumatra, Indonesia. Hydrochloric acid (HCl), dioxane, sodium hydroxide (NaOH), acetic acid anhydride, dimethyl sulfoxide (DMSO), and tetrahydrofuran (THF) were purchased from Merck (Darmstadt, Germany), while pyridine was obtained from Wako Pure Chemical Industries (Osaka, Japan). Lignin alkali (kraft) from Sigma-Aldrich (Saint Louis, MO, USA) was used for the lignin standard. All chemical materials used in this study were analytical grade without any purification.

2.2. Lignin Isolation

Dilute acid precipitation was used to isolate lignin from BL through a single-step and fractionated-step based on Hermiati et al. [20] with one major modification. For the single step, HCl 1 M was poured into BL until pH 2. The solution was kept overnight and the residue was separated by decantation. The deionized (DI) water was added to the residue with similar acid volume, and the decantation process was conducted again after 24 h. This process was repeated six times. The residue was kept in the refrigerator overnight and separated by vacuum filtration. Wet lignin on the filter paper was washed with DI water and dried in an oven at 45 °C for 24 h. The lignin yield percentage was measured by dividing the dry weight of lignin (g) by the dry weight of BL. Dried lignin was kept in sealed plastic for further analysis.

For fractionated-step, HCl 1 M was added into BL until pH 7. Ethanol, as much as four times the volume of the acid, was added to the solution. Non-lignin components such as sugar and carbohydrates were filtrated as a residue. The filtrate was evaporated until the ethanol completely dried up. The acidification was continued by adding acid until pH 2. Lignin precipitate was separated without decantation six times by water, unlike the single-step. The following step is similar to the procedure from the single-step method.

2.3. Chemical Features Measurement: Chemical Component, Total Phenolic Hydroxyl, and Solubility

The water content of lignin was determined according to TAPPI T 264 cm-97 [21], and ash content was calculated following the TAPPI T211 om-02 method [22]. Contents of acid-insoluble lignin (AIL) and acid-soluble lignin (ASL) were analyzed based on the method of Sluiter et al. (NREL/TP-510-42618) [23]. Triplicate analysis was performed for all chemical features measurements. Lignin alkali (kraft) from Sigma-Aldrich (370958) (Saint Louis, MO, USA) was used as a reference.

Total phenolic hydroxyl (phOH) was determined by the UV-Vis method [24]. A total of 1 mg/mL of lignin was diluted in dioxane: 0.2 M NaOH (1:1), and the mixture was filtered by microfiltration (0.45 µm). The filtrate was diluted in 0.2 M NaOH until the 0.08 mg/mL concentration was reached. The UV spectrum was recorded in a 200–600 nm range by Shimadzu UV vis-1800 spectrophotometer, where lignin in pH6 was used as reference. The absorbance of maximum spectra at 300 and 350 nm was used to calculate total phOH by the following equation:

$$Total\ phOH\ (\text{mmol/g}) = (0.425 \times A_{300\text{ nm}}) + (0.812 \times A_{350\text{ nm}}) \times \frac{1}{c \times a} \times \frac{10}{17}$$

where A is absorbance, c stands for lignin concentration, and a is path length (1 cm) [24].

The solubility test of lignin in the base and the organic solvent was conducted according to the method by Hermiati et al. Lignin 7 mg/5 mL was dissolved in NaOH pH 12 as the alkali solvent and mixture of dioxane water (9:1). Each solution was diluted 50 times by DI water. The UV spectrum was measured in the range of wavelength 200–400 nm [20].

2.4. Morphological Assessment by SEM

A scanning electron microscope (JSM-IT200, JEOL, Tokyo, Japan) was used to observe morphological surfaces and particle size of the reference lignin and the isolated lignins. Lignin samples were placed on the carbon tube, and the surface was coated with gold using Ion Coater iB2. The micrograph of the sample was recorded at 200 and 5000 magnifications under a high vacuum and working distance of 11 mm with 5.0 kV accelerating voltage.

2.5. Functional Group Analysis by UATR-FTIR

Attenuated total reflection Fourier-transform infrared (ATR-FTIR) spectroscopic equipped with UATR unit cell from PerkinElmer (spectrum two) (PerkinElmer Corporation, Waltham, MA, USA) was employed to investigate the functional group of lignin. The sample was placed on the diamond crystal, and the spectrum at a wavelength of 400–500 cm^{-1} was taken by pressing the torque knob with the same pressure. An average of 32 scans with 4 cm^{-1} resolution were used to acquire the spectrum. The same average scanning was carried out for background correction and scanning before analysis.

2.6. Fingerprint Observation by 1H and 1C NMR

Lignin samples were acetylated before undergoing nuclear magnetic resonance (NMR) and molecular weight distribution test based on Wen et al. [25] method with a minor adjustment. A total of 200 mg lignin was dissolved in an 8 mL mixture of acetic acid anhydride: pyridine (1:1) for 72 h in the dark bottle. Ethanol was added until the mixture was concentrated. Acetylated lignin was obtained by slowly dropping the mixture into ice acid (pH 2) and separated through centrifugation. The wet acetylated lignin was washed with 50 mL DI water three times and freeze-dried until dry acetylated-lignin (AL) was obtained.

Each 20 mg AL sample was diluted in 0.7 mL DMSO. The solution was transferred to a 3 mm tube. ^{1}H NMR (JEOL JNM-ECZR 500, Tokyo, Japan) data points were acquired with an acquisition time of 1.75 s, a relaxation time of 5.0 s, and 24 scans. For typical ^{13}C NMR, 20,480 spectra scanning were averaged to increase the signal-to-noise ratio with 2.0 s delayed relaxation and 0.9 s acquisition time.

2.7. Thermal Investigation by TGA and DSC

The thermal investigation of the isolated lignin was conducted using a thermogravimetric analyzer (TGA 4000, PerkinElmer, Waltham, MA, USA) and differential scanning calorimetry (DSC) (DSC 4000, PerkinElmer, Waltham, MA, USA). For TGA analysis, about 4 mg lignin sample was placed on the crucible ceramics sample holder, and the measurement was conducted under argon atmosphere with the flow of 20 mL/min. The sample was heated from 25 °C to 750 °C at a 10 °C/min rate. The automatic curve of weight loss versus temperature was generated from the instrument.

DSC analysis was carried out with ~4 mg lignin samples on a standard aluminium pan to determine the glass transition temperature (Tg) and curing properties of lignin. Each sample was heated until 300 °C with a 10 °C/min heating rate under a nitrogen atmosphere (flowrate = 20 mL/min). Tg value was automatically calculated by DSC 4000 pyris 1 PerkinElmer software (Pyris 11 software Version 11.1.1.0492, PerkinElmer, Shelton CT, USA).

2.8. Chemical Elucidation by Py-GCMS

Chemical elucidation analysis of lignin was studied by pyrolysis-gas chromatography mass spectrometry (PyGC/MS) (Shimadzu GC/MS system QP-2020 NX, Shimadzu, Kyoto Japan) equipped with multi-shot pyrolyzer EGA/PY-3030D. Between 500–600 µg of lignin was placed in eco-cup SF PY1-EC50F, and the cup was sealed by glass wool. The eco cup was pyrolyzed at 500 °C for 0.1 min using helium as carrier gas and SH-Rxi-5Sil MS column (30 m × 0.25 mm i.d. film thickness. 0.25 µm). The PyGC/MS temperature was programmed as follows: 50 °C for 1 min, 5 °C/min to 280 °C, and 13 min at 280 °C. The mass spectrum was taken at 70 eV with a pressure of 20.0 kPa (15.9 mL/min, column flow 0.61 mL/min). The obtained pyrolysis product was identified by approaching mass spectra and retention times using the data library in NIST LIBRARY 2017.

2.9. Molecular Weight by GPC

Gel permeation chromatography (GPC) is a rapid and versatile tool to provide information on the molecular weight of lignin. Lignin was dissolved in the THF, and Shimadzu LC-20 (Shimadzu, Kyoto, Japan) equipped with a UV-RID detector was used to quantify the molecular weight distribution acetylated lignin. The analysis was employed using the LF-800 column with an injection volume of 20 uL. Polystyrene standard was used to create a calibration curve and GPC system calibration.

3. Result and Discussion

3.1. Chemical Composition and Lignin Solubility

Lignin recovery was one critical factor for selecting the lignin isolation method that is economically feasible. Lignin yield recovery from BL by single and fractionated-step (oven dry based) was 35.39% and 16.34%. Both isolation methods reported lignin recovery yield at the expected range of 20–40% [5]. Ethanol fractionation resulted in lignin depolymerization in the liquid solution, decreasing the solid lignin residue by acid. Lignin yield recovery is related to the larger size of the fractionated-step lignin based on the SEM micrograph. Large lignin particle size results in a smaller reaction surface area for precipitation, reducing the amount of lignin recovered after acid precipitation. This suggestion was in correlation with the ASL content. This finding agreed with Hamzah et al. (2020), where lignin recovery from *Miscanthus x giganteus* decreased from 75% to 25% with the increased ethanol concentration from 0% to 50% [26].

The chemical composition of lignin Is presented in Table 1, where the data is the average value from a triplicate experiment with a deviation standard less than ±5%. Ethanol was added in fractionated-step to precipitate non-lignin components such as sugar and carbohydrates, theoretically increasing lignin purity. However, the total lignin from single-step lignin (~99%) was slightly higher than from the fractionated step. Besides, the impurities component in the single-step lignin, represented as ash content, was lower than the fractionated step. It suggests that the single-step isolation method effectively isolates lignin with high purity.

Table 1. Chemical composition of lignin.

	Water Content (%)	Ash Content (%)	AIL (%)	ASL (%)	Total phOH (mmol/g)
Lignin reference	2.60 ± 0.27	2.44 ± 0.00	96.02 ± 0.50	1.54 ± 0.06	6.00 ± 0.50
Lignin single method	5.65 ± 1.14	0.53 ± 0.07	77.45 ± 0.48	22.02 ± 0.83	7.40 ± 0.71
Lignin fraction method	15.79 ± 0.74	1.94 ± 0.08	69.94 ± 5.55	28.12 ± 0.94	7.31 ± 0.78

Interestingly, lignin isolation from BL by dilute hydrochloric acid obtained high ASL content while reference lignin had low ASL content. Different isolation methods likely obtained the different proportions of AIL and ASL content. Similar results were reported by Sameni et al. [8] where isolation lignin from BL by using dilute sulfuric acid resulted in low ASL (<4%) and high AIL ~91. Sulfuric acid is popular acid to isolate lignin and obtain a high concentration of AIL. Consequently, it will increase sulfuric and ash content [5].

In this study, the highest free phOH content was obtained from the isolated lignin, where the lowest was from the reference lignin. This finding agreed with the previous report where Kraft pulping process and precipitation lignin by HCl enhanced condensed structure and the phenolic hydroxyl group [10,27]. Unfortunately, we could not find the source and isolation process of reference lignin from Sigma-Aldrich. The total phOH content is correlated to the Tg value because a higher condensed structure in polymer created a high char amount in high temperature. Eventually, the combustion rate can be reduced by the presence of char [11,28,29].

UV spectroscopy is used to monitor the lignin purity and molecular distribution. A similar pattern of UV-Vis spectra from both commercial and isolated lignin is observed in Figure 1a. The distinct absorption at 215–222 nm corresponded to non-conjugated phenolic groups (excitation of π-π^*) that appear due to shifting band is an effect of hypsochromic NaOH. The spectrum of single-step lignin is slightly higher than others which may correlate to the total phOH (Table 1). Another intensive peak was observed in the range of 296–303 nm, originating from the conjugated phenolic group due to n-π^* excitation [30].

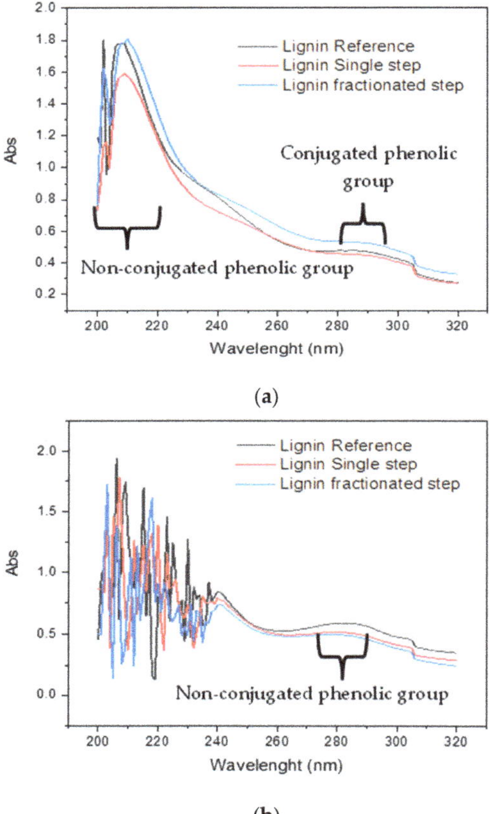

Figure 1. Lignin solubility in base (**a**) and organic solvent (**b**) determined by UV-Vis.

A glance at Figure 1b reveals an identical UV-Vis spectrum among three lignin samples when lignin is diluted in dioxane/water. Unlike the lignin in the base solution, solubilization lignin in dioxane/water is limited to wavelengths above 250 nm seen in the spectra Figure 1b. This finding is similar to lignin Alfa grass kraft from industrial waste [31] and Kraft-anthraquinone (AQ) lignin [32]. According to Ammar et al.'s report, the large absorbance of lignin in dioxane/water at 280 nm corresponded to non-conjugated phenolic hydroxyl groups. In comparison, the presence of both ferulic acids and p-coumaric acids could be attributed to the presence of the second type region of lignin absorption at about 300 nm [31]. Lignin reference has slightly higher absorbance than isolated lignin regarding purity.

3.2. SEM Micrograph of Lignin

SEM micrograph in Figure 2a–c, shown in 200-times magnification, described irregular and not uniform particles in terms of size from lignin samples. The reference lignin had a smaller particle of 59 μm than lignin from fractionated-step isolation (~72 μm). Interestingly, the single-step lignin had the lowest particle size, ~51 μm, indicating that ethanol impacted the behavior of lignin aggregates. At 5000-times magnification (Figure 2d–f), the morphological image of reference lignin (Figure 2d) and lignin from a single-step (Figure 2e) showed more rupture and pores on the lignin surface. Lignin from fractionated-step (Figure 2f) depicted smooth and rigid surfaces. This finding was following lignin isolation from *Miscanthus x giganteus* where the particle size of lignin increased with higher ethanol concentration, from 306 to 2050 nm. Besides, more crystalline structures and pores on the lignin surface were observed in more concentrated ethanol [26].

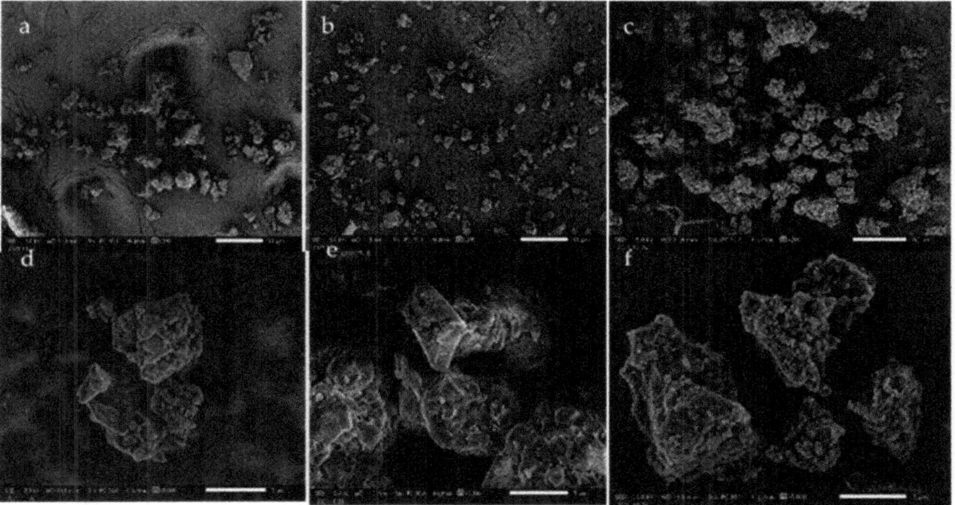

Figure 2. SEM micrographs show the morphological surface of reference lignin (**a**), single-step (**b**), fractionated-step (**c**) at 200× magnification and reference lignin (**d**), single-step (**e**), and fractionated-step (**f**), at 5000× magnification.

3.3. Functional Group of Lignin

FTIR is a versatile analytical tool to investigate functional groups and the general structure in lignin. The functional group of lignin may vary depending on the source of lignin. The main functional group in lignin is hydroxyl, methoxyl, carboxylic acid, and carbonyl. Figure 3 shows the FTIR spectra of reference lignin, and two isolated lignin, while a summary of peaks interpretation is available in Table 2. As shown in Figure 3, most of the peaks were similar between three samples, such as the broadband corresponding to hydroxyl group stretching (O-H) from aliphatic and aromatic in lignin structure detected at the wavelength 3500–3400 cm^{-1} (a), while the sharp peak at 2918 (b) and 2854 (c) cm^{-1}, respectively, were attributed to C-H stretching in methylene from side chain and aromatic methoxyl groups [33]. A small band assigned to carbonyl (C=O) stretching in unconjugated aldehyde and ketone in the ester group at 1716–1704 cm^{-1} (d) was found in the reference lignin. Still, a pronounced peak was obtained from both isolated lignin due to different lignin structures. Noticeable peaks attributed to vibration of the aromatic skeleton in all types of lignin appeared in the range of 1590–1460 cm^{-1} (e–g) [34]. An intense peak at 1430–1420 cm^{-1} (h) referred to aromatic skeletal vibration with deformation of C-H asymmetric in a methyl group [4].

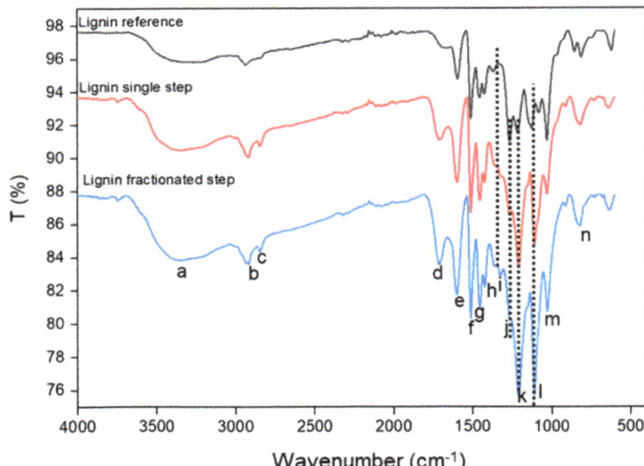

Figure 3. Functional group peaks of reference lignin, isolated lignin from single-step and fractionated-step by UATR-FTIR.

Table 2. Interpretation bands of UATR-FTIR spectra.

Code	Wavelength (cm^{-1})	Functional Group
a	3359	Hydroxyl group stretching (O-H) from aliphatic and aromatic [36]
b	2918	C-H stretching in methylene [33]
c	2854	C-H stretching in methoxy [33]
d	1710	Carbonyl (C=O) stretching in unconjugated aldehyde and ketone [34]
e	1590	C=C (aromatic rings) [34]
f	1511	C=C (aromatic rings) [34]
g	1470	Aromatic ring vibration with C-O [34]
h	1430	Deformation C-H in methyl group [4]
i	1326	C-O breathing (syringyl) [37]
j	1266	C-O(H) (phenolic OH guaiacyl) [36]
k	1213	C-O(Ar) in guaicyl ring [35]
l	1111	Deformation Ar-CH in syringyl ring [35]
m	1030	Unconjugated C-O in guaicyl [35]
n	855	CH out of plane bending in guaicyl [35]

However, an obvious difference in lignin structure between reference and isolated lignin was observed in adsorption at 1326 cm^{-1} (i) and 1111 cm^{-1} (l) as the breathing of C-O and deformation C-H in syringyl rings. The band was absent in the reference lignin, but it appeared in two isolated lignins. Conversely, stronger C-O stretching in the guaicyl unit at 1266 cm^{-1} (j) and 1213 cm^{-1} (k) was recorded in reference lignin but not in isolated lignin since the lignin was extracted from *Acacia mangium* (hardwood). This finding suggests that the reference lignin may be derived from softwood. This result correlates with Sameni et al. [35] finding where syringyl unit portion was absent in lignin from softwood and abundant in lignin from hardwood. Furthermore, higher peak absorption of unconjugated C-O (at 1030 cm^{-1} (m)) and CH out-of-plane bending (at 855 cm^{-1} (n)) in the guaicyl ring in reference lignin suggests a higher concentration of guiacyl in softwood than hardwood. Besides, these peaks were also slightly sharper in the lignin fractionated step than in the single-step. Further semi-quantitative analysis of syringyl versus guaicyl percentage is described in the Py-GCMS section.

3.4. ^{1}H and ^{13}C NMR

NMR analysis is frequently used to predict lignin's structural details concerning its molecular characteristics, reactivity, and composition. The acetylation of lignin before NMR analysis aims to decrease the impurities in lignin that may interfere with the spectrum [25]. Due to the complex structure of polymer lignin, typically simple proton ^{1}H NMR resulted in overlapping spectra which is difficult to justify the structure. Hence, ^{13}C NMR is needed to support the hypothesis of ^{1}H NMR. The presence of condensed and uncondensed aliphatic and aromatic carbon and aryl ethers can be detected by natural ^{13}C isotope NMR. However, longer scanning and acquisition times are required to improve signal sensitivity due to the low abundance of carbon isotope in the lignin molecule [4]. Still, quantitative ^{13}C-NMR can be a useful technique for lignin structural investigation, particularly in determining molecular alterations caused by different isolation procedures and biomass sources [8,25,35].

The ^{1}H NMR spectrum in Figure 4 shows similar peaks among the three lignin samples. The small peak at 0.8 and 1.23 ppm occurred because of saturated aliphatic lignin protons in the methyl and methylene chain. The intense signal at 1.98 ppm indicates the presence of an aliphatic acetate group. The strong signal at 2.5 ppm and 3.3 is because of protons in water and DMSO. A pronounced peak at 3.76 ppm corresponds to methoxyl protons (-OCH_3). A sharper signal at 6.7–6.9 in isolated lignin spectrum suggests more syringyl units than reference lignin.

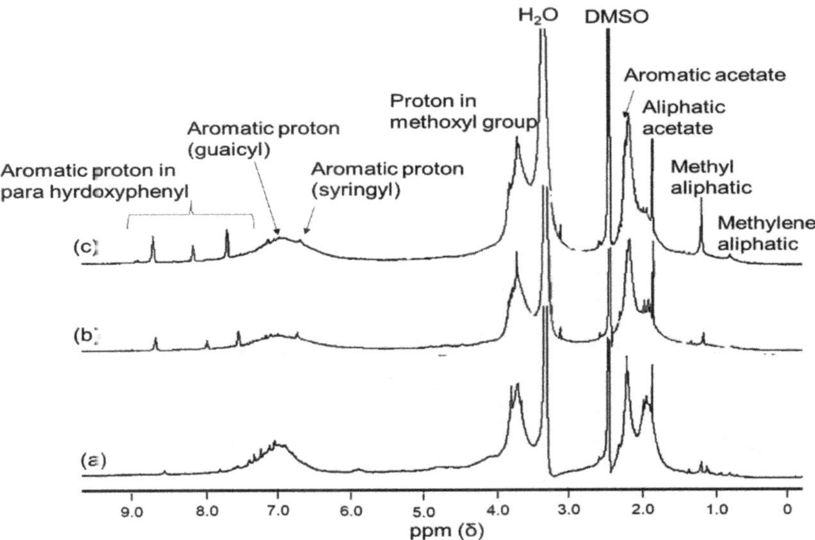

Figure 4. ^{1}H NMR signal of standard lignin (**a**), single-step lignin (**b**), and fractionated-step lignin (**c**).

Conversely, a more intensive peak at 7 ppm was found in reference lignin due to the higher guaicyl content of reference lignin than isolated lignin. This trend agrees with the FTIR and Pyr-GC/MS results, where isolated lignin has more syringyl units than reference lignin. The other strong signal in the range of 7.5–8.5 ppm reveals an aromatic presence in p-hydroxyphenyl proton positions 2 and 6 [38]. The obtained signal in the NMR spectra of isolated lignin was similar to lignin from sweet sorghum stem (SST) [34] and lignin kraft [8]. Nevertheless, the peak spectrum was slightly different with lignin from Ginkgo shells, where the Hα signals at 5.5–5.9 ppm related to linkages of β-O-4′ and β-5′ were produced [39]. The results may reveal that different sources of lignin generated different structures.

The ^{13}C NMR spectra (Figure 5) shows a unique signal related to the lignins in this study. All the lignin samples show a similar trend of peaks where five typical lignin signals show strong resonance such as aliphatic chains, solvent (DMSO), methoxy, C3/C5, and ester. The signal in region between 20–30 ppm represents an aliphatic chain structure in lignin where isolated lignin (b–c) has a stronger signal than commercial lignin (a), which corresponds with FTIR spectra (2918 and 2854 cm^{-1}) and ^1H NMR (0.8 and 1.23 ppm). Meanwhile, the strong peak at 40 ppm belongs to DMSO as a solvent. The intensive signal at 55.9 ppm is attributed to the methoxy group in the G and S units. A signal related to esterified syringyl unit in C3/C5 is observed at 152 ppm in isolated lignin (b–c) but is not detected in reference lignin (a). Repeatedly, this result agrees with FTIR, Py-GC/MS, and ^1H NMR. A strong signal at a 170–160 ppm range implies ester linkage (-COO^-) at γ position [34].

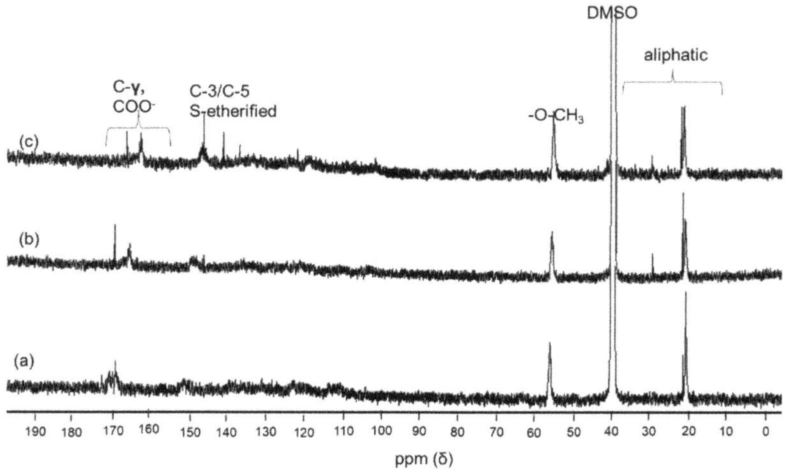

Figure 5. ^{13}C NMR spectra of standard lignin (**a**), single-step lignin (**b**), and fractionated-step lignin (**c**).

In general, the obtained signal in this study was also detected in lignin from sweet sorghum stem SST [34] and lignin kraft from the industrial residue [32]. However, the assigned peak correlated to the G unit is not detected in this current spectrum. it is likely that this is caused by either the lignin concentration being too dilute or the presence of the Nuclear Overhauser Effect (NOE). According to a report by Wen et al. [40], an important aspect of lignin characterization fulfills three criteria. First, lignin should be free from impurities; for this case, acetylation improved the purity of lignin, which was proved by residual carbohydrate's absence signal at 62, 73–75, and 100–102 ppm [25]. Second, lignin solution must be concentrated to minimize baselined phasing distortion and increase the signal-to-noise ratio, yet this requirement negatively affects the LC column. Third, to avoid the NOE, the inverse-gated decoupling sequence (i.e., C13IG pulse) should be utilized, which entails turning off the proton decoupling during the recovery between pulses [40].

3.5. Thermal Behavior of Lignin

Mass loss (TG) and mass loss rate (DTG) curves of *A. mangium* lignin are shown in Figure 6a to indicate the similar thermal characteristics between isolated lignin and reference lignin. The reaction region of all stages shifts toward a higher temperature by increasing the heating rate for isolated and reference lignin. The primary loss stage of two isolated lignins and their reference was located in a broad temperature range (between 100 °C and 700 °C), representing a complex structure consisting of phenolic hydroxyl, carbonyl benzylic hydroxyl functionalities [41]. The decomposition of lignin by temperature can be divided into three stages. The initial pyrolysis stage at around 100 °C with a higher mass loss was represented by the fractionated-step lignin. This first stage, up to 200 °C, is mainly attributed to the moisture evaporation in lignin and releasing of volatile products such as carbon dioxide and carbon monoxide [30]. A similar degradation study of lignin reported that an endothermic peak ranges from 100–180 °C, corresponding to the elimination of humidity [42].

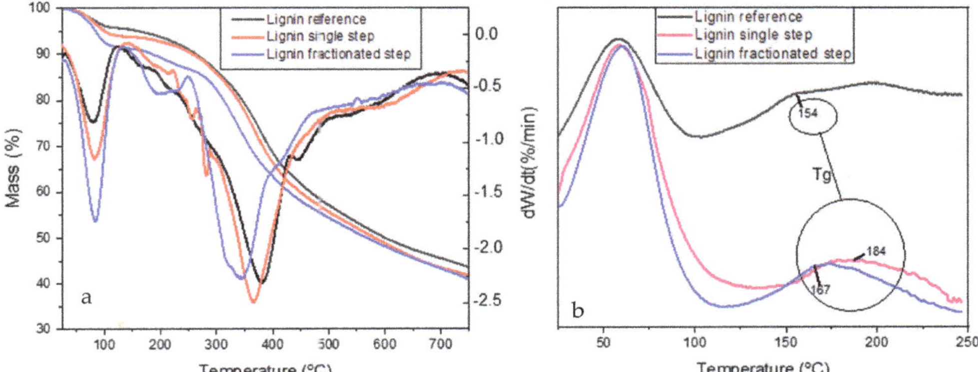

Figure 6. TGA (**a**) and DSC (**b**) thermogram of lignin standard, single-step lignin, and fractionated-step lignin.

The second pyrolysis stage, between 120 °C and 270 °C, indicated the decomposition of lignin into some possible degradation products and the removal of carbohydrates from lignin. The peak of the stage was around 200 °C, and below this peak, lignin is thermally stable. The losses in this stage were derived by aromatic decomposition as a phenolic compounds, such as the cleavage of ether linkages among the C9 units [43]. Thermal degradation of lignin is followed by condensation processes, leading to unsaturated C=C bonds occurring in the temperature range of 160 to 270 °C. Afterward, the production of vinyl guaiacol, ethyl, and methyl byproducts is usually obtained at 230 and 260 °C with the degradation of the propanoic side chains of lignin [41].

The third pyrolysis stage had a temperature range of 270 °C to 700 °C, with the prominent peak being around 350 °C and 380 °C for the fractionated-step lignin and single-step, respectively. The reference lignin reached the highest peak, indicating it as the most stable. The lignin structure is decomposed majorly at a temperature of 260–478 °C. At temperatures below 310 °C, aryl ether links tend to cleave, caused by low thermal stability [44]. At higher temperatures (>500 °C), aromatic structures rearranged and condensed the lead into char [44,45] and released volatile products. The high capacity to produce char by lignin makes it an efficient alternative to improve the flame retardancy of polymers [7]. Similar tendencies were also observed in other studies concerning the mass loss and the evolution of the volatiles against the origin and pyrolysis temperature [46]. Hu et al. [47] studied the isolated lignins extracted by different solvents and reported that CH_4, CO, and phenols are lignin's main mass loss stage. Based on TGA analysis in the third stage, the fractionated-step lignin was decomposed at a lower temperature with moderate mass loss than the single-step lignin due to the extraction method used. This result indicates that fractionated-step lignin yields better purity than single-step lignin. The mass loss in this third stage was remarkable over in the second stage for isolated lignin and reference lignin, attributed to volatiles' intensive evolution in the third stage [43].

The glass transition temperature (T_g) of the lignin fractionated-step and single-step was higher than that of reference lignin, with T_g temperatures of 184, 167, and 154 °C, respectively (Figure 6b). Furthermore, the first peak at 50 °C indicated an endothermic process in which the lignin absorbed heat energy to evaporate water and other volatile substances [48]. T_g represents the end of an endothermic process in which the lignin structure changes from a glassy state into a rubbery (plasticized) state. The wide range of T_g values indicated the flexibility and stiffness at higher temperatures, beneficial in industrial applications [35]. The T_g value varies widely depending on the method of lignin isolation, adsorbed water, molecular weight, and thermal history [49]. Since single-step lignins with higher T_g values are more stable at high temperatures, the process requires higher temperature operation.

The higher Tg value of isolated lignin compared to reference lignin was due to the higher amount of phOH content in the isolated lignin (Table 1). Intramolecular hydrogen bonds between phOH groups in the main back bonds of lignin contributed to the higher T_g. The bonds created a physically cross-linked structure [50]. The T_g value is influenced by the solubility of the organic solvents, where higher solubility is obtained with a lower T_g value [51,52]. The reference lignin had a higher absorbance in dioxane, the organic solvent, than isolated lignin, as presented in Figure 1b (solubility lignin in dioxane). This finding is supported by the lower T_g value of reference lignin than the isolated lignin and agrees with the results reported by Dastpak et al. [52] where Kraft lignin had higher T_g and lower solubility in the organic solvent than organosolv lignin. The T_g value corresponds positively with the molecular mass of lignin [53–55]. The T_g shifts to higher temperatures by increasing the average molar mass [56]. Based on the T_g value, the fractionated-step lignin had a lower molecular mass than the single-step lignin. This might be caused by the more extended process obtained by using acid precipitation to condense lignin, continued by ethanol addition to adsorbing the carbohydrate attached to lignin during the kraft pulping process. This suggestion was supported by GPC analysis in the next section.

Impurities influenced the T_g value in the lignin sample represented by ash content. The fractionated-step lignin and reference lignin had higher ash content (1.96% and 2.58%, respectively) compared to single-step lignin (0.48%), resulting in a lower T_g value. Sameni et al. (2013) also reported a higher percentage of impurities obtained with a lower T_g value. The abundance of aromatic rings in the main backbone of lignin can also contribute to the higher T_g value of isolated lignin. The varied T_g values were due to the heterogeneous structures and the broad molecular weight of isolated lignin samples [57]. These factors also affect interchain hydrogen bonding, cross-linking density, and rigid phenyl groups [58]. Although several studies reported an increase of char residue with higher Tg values, the results may be inconsistent due to the plant sources and extraction conditions [35]. The two isolated lignins from *A. mangium* in this study had higher T_g values in comparison to the others hardwood lignin, for instance T_g value from *Eucalyptus Grandis* was 161 °C [59], while eucalyptus kraft lignin was 133 °C [60], and other hardwood kraft lignin showed values such as 108 °C [61] and 138 °C [62]. pH solution conditions also influenced the T_g during lignin precipitation that varied from 106.12–131.81 °C at pH 1–5 [55].

3.6. Chemical Elucidation by Mass Spectrometry

Py-GCMS helps determine the lignin degradation and the existence of carbohydrates and other additives [63]. The PyGCMS method can help unravel the nature of lignins elemental composition, number of formed products, and the isolation method [64]. The monomer unit in hardwood lignin consists of syringyl (S) and guaiacyl (G) units with varying proportions, while softwood is dominated by a high proportion of G units with less p-hydroxyl phenyl (H) units [65]. In native wood, higher S units are easier to delignify than lower S units contributed by the erythro-rich and S unit rich in β-O-4 structure [66]. Some kinds of pyrolysis products can be found in pyrogram (Figure 7)-derived S unit, G unit, and H unit with different relative abundance between lignin samples. The higher retention time indicated that the compound was degraded at a higher temperature. At a pyrolysis temperature of 350 °C, phenolic compounds such as eugenol (G6), aldehydes, ketones, or alcohol group from G- and S-unit were released. Other compounds such as vanillin (G7) and acetoguaiacone from G-unit were degraded at 200–400 °C, in which the β-ether was separated [67].

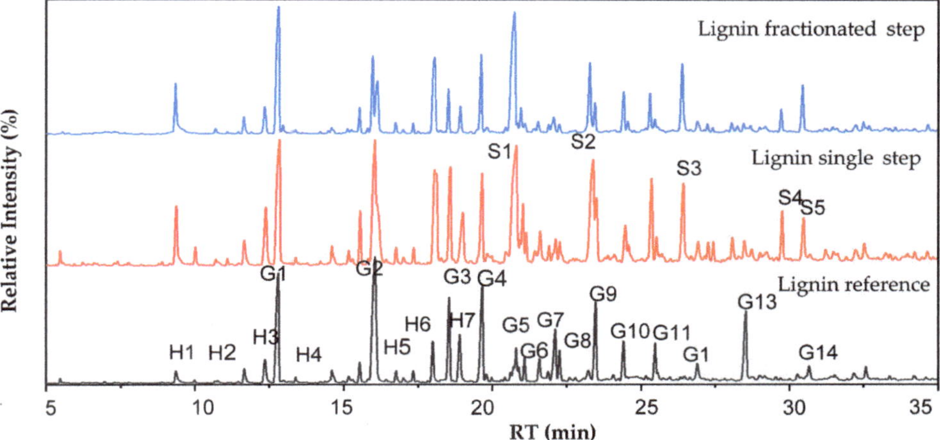

Figure 7. Pyrogram of single-step lignin and fractionated-step lignin compared to reference lignin.

Table 3 shows pyrolysis products of reference, single-step, and fractionated-step lignin. The compounds detected by Py-GCMS are classified into five categories: aliphatic oxygen compounds and hydrocarbons, aromatic hydrocarbons, furan and phenol derivatives [68]. Based on pyrolysis products, single-step lignin had the highest total relative abundance of H unit followed by fractionated-step lignin and reference lignin. G unit presented the highest portion compared to the S and H unit in both isolated and reference lignin. The result indicated that the primarily pyrolysis composition of *A. mangium* kraft lignin from kraft black liquor of industrial pulp and paper was an S unit. This result is similar to wheat straw and pine sawdust lignin, mainly G unit. However, a different finding was reported in palm kernel shell (PKS) lignin, which was dominated by the H unit [69]. The classification of reference lignin was G-lignin, while single-step and fractionated-step lignin are classified as SGH-lignin. The total relative abundance of H and G units of single-step lignin was higher than that of fractionated-step. Inversely, the total relative abundance of the S unit of the single-step lignin was lower than fractionated-step lignin. H and G units are easier formed in terms of the possibility of forming condensation in the acid precipitation process. Based on this classification, the biphenyl bond of single-step lignin is relatively higher than fractionated-step lignin. The H unit does not have a methoxy group, and the G unit only has one methoxy group, leading to the formation of C-C biphenyl linkages. Thus, there is a higher yield of single-step lignin compared to fractionated-step lignin. The biphenyl linkages are included as covalent linkages, which are relatively more stable than β-O-4 ether linkages [68], so single-step lignin is relatively more stable and resistant to thermal treatment and biodegradation. The fractionated-step lignin had a total relative abundance of S unit higher than the single-step lignin, which showed that the fractionated-step lignin had more β-O-4 ether linkages and was more reactive than the single-step lignin.

Table 3. The list of pyrolysis product reference lignin, single-step lignin, and fractionated-step lignin.

Unit	Pyrolysis Product	Relative Abundance (%)			Fragmentation (m/z)
		Reference Lignin	Single-Step Lignin	Fractionated-Step Lignin	
H1	Phenol	1.02	2.76	3.58	94, 66, 45
H2	Phenol, 2-methyl-	1.18	1.39	1.32	108, 90, 79
H3	Phenol, 3 + 4-methyll	2.24	3.25	2.26	107, 90, 79
H4	Phenol, 2,4-dimethyl-	1.00	0.40	0.00	122, 107, 77
H5	Phenol, 4-vinyl	0.20	0.48	0.66	120, 91, 65, 40
H6	Catechol, 3-methyl	3.63	9.70	9.66	124, 78
H7	Catechol, 4-methyl	3.90	3.81	1.99	124, 78
Total relative abundance of H unit		**13.17**	**21.80**	**19.47**	
G1	Guaiacol	12.36	9.99	15.23	124, 109, 81
G2	Guaiacol-4-methyl-	18.62	13.49	5.96	138, 123, 95
G3	Guaiacol, 4-ethyl	7.02	5.19	3.13	152, 137
G4	Guaiacol, 4-vinyl	10.59	4.87	6.22	150, 135, 107, 77
G5	Guaiacol, 4-propyl	1.70	3.45	0.00	166, 137
G6	Eugenol	1.08	1.25	0.33	164, 149, 77
G7	Vanillin	5.57	1.36	1.80	151, 123, 109
G8	Isoeugenol (cis)	2.33	0.50	2.50	164, 149
G9	Isoeugenol (trans)	7.08	3.39	1.98	164, 149
G10	Acetoguaiacone	3.15	2.06	3.39	166, 151, 123
G11	Guaiacyl acetone	3.47	0.62	0.48	180, 137
G12	Propioguaiacone	1.70	0.00	1.23	180, 151, 123
G13	Dihydroconiferyl alcohol	7.30	0.00	0.00	182, 137
G14	Coniferyl alcohol	1.95	0.00	0.00	180, 137, 124, 91
Total relative abundance of G unit		**83.93**	**46.17**	**42.26**	
S1	Syringol	1.70	13.70	19.41	154, 139, 111, 96
S2	Syringol, 4-methyl	0.96	9.47	6.45	168, 153, 125
S3	Syringol, 4-vinyl	0.14	4.79	6.68	180, 165, 137
S4	Syringol, 4-propenyl (trans)	0.11	2.32	1.62	194, 179, 91
S5	Acetosyringone	0.00	2.24	4.12	196, 181, 153
Total relative abundance of S unit		**2.90**	**32.52**	**34.15**	
S/G ratio		**0.03**	**0.70**	**0.81**	

3.7. Molecular Weights

The reactivity and physicochemical property are indicated by critical parameters such as molecular mass. The GPC curve in Figure 7 depicts the molecular weight of number-average (Mn), weight-average (Mw), and polydispersity index (PDI, Mw/Mn) of reference and isolated lignin from BL. The obtained molar mass distribution of lignin in this study is in the range of the Kraft lignin in a THF-based system reported by Baumberger et al. [70] (Mn = 200–2000 Da, Mw = 1500–50,000 Da). The number of Mw, Mn, and PDI depends on the isolation method, biomass source, and purification [71]. Figure 8 also shows that the trend of both Mw and Mn is reference lignin > single-step lignin > fractionated-step lignin. Markedly, the trend was similar to AIL content and lignin solubility in an organic solvent. However, the opposite trend is seen against the S/G ratio and total phOH. Reference lignin has a higher molecular mass, AIL content, and solubility in an organic solvent, yet it has a lower S/G ratio, Tg value, and total phOH than isolated lignin. This result contrasts with Gordobil et al. (2018), where a positive correlation between molecular weight, S/G ratio, and total phOH was observed. Different analytical methods may affect the S/G ratio and total phOH, ^{31}P NMR vs Py-GCMS (S/G ratio) and UV-Vis (total phOH). Higher AIL content resulted in high molecular weight and ASL content due to different governing mechanisms of cleavage bonds and functional groups in lignin. This finding was similar to the results reported by Stiefel et al. [72], where the insoluble acid is slightly correlated (r^2 = 0.739) with the molecular weight in lignin from different treatments [72].

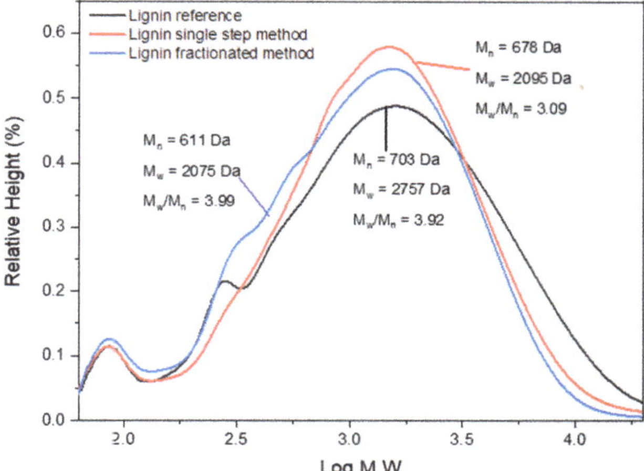

Figure 8. GPC curve including number-average (Mn), weight-average (Mw), and polydispersity index (PDI, Mw/Mn) of reference lignin, single-step lignin, and fractionated-step lignin.

The higher molecular weight of reference lignins was attributed to a higher percentage of the G unit [73]. This finding substantiated that reference lignins are derived from softwood. Higher PDI of fractionated-step lignin indicated wider molecular weight distribution as well as the existence of the impurities that positively correlated to ash content (Table 1) [52].

3.8. Future Potential of A. mangium Lignin from BL in Adhesive Applications for Wood-Based Composites

The structural features of *A. mangium* lignin extracted from BL exhibited a strong link to many alternative ways in its possible applications, according to the findings of this study. The fingerprint result (UATR-FTIR, ^{1}H, and ^{13}C NMR) and elucidation structure by Py-GCMS showed a higher abundance of G-unit which is the most active unit in phenolic resin polymerization. Besides, the high Tg value of isolated lignin is suitable for wood adhesive applications. The result was similar to lignin from coconut husk that was examined by Abd Latif et al. [74] as an alternative material for lignin-phenol-glyoxal adhesives. Markedly, high MW indicated higher content of aromatic protons which has a better chance of polymerization. Hence, lignin from the single-step method would be more suitable for wood adhesive applications. Another consideration is the presence of large amounts of phenolic hydroxyls in the isolated lignin structure which means making them reactive to create linkage with aldehyde [18].

4. Conclusions

This study investigated the chemical and physical properties of lignin derived from pulp mill factory residue (*Acacia mangium*) using diverse techniques. Lignin was successfully isolated through single-step and fractionated-step dilute acid precipitation. According to fingerprint analysis by FTIR, ^{1}H, and ^{13}C NMR, unique lignin peaks such as aromatic unit guaiacyl (G) and syringyl (S) were observed. The results were confirmed by the commercial lignin used as a reference. Dilute hydrochloric acid obtained high acid-soluble lignin (ASL) content. In contrast, fractionated-step lignin had lower lignin content than single-step. Still, it had a linear correlation against total phenolic hydroxyl (phOH) content, thermal stability, G-unit, and molecular weight distribution. More condensed G-unit in single-step lignin induced higher molecular weight distribution (Mw and Mn) and Tg value and total phOH. Single-step precipitation obtained the highest lignin yield, ~35.39 %. Comprehensive analysis of technical lignin aided in gathering knowledge of the structure and properties of lignin in suggesting better valorization strategies and enhanced future potential for wider industrial application of lignin as a renewable raw material.

Author Contributions: Conceptualization, N.N.S., E.B.S. and W.F.; methodology, N.N.S., E.B.S., W.S. and W.F.; formal analysis, N.N.S., W.F., A.K., E.W.M. and A.H.I.; investigation, N.N.S., E.B.S., F.F., F.P.S., M.I. and M.A.R.L.; resources, N.N.S. and W.F.; data curation, N.N.S., E.B.S., M.I. and M.A.R.L.; writing—original draft preparation, review and editing, N.N.S., W.F., P.A., V.S. and M.G.; supervision, N.N.S., W.S. and W.F.; project administration, N.N.S. All authors have read and agreed to the published version of the manuscript.

Funding: This research was supported by project from Research Organization for Engineering Science, National Research and Innovation Agency (BRIN) Indonesia No. 26/A/DT/2021 "Valorization of Black Liquor from Pulp Mill by Product as Antimicrobial Agent for Textile". This research was also supported by project No. НИС-Б-1145/04.2021 "Development, Properties, and Application of Eco-Friendly Wood-Based Composites" carried out at the University of Forestry, Sofia, Bulgaria. This research was also supported by the Slovak Research and Development Agency under contracts No. APVV-18-0378 and APVV-19-0269.

Institutional Review Board Statement: Not applicable.

Informed Consent Statement: Not applicable.

Data Availability Statement: The data presented in this study are available on request from the corresponding author.

Acknowledgments: Authors acknowledge the Research Organization for Engineering Science National Research and Innovation Agency (BRIN) for their funding support of this project. The authors also wish to thank Advanced Characterization Laboratories Cibinong—Integrated Laboratory of Bioproduct, National Research and Innovation Agency through E- Layanan Sains, Badan Riset dan Inovasi Nasional for the facilities, scientific and technical support form.

Conflicts of Interest: The authors declare no conflict of interest.

References

1. IDNFinancial. Indonesia's Pulp and Paper Are in the Top 10 in the World. Available online: https://www.idnfinancials.com/archive/news/22291/Indonesias-pulp-and-paper-are-in-the-top-10-in-the-world (accessed on 25 October 2021).
2. Bajpai, P. Chapter 12—Pulping Fundamentals. In *Biermann's Handbook of Pulp and Paper*, 3rd ed.; Bajpai, P., Ed.; Elsevier: Amsterdam, The Netherlands, 2018; pp. 295–351.
3. Speight, J.G. Chapter 13—Upgrading by Gasification. In *Heavy Oil Recovery and Upgrading*; Speight, J.G., Ed.; Gulf Professional Publishing: Houston, TX, USA, 2019; pp. 559–614.
4. Melro, E.; Filipe, A.; Sousa, D.; Medronho, B.; Romano, A. Revisiting lignin: A tour through its structural features, characterization methods and applications. *New J. Chem.* **2021**, *45*, 6986–7013. [CrossRef]
5. Vishtal, A.; Kraslawski, A. Challenges in industrial applications of technical lignins. *BioRes* **2011**, *6*, 3547–3568. [CrossRef]
6. Solihat, N.N.; Sari, F.P.; Falah, F.; Ismayati, M.; Lubis, M.A.R.; Fatriasari, W.; Santoso, E.B.; Syafii, W. Lignin as an Active Biomaterial: A Review. *J. Sylva Lestari* **2021**, *9*, 1–22. [CrossRef]
7. Vahabi, H.; Brosse, N.; Latif, N.A.; Fatriasari, W.; Solihat, N.; Hashimd, R.; Hussin, M.; Laoutid, F.; Saeb, M. Chapter 24-Nanolignin in materials science and technology-does flame retardancy matter? In *Biopolymeric Nanomaterials: Fundamental and Applications*; Kanwar, S., Kumar, A., Nguyen, T.A., Sharma, S., Slimani, Y., Eds.; Elsevier: Amsterdam, The Netherlands, 2022; Volume 1; pp. 515–550.
8. Sameni, J.; Krigstin, S.; Sain, M. Characterization of lignins isolated from industrial residues and their beneficial uses. *BioRes* **2016**, *11*, 8435–8456. [CrossRef]
9. Ház, A.; Jablonský, M.; Šurina, I.; Kačík, F.; Bubeníková, T.; Ďurkovič, J. Chemical Composition and Thermal Behavior of Kraft Lignins. *Forests* **2019**, *10*, 483. [CrossRef]
10. Haz, A.; Strizincova, P.; Majova, V.; Sskulcova, A.; Surina, I.; Jablonsky, M. Content of Phenolic Hydroxyl Groups In Lignin Characterisation of 23 Isolated Non-Wood Lignin With Various Acids. *Int. J. Recent Sci. Res.* **2016**, *7*, 11547–11551.
11. Handika, S.O.; Lubis, M.A.R.; Sari, R.K.; Laksana, R.P.B.; Antov, P.; Savov, V.; Gajtanska, M.; Iswanto, A.H. Enhancing Thermal and Mechanical Properties of Ramie Fiber via Impregnation by Lignin-Based Polyurethane Resin. *Materials* **2021**, *14*, 6850. [CrossRef]

12. Ang, A.F.; Ashaari, Z.; Lee, S.H.; Md Tahir, P.; Halis, R. Lignin-based copolymer adhesives for composite wood panels—A review. *Int. J. Adhes. Adhes.* **2019**, *95*, 102408. [CrossRef]
13. Çetin, N.S.; Özmen, N. Use of organosolv lignin in phenol-formaldehyde resins for particleboard production II. Particleboard production and properties. *Int. J. Adhes. Adhes.* **2002**, *22*, 477–480. [CrossRef]
14. Donmez Cavdar, A.; Kalaycioglu, H.; Hiziroglu, S. Some of the properties of oriented strandboard manufactured using kraft lignin phenolic resin. *J. Mater. Processing Technol.* **2008**, *202*, 559–563. [CrossRef]
15. Madyaratri, E.W.; Ridho, M.R.; Aristri, M.A.; Lubis, M.A.R.; Iswanto, A.H.; Nawawi, D.S.; Antov, P.; Kristak, L.; Majlingová, A.; Fatriasari, W. Recent Advances in the Development of Fire-Resistant Biocomposites—A Review. *Polymers* **2022**, *14*, 362. [CrossRef]
16. Ahmad Saffian, H.; Hyun-Joong, K.; Md Tahir, P.; Ibrahim, N.A.; Lee, S.H.; Lee, C.H. Effect of Lignin Modification on Properties of Kenaf Core Fiber Reinforced Poly(Butylene Succinate) Biocomposites. *Materials* **2019**, *12*, 4043. [CrossRef] [PubMed]
17. Ahmad Saffian, H.; Talib, M.A.; Lee, S.H.; Md Tahir, P.; Lee, C.H.; Ariffin, H.; Asa'ari, A.Z.M. Mechanical Strength, Thermal Conductivity and Electrical Breakdown of Kenaf Core Fiber/Lignin/Polypropylene Biocomposite. *Polymers* **2020**, *12*, 1833. [CrossRef]
18. Tejado, A.; Pena, C.; Labidi, J.; Echeverria, J.M.; Mondragon, I. Physico-chemical characterization of lignins from different sources for use in phenol-formaldehyde resin synthesis. *Bioresour. Technol.* **2007**, *98*, 1655–1663. [CrossRef] [PubMed]
19. Zinovyev, G.; Sumerskii, I.; Korntner, P.; Sulaeva, I.; Rosenau, T.; Potthast, A. Molar mass-dependent profiles of functional groups and carbohydrates in kraft lignin. *J. Wood Chem. Technol.* **2016**, *37*, 171–183. [CrossRef]
20. Hermiati, E.; Risanto, L.; Lubis, M.A.R.; Laksana, R.P.B.; Dewi, A.R. Chemical characterization of lignin from kraft pulping black liquor of *Acacia mangium*. *AIP Conf. Proc.* **2017**, 020005. [CrossRef]
21. TAPPI. TAPPI Test Method T 264 cm-97 Preparation of Wood for Chemical Analysis. 1997. Available online: https://tappi.micronexx.com/CD/TESTMETHODS/T264.pdf (accessed on 26 December 2021).
22. TAPPI. TAPPI Test Method T 211 om-02 Ash in Wood, Pulp, Paper and Paperboard: Combustion at 525 °C. 2002. Available online: https://www.tappi.org/content/sarg/t211.pdf (accessed on 26 December 2021).
23. Sluiter, A.; Hames, B.; Ruiz, R.; Scarlata, C.; Sluiter, J.; Templeton, D.; Crocker, D. *Determination of Structural Carbohydrates and Lignin in Biomass—Laboratory Analytical Procedure (LAP)*; Technical Report NREL/TP-510-42618; National Renewable Energy Laboratory: Golden, CO, USA, 2012.
24. Serrano, L.; Esakkimuthu, E.S.; Marlin, N.; Brochier-Salon, M.-C.; Mortha, G.; Bertaud, F. Fast, Easy, and Economical Quantification of Lignin Phenolic Hydroxyl Groups: Comparison with Classical Techniques. *Energy Fuels* **2018**, *32*, 5969–5977. [CrossRef]
25. Wen, J.-L.; Sun, S.-L.; Xue, B.-L.; Sun, R.-C. Quantitative structural characterization of the lignins from the stem and pith of bamboo (Phyllostachys pubescens). *Holzforschung* **2013**, *67*, 613–627. [CrossRef]
26. Hamzah, M.H.; Bowra, S.; Cox, P. Effects of Ethanol Concentration on Organosolv Lignin Precipitation and Aggregation from *Miscanthus* x *giganteus*. *Processes* **2020**, *8*, 845. [CrossRef]
27. Ponnuchamy, V.; Gordobil, O.; Diaz, R.H.; Sandak, A.; Sandak, J. Fractionation of lignin using organic solvents: A combined experimental and theoretical study. *Int. J. Biol. Macromol.* **2021**, *168*, 792–805. [CrossRef]
28. Gordobil, O.; Herrera, R.; Yahyaoui, M.; İlk, S.; Kaya, M.; Labidi, J. Potential use of kraft and organosolv lignins as a natural additive for healthcare products. *RSC Adv.* **2018**, *8*, 24525–24533. [CrossRef]
29. Sadeghifar, H.; Argyropoulos, D.S. Correlations of the Antioxidant Properties of Softwood Kraft Lignin Fractions with the Thermal Stability of Its Blends with Polyethylene. *ACS Sustain. Chem. Eng.* **2015**, *3*, 349–356. [CrossRef]
30. Alzagameem, A.; Khaldi-Hansen, B.E.; Buchner, D.; Larkins, M.; Kamm, B.; Witzleben, S.; Schulze, M. Lignocellulosic Biomass as Source for Lignin-Based Environmentally Benign Antioxidants. *Molecules* **2018**, *23*, 2664. [CrossRef] [PubMed]
31. Ammar, M.; Mechi, N.; Slimi, H.; Elaloui, E. Isolation and Purification of Alfa Grass Kraft Lignin from Industrial Waste. *Curr. Trends Biomed. Eng. Biosci.* **2017**, *6*, 31–35. [CrossRef]
32. Wang, K.; Xu, F.; Sun, R. Molecular characteristics of Kraft-AQ pulping lignin fractionated by sequential organic solvent extraction. *Int. J. Mol. Sci.* **2010**, *11*, 2988–3001. [CrossRef]
33. Abdelaziz, O.Y.; Hulteberg, C.P. Physicochemical Characterisation of Technical Lignins for Their Potential Valorisation. *Waste Biomass Valorization* **2016**, *8*, 859–869. [CrossRef]
34. She, D.; Xu, F.; Geng, Z.; Sun, R.; Jones, G.L.; Baird, M.S. Physicochemical characterization of extracted lignin from sweet sorghum stem. *Ind. Crops Prod.* **2010**, *32*, 21–28. [CrossRef]
35. Sameni, J.; Krigstin, S.; Santos Rosa, D.D.; Leao, A.; Sain, M. Thermal Characteristics of Lignin Residue from Industrial Processes. *BioResources* **2013**, *9*, 725–737. [CrossRef]
36. Solihat, N.N.; Sari, F.P.; Risanto, L.; Anita, S.H.; Fitria, F.; Fatriasari, W.; Hermiati, E. Disruption of Oil Palm Empty Fruit Bunches by Microwave-assisted Oxalic Acid Pretreatment. *J. Math. Fund. Sci.* **2017**, *49*, 244–257. [CrossRef]
37. Solihat, N.N.; Fajriutami, T.; Adi, D.T.N.; Fatriasari, W.; Hermiati, E. Reducing sugar production of sweet sorghum bagasse kraft pulp. *AIP Conf. Proc.* **2017**, *1803*, 020012 [CrossRef]
38. Lu, Y.; Lu, Y.-C.; Hu, H.-Q.; Xie, F.-J.; Wei, X.-Y.; Fan, X. Structural Characterization of Lignin and Its Degradation Products with Spectroscopic Methods. *J. Spectrosc.* **2017**, *2017*, 1–15. [CrossRef]
39. Jiang, B.; Zhang, Y.; Guo, T.; Zhao, H.; Jin, Y. Structural Characterization of Lignin and Lignin-Carbohydrate Complex (LCC) from Ginkgo Shells (*Ginkgo biloba* L.) by Comprehensive NMR Spectroscopy. *Polymers* **2018**, *10*, 736. [CrossRef] [PubMed]

40. Wen, J.L.; Sun, S.L.; Xue, B.L.; Sun, R.C. Recent Advances in Characterization of Lignin Polymer by Solution-State Nuclear Magnetic Resonance (NMR) Methodology. *Materials* **2013**, *6*, 359. [CrossRef] [PubMed]
41. Hansen, B.; Kusch, P.; Schulze, M.; Kamm, B. Qualitative and Quantitative Analysis of Lignin Produced from Beech Wood by Different Conditions of the Organosolv Process. *J. Polym. Environ.* **2016**, *24*, 85–97. [CrossRef]
42. Brebu, M.; Vasile, C. Thermal degradation of lignin—A Review. *Cellul. Chem. Technol.* **2010**, *44*, 353–363.
43. Zhao, J.; Xiuwen, W.; Hu, J.; Liu, Q.; Shen, D.; Xiao, R. Thermal degradation of softwood lignin and hardwood lignin by TG-FTIR and Py-GC/MS. *Polym. Degrad. Stab.* **2014**, *108*, 133–138. [CrossRef]
44. Ramezani, N.; Sain, M. Thermal and Physiochemical Characterization of Lignin Extracted from Wheat Straw by Organosolv Process. *J. Polym. Environ.* **2018**, *26*, 3109–3116. [CrossRef]
45. Liu, S.-M.; Huang, J.-Y.; Jiang, Z.-J.; Zhang, C.; Zhao, J.-Q.; Chen, J. Flame retardance and mechanical properties of a polyamide 6/polyethylene/surface-modified metal hydroxide ternary composite via a master-batch method. *J. Appl. Polym. Sci.* **2010**, 3370–3378. [CrossRef]
46. Pourjafar, S. An Investigation of the Thermal Degradation of Lignin. Ph.D. Thesis, University of North Dakota, Grand Forks, ND, USA, 2017.
47. Hu, J.; Shen, D.; Xiao, R.; Wu, S.; Zhang, H. Free-Radical Analysis on Thermochemical Transformation of Lignin to Phenolic Compounds. *Energy Fuels* **2013**, *27*, 285–293. [CrossRef]
48. Aristri, M.A.; Lubis, M.A.R.; Laksana, R.P.B.; Falah, F.; Fatriasari, W.; Ismayati, M.; Wulandari, A.P.; Nurindah, N.; Ridho, M.R. Bio-Polyurethane Resins Derived from Liquid Fractions of Lignin for the Modification of Ramie Fibers. *J. Sylva Lestari* **2021**, *9*, 223–238. [CrossRef]
49. Erdtman, H. Lignins: Occurrence, formation, structure and reactions, K.V. Sarkanen and C.H. Ludwig, Eds., John Wiley & Sons, Inc., New York, 1971. 916 pp. $35.00. *J. Polym. Sci. Part B Polym. Lett.* **1972**, *10*, 228–230. [CrossRef]
50. Naseem, A.; Tabasum, S.; Zia, K.M.; Zuber, M.; Ali, M.; Noreen, A. Lignin-derivatives based polymers, blends and composites: A review. *Int. J. Biol. Macromol.* **2016**, *93*, 296–313. [CrossRef] [PubMed]
51. Lora, J.H.; Glasser, W.G. Recent Industrial Applications of Lignin A Sustainable Alternative to Nonrenewable Materials. *J. Polym. Environ.* **2002**, *10*, 39–48. [CrossRef]
52. Dastpak, A.; Lourençon, T.V.; Balakshin, M.; Farhan Hashmi, S.; Lundström, M.; Wilson, B.P. Solubility study of lignin in industrial organic solvents and investigation of electrochemical properties of spray-coated solutions. *Ind. Crops Prod.* **2020**, *148*, 112310. [CrossRef]
53. Pang, T.; Wang, G.; Sun, H.; Sui, W.; Si, C. Lignin fractionation: Effective strategy to reduce molecule weight dependent heterogeneity for upgraded lignin valorization. *Ind. Crops Prod.* **2021**, *165*, 113442. [CrossRef]
54. Gregorova, A. *Application of Differential Scanning Calorimetry to the Characterization of Biopolymers*; IntechOpen: London, UK, 2013.
55. Sathawong, S.; Sridach, W.; Techato, K.-a. Lignin: Isolation and preparing the lignin based hydrogel. *J. Environ. Chem. Eng.* **2018**, *6*, 5879–5888. [CrossRef]
56. Laurichesse, S.; Avérous, L. Chemical modification of lignins: Towards biobased polymers. *Prog. Polym. Sci.* **2014**, *39*, 1266–1290. [CrossRef]
57. Jardim, J.M.; Hart, P.W.; Lucia, L.; Jameel, H. Insights into the Potential of Hardwood Kraft Lignin to Be a Green Platform Material for Emergence of the Biorefinery. *Polymers* **2020**, *12*, 1795. [CrossRef]
58. Heitner, C.; Dimmel, D.; Schmidt, J. *Lignin and Lignans: Advances in Chemistry*, 1st ed.; CRC Press, Taylor & Francis Group: Boca Raton, FL, USA, 2010.
59. Poletto, M. Assessment of the thermal behavior of lignins from softwood and hardwood species. *Maderas. Cienc. y Tecnol.* **2017**, *19*, 63–74. [CrossRef]
60. Torrezan, T. *Avaliação do Comportamento Reológico, Térmico e Mecânico de Misturas de PBAT com Elevados Teores de Lignina*; Universidade Federal de São Carlos: São Paulo, Brazil, 2019; Available online: https://repositorio.ufscar.br/handle/ufscar/12413 (accessed on 16 November 2021).
61. Kadla, J.F.; Kubo, S. Lignin-based polymer blends: Analysis of intermolecular interactions in lignin–synthetic polymer blends. *Compos. Part A Appl. Sci. Manuf.* **2004**, *35*, 395–400. [CrossRef]
62. Ropponen, J.; Räsänen, L.; Rovio, S.; Ohra-aho, T.; Liitiä, T.; Mikkonen, H.; van de Pas, D.; Tamminen, T. Solvent extraction as a means of preparing homogeneous lignin fractions. *Holzforschung* **2011**, *65*, 543–549. [CrossRef]
63. Ponomarenko, J.; Dizhbite, T.; Lauberts, M.; Viksna, A.; Dobele, G.; Bikovens, O.; Telysheva, G. Characterization of Softwood and Hardwood LignoBoost Kraft Lignins with Emphasis on their Antioxidant Activity. *BioResources* **2014**, *9*, 2051–2068. [CrossRef]
64. Haz, A.; Jablonský, M.; Orságová, A.; Surina, I. Characterization of lignins by Py-GC/MS. In Proceedings of the 4nd International Conference Renewable Energy Sources 2013, High Tatras, Slovak Republic, 21–23 May 2013.
65. Nawawi, D.; Rahayu, I.; Wistara, N.; Sari, R.; Syafii, W. Distribusi sel pori pada kayu tarik dan korelasinya dengan komposisi lignin. *J. Ilmu Kehutan.* **2019**, *13*, 70–76.
66. Nawawi, D.S.; Syafii, W.; Tomoda, I.; Uchida, Y.; Akiyama, T.; Yokoyama, T.; Matsumoto, Y. Characteristics and Reactivity of Lignin in Acacia and Eucalyptus Woods. *J. Wood Chem. Technol.* **2017**, *37*, 273–282. [CrossRef]
67. González Martínez, M.; Ohra-aho, T.; da Silva Perez, D.; Tamminen, T.; Dupont, C. Influence of step duration in fractionated Py-GC/MS of lignocellulosic biomass. *J. Anal. Appl. Pyrolysis* **2019**, *137*, 195–202. [CrossRef]

68. Wądrzyk, M.; Janus, R.; Lewandowski, M.; Magdziarz, A. On mechanism of lignin decomposition—Investigation using microscale techniques: Py-GC-MS, Py-FT-IR and TGA. *Renew. Energy* **2021**, *177*, 942–952. [CrossRef]
69. Chang, G.; Huang, Y.; Xie, J.; Yang, H.; Liu, H.; Yin, X.; Wu, C. The lignin pyrolysis composition and pyrolysis products of palm kernel shell, wheat straw, and pine sawdust. *Energy Convers. Manag.* **2016**, *124*, 587–597. [CrossRef]
70. Baumberger, S.; Abaecherli, A.; Fasching, M.; Gellerstedt, G.; Gosselink, R.; Hortling, B.; Li, J.; Saake, B.; Jong, E.d. Molar mass determination of lignins by size-exclusion chromatography: Towards standardisation of the method. *Holzforschung* **2007**, *61*, 459–468. [CrossRef]
71. Tolbert, A.; Akinosho, H.; Khunsupat, R.; Naskar, A.K.; Ragauskas, A.J. Characterization and analysis of the molecular weight of lignin for biorefining studies. *Biofuels Bioprod. Biorefining* **2014**, *8*, 836–856. [CrossRef]
72. Stiefel, S.; Marks, C.; Schmidt, T.; Hanisch, S.; Spalding, G.; Wessling, M. Overcoming lignin heterogeneity: Reliably characterizing the cleavage of technical lignin. *Green Chem.* **2016**, *18*, 531–540. [CrossRef]
73. Ahmad, Z.; Dajani, W.W.A.; Paleologou, M.; Xu, C.C. Sustainable Process for the Depolymerization/Oxidation of Softwood and Hardwood Kraft Lignins Using Hydrogen Peroxide under Ambient Conditions. *Molecules* **2020**, *25*, 2329. [CrossRef] [PubMed]
74. Abd Latif, N.H.; Brosse, N.; Ziegler-Devin, I.; Chrusiel, L.; Hashim, R.; Hussin, M.H. A Comparison of Alkaline and Organosolv Lignin Extraction Methods from Coconut Husks as an Alternative Material for Green Application. *BioRes* **2022**, *17*, 469–491. [CrossRef]

 *polymers*

Article

Novel In Situ Modification for Thermoplastic Starch Preparation based on *Arenga pinnata* Palm Starch

Muhammad Ghozali [1,2], Yenny Meliana [2] and Mochamad Chalid [1,*]

[1] Green Polymer Technology Group, Department of Metallurgical and Material Engineering, Faculty of Engineering, Universitas Indonesia, Depok 16424, Indonesia

[2] Research Center for Chemistry, National Research and Innovation Agency (BRIN), Tangerang Selatan 15314, Indonesia

* Correspondence: chalid@metal.ui.ac.id or m.chalid@ui.ac.id

Abstract: Thermoplastic starch (TPS) has three main disadvantages, i.e., poor mechanical properties, low thermal stability and water sensibility. To overcome these disadvantages, TPS properties can be improved by starch modification, adding reinforcements and blending with other polymers. In this research, to prepare modified TPS, starch modification was carried out by in situ modification. The modified TPS was prepared by adding *Arenga pinnata* palm starch (APPS), glycerol and benzoyl peroxide simultaneously in the twin-screw extruder. Morphology analysis of TPS revealed that the starch granules were damaged and gelatinized in the extrusion process. No phase separation is observed in TPS, which exhibits that starch granules with and without benzoyl peroxide were uniformly dispersed in the matrix. The addition of benzoyl peroxide resulted in increased density of TPS from 1.37 to 1.39 g·cm^{-3}, tensile strength from 7.19 to 8.61 MPa and viscosity from 2482.19 to 2604.60 Pa.s. However, it decreased the elongation at break of TPS from 33.95 to 30.16%, melt flow rate from 7.13 to 5.73 gr/10 min and glass transition temperature from 65 to 52 °C. In addition, the thermal analysis showed that the addition of benzoyl peroxide increased the thermal stability of TPS and extended the temperature range of thermal degradation.

Keywords: thermoplastic starch; *Arenga pinnata*; modification; benzoyl peroxide; twin-screw extruder

Citation: Ghozali, M.; Meliana, Y.; Chalid, M. Novel In Situ Modification for Thermoplastic Starch Preparation based on *Arenga pinnata* Palm Starch. *Polymers* **2022**, *14*, 4813. https://doi.org/10.3390/polym14224813

Academic Editor: Nathanael Guigo

Received: 12 October 2022
Accepted: 3 November 2022
Published: 9 November 2022

1. Introduction

Thermoplastic starch (TPS) is considered one of the most promising alternatives to fossil-based ones for disposable packaging material applications [1], mainly because of its low price, biodegradability [2] and renewability [1]. TPS can be obtained from native starch granules found in numerous plants, such as rice, corn, wheat, cassava, potato [1], or sago [2]. However, the use of those starches will intersect with food sources. Therefore, other starch sources are needed in order to avoid the debate and criticism regarding the use of food sources [3]. Some fruit wastes can be extracted and considered as alternative sources of starch, including kiwifruit, pineapple stems, mango kernels, apple pulp, banana peel, litchi, tamarind, longan and loquat, annatto, jackfruit and avocado seeds [4]. However, the availability of these fruit wastes is limited. *Arenga pinnata* palm starch (APPS) is also considered an agro-industrial residue in the agricultural industry [5]. *A. pinnata* palm starch can be obtained from the core of an unproductive (in terms of sugar and fruit) *A. pinnata* palm tree's trunk [6,7]. *A. pinnata* tree grows in more humid parts of subtropical and tropical areas [8]. It is widespread from South Asia to Southeast Asia and from the east of India [8] and Taiwan to Philippines, Indonesia, Papua New Guinea, India, North Australia, Malaysia, Thailand, Burma, Vietnam [3,7] and southwest of China [9]. Therefore, the abundant availability of APPS can be considered as a potential source of starch. A tree of *A. pinnata* can produce about 50–100 kg of APPS [8,10]. The starch content of APPS is approximately 10.5–36.7% [6] with 36.6 [6]–59.2% [8] amylose content. The density of APPS is around 1.54 g·cm^{-3} [3,11]. The gelatinization temperature of APPS was around

67 °C [6]. APPS was characterized by a C-type pattern crystalline structure [5,8]. In addition, *A. pinnata* palm starch has been used as a material for the preparation of TPS [5,9–11]. The density, tensile strength and elongation at break values of the TPS prepared from APPS are 1.41 g·cm^{-3}, 4.8 MPa and 38.10%, respectively [5], while other studies reported a density value of TPS prepared from APPS of 1.40 g·cm^{-3} [11], tensile strength and elongation at break of 2.42 MPa and 8.03%, respectively [10].

TPS can be prepared by starch into TPS [12] under heat and shear in the presence of plasticizers [13,14]. The starch granules are destructurized, plasticized and melted, forming that has similar characteristics to that of thermoplastics [13]. Generally, the preparation process of TPS can be divided into two processes, the wet and dry process. The wet process is commonly used in the laboratory by solvent casting [15–17]. Solvent casting is a batch process, so time consuming [15], with factors of the need to evaporate large amounts of solvents [16], low in efficiency, high in cost [18] and not suitable for industrial production, while the dry process, usually via an extrusion process, is a continuous process, efficient and more suitable for industrial production. Extrusion is the most widely used for plastic films because of its advantages, such as simple production equipment, low investment, continuous production [15], easier handling, a broad range of processing conditions, good mixing [17] and easy scale-up [13,18,19]. Thus, the extrusion process is a promising approach to producing bio-based plastics [16] for industrial production.

However, compared to conventional plastics, TPS has three main drawbacks, i.e., low mechanical properties, low thermal stability as well as water and humidity sensibility [13,17,20]. To overcome these drawbacks, several solutions have been studied, such as chemical modification of starch, mixing with other polymers and incorporation of reinforcing materials [2,12,13,18,21]. Cellulose is one of promising reinforcing materials for enhancing mechanical properties of TPS [13,18]. Digestate sludge from an agricultural biogas plant is also considered as a promising reinforcing material for improving the mechanical properties of TPS biocomposites [22], while the modification of starch is generally modified physically, chemically, enzymatically or by combinations. Usually, chemical modification occurs via etherification, esterification and an oxidation process. Preparation of starches by introducing functional groups shows its helpfulness in reducing TPS's hydrophilic performance and improving its compatibility. Starches are generally modified chemically to promote the hydrophobic, mechanical and thermal characteristics to increase TPS applications [21]. Oxidation is one chemical method for starch modification to obtain oxidized starch [23]. Oxidized starches also have the potential to be helpful in the preparation of biodegradable food packaging [21].

In the oxidized starch, the purpose of oxidation is to generate more functional groups, i.e., carbonyl and carboxyl, thereby increasing the functionality and reactivity of native starch. The preferred oxidants are hydrogen peroxide, sodium hypochlorite, potassium permanganate, chromic acid, nitrogen dioxide [21], ozone and sodium periodat [24]. The stage of oxidation in starch means the hydroxyl groups are oxidized to carbonyl groups first and then to carboxyl groups [24]. Oxidation reduced the relative crystallinity and viscosity of starches [21]. Starch oxidation improves moisture resistance with hydrophobic carbonyl groups, replacing the hydrophilic hydroxyl groups in starch, resulting in improved mechanical and thermal properties and lower humidity absorption compared to the TPS control film [17]. The use of oxidized starch improved the toughness, elongation at break, compatibility, thermal stability and rheological properties. However, it lowered the storage modulus and glass transition temperature of TPS [25].

Modifying starch by oxidation to obtain modified TPS has disadvantages, i.e., it requires time and cost and the process steps become longer because the starch is modified first and then the oxidized starch produced is used as raw material for TPS. The objective of this study is to overcome these disadvantages, in particular shortening the modified TPS preparation time. Therefore, in this study, starch modification was carried out via an in situ process simultaneously with the preparation of TPS with the extrusion process in the twin-screw extruder.

2. Materials and Methods

2.1. Materials

In this research, *Arenga pinnata* palm starch (APPS) with an amylose content of 23.19% was obtained from the local industry located at Klaten, Central Java, Indonesia. Glycerol (CAS 56-61-5) and benzoyl peroxide (CAS 94-36-0) were purchased from Merck, Darmstadt, Germany.

2.2. Preparation of Thermoplastic Starch (TPS)

The preparation of thermoplastic starch (TPS) was conducted by a twin-screw extruder [2,12–19,25–31]. APPS (70%w), glycerol (30%w) and benzoyl peroxide (0.1 phr of the weight of APPS + glycerol) were mixed and stirred using a mixer until well distributed. Then the mixture was stored overnight (24 h) to diffuse the glycerol into the APPS granules completely. Furthermore, the mixture was fed manually into a co-rotating twin-screw extruder (Compounder ZK 16 T x 36 L/D, Collin, Germany) at a screw speed of 90 rpm with a temperature extruder barrel as the following profile 40/80/120/150/150/150/150/150 °C from zones 1–8. Then the extruded strips were cut into pellets (diameter 2–3 mm) by a pelletizer. In this study, when only glycerol is added, the thermoplastic starch is abbreviated as TPS and when glycerol and benzoyl peroxide are added, the thermoplastic starch is abbreviated as TPSB.

2.3. Fourier-Transform Infrared (FTIR)

The functional groups of the APPS, TPS and TPSB were obtained using Fourier-Transform Infrared Spectroscopy (Bruker Tensor II, Etlingen, Germany). Thus, 32 scans were recorded for each sample while it was set in attenuated total reflectance (ATR) mode with a diamond ATR crystal at a wave number range between 500 and 4000 cm^{-1}.

2.4. Density Measurements

The density of the APPS, TPS and TPSB was measured (five replicates) in accordance with the ASTM D 792 standard at 23 °C and 50% relative humidity.

2.5. Scanning Electron Microscopy (SEM)

The surface morphology of APPS and the cross-section morphology of TPS and TPSB were observed using scanning electron microscopy (JEOL, JSM-IT200, Tokyo, Japan) at 3 kV for APPS and 10 kV for TPS and TPSB. After submersion in liquid nitrogen, the TPS and TPSB samples was broken. The surfaces had a thin gold layer applied to the sample before analysis.

2.6. X-ray Diffraction (XRD)

The crystallinity of APPS, TPS and TPSB was studied by X-ray diffractometer (Malvern Panalytical's Aeris, Eindhoven, Netherlands). The X-ray diffraction pattern observed the 1.54 wavelength, 40 kV voltage and 15 mA filament emission. The radiation reflection of APPS, TPS and TPSB was measured at the 2θ angle of 5–50°.

2.7. Mechanical Properties

The mechanical properties of TPS and TPSB were determined using an electronic universal testing machine (UTM) (Shimadzu AG-X plus 50 kN, Kyoto, Japan), in accordance with ASTM D 638-14 standard, at 21 °C and relative humidity around 50%. Thermo Scientific Haake MiniJet Pro is used to prepare a dumbbell (dogbone) tensile test sample specimen according to ASTM D 638-14 type V. Five specimens of each type of TPS and TPSB film were calculated at tensile rates of 10 mm/min. Each sample was randomly measured three times at various locations using a digital thickness gauge (Preisser Digimet, Gammertingen, Germany) to determine the film thickness.

2.8. Rheological Properties

The rheological properties of TPS and TPSB were studied using the Melt Flow Index (MFI) Instrument Ceast Model 7026, in accordance with ASTM E 1238. Measurement of melt flow rate (MFR), viscosity and shear rate value was carried out at 190 °C with load 2.16 kg.

2.9. Differential Scanning Calorimeter (DSC)

The thermal properties of APPS, TPS and TPSB were characterized using differential scanning calorimetry (DSC) (DSC 4000, PerkinElmer, Waltham, MA, USA. DSC analysis was carried out with approximately 5 mg APPS, TPS and TPSB samples on a standard aluminum pan to determine the samples' glass transition temperature (Tg). Each sample was heated until 200 °C with a 10 °C/min heating rate under a nitrogen atmosphere (flow rate = 20 mL/min).

2.10. Thermogravimetric Analysis (TGA)

The stability of APPS, TPS and TPSB was studied using a thermogravimetric analyzer (TGA 4000, PerkinElmer, Waltham, MA, USA). A crucible pan containing 5 mg of sample was heated from 25 °C to 600 °C at a rate of 10 °C/min. A flow of 20 mL/min was used to purge the nitrogen gas.

3. Results and Discussions

3.1. FTIR Analysis

The FTIR spectrum of APPS, TPS and TPSB are shown in Figure 1. The FTIR spectrum of APPS, TPS and TPSB shows absorption peaks at similar wave numbers. The absorption peak widened at a wave number of about 3265 cm^{-1}, associated with the hydroxyl group from APPS and glycerol, while the absorption peak at a wave number of about 2921 and 2884 cm^{-1} was associated with the C–H group in starch [18,26]. In addition, the absorption peak at wave number 1645 cm^{-1} is the H–O–H vibration of water molecules in the amorphous region, which may be absorbed in the samples [16,18]. The absorption peaks around 1450–1330 cm^{-1} are associated with CH_2 bending and wagging (out of plane bending) of CH_2. The absorption peaks between 1500 cm^{-1} and 1200 cm^{-1} overlap each other between C–H stretching and O-H bending, making it difficult to distinguish the difference in the absorption peaks in this part of the spectrum [17]. Fuente et al., 2022 [16] reported the absorption peak at a wave number between 1200 cm^{-1} and 900 cm^{-1} is the vibration of a functional group of C–O, C–C and C–O–H. Similar absorption peaks of APPS were also reported by previous studies [5,8,10], while the FTIR spectrum result of TPS was verified by [5,10]. Zhang et al., 2013 [25] reported the hydroxyl groups of starch were changed to carbonyl and carboxyl groups during oxidation. Furthermore, these carbonyl and carboxyl groups of oxidized starch form strong hydrogen bonds with the hydroxyl groups on starch. The anhydroglucose ring of oxidized starch is still maintained, so its chemical structure looks like starch and TPS [25]. In addition, the lack of a significant difference in the absorption peaks of TPS and TPSB, possibly caused by the low concentration of benzoyl peroxide used, causes changes in the functional groups of starch molecules that are not drastic enough to be identified by this technique.

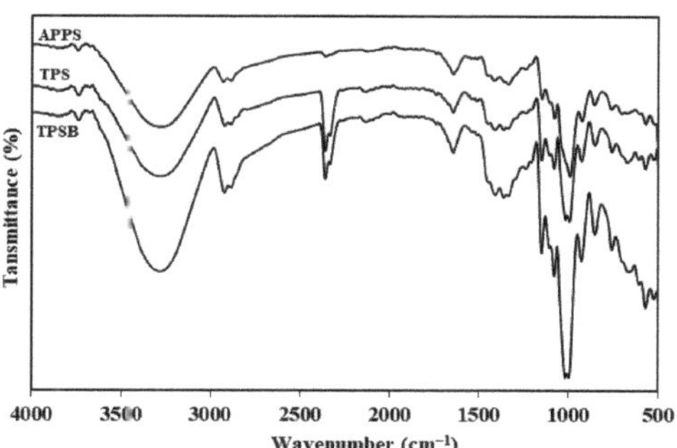

Figure 1. FTIR spectrum of APPS, TPS and TPSB.

3.2. Physical Properties

The density of APPS, TPS and TPSB is shown in Figure 2. The density values of APPS, TPS and TPSB were 1.79, 1.37 and 1.39 g·cm^{-3}, respectively. The presence of glycerol as a plasticizer destroys and weakens the inter- and intra-molecular hydrogen bonding between starch molecules, thereby increasing the free volume and mobility between molecular chains and decreasing the density of TPS [5,11,25]. Previous solvent-casting methods reported that TPS's density values varied from 1.40 g·cm^{-3} [11] to 1.41 g·cm^{-3} [5]. Compared to the solvent-casting method, the density of TPS in this research was lower, which proved that the extrusion method could potentially increase the interaction between the glycerol and APPS, leading to an increase in free volume and mobility between molecular chains. The use of benzoyl peroxide as an oxidizing agent on in situ modified TPS preparation increases the TPS density, although not very significantly, from 1.37 g·cm^{-3} to 1.39 g·cm^{-3}. The higher density value of TPSB compared to TPS is, presumably, because the TPSB molecules are arranged more neatly and orderly so that the density value is greater than the TPS density value. The extrusion process is expected to produce thermoplastic materials with more neatly and consistently ordered molecules [16]. In addition, during the extrusion process, oxidation of the starch molecule may lead to the formation of carbonyl or carboxyl groups. The carbonyl groups that may be formed can form strong hydrogen bonds with the hydroxyl groups of starch, resulting in a stiffer film and increase in density. However, because the use of benzoyl peroxide is very small, the result changes are also not significant. The insignificant changes were also confirmed by the absence of new functional groups in the results of the FTIR analysis.

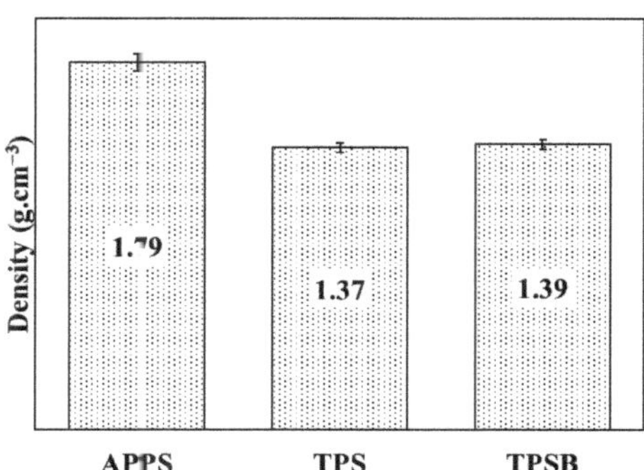

Figure 2. Density of APPS, TPS and TPSB.

3.3. Morphology

The morphology of APPS, TPS and TPSB was examined by SEM. The samples were fractured in liquid nitrogen before testing. Figure 3 shows the surface morphology of APPS and the cross-section morphology of TPS and TPSB. Figure 3 shows that the APPS used in this study is granular and has inhomogeneous shapes and sizes. Some are spherical, oval and irregular, with diameters between 13 and 52 mm. Similar results on the morphology of APPS were also reported [8]. The TPS cross-section surface is rather smooth, meaning the APPS granules changed phase during the process in the extruder. The starch granules were physically broken into small fragments and melted due to the continuous interaction of the plasticizer, heat and shear rate in the twin-screw extruder, which resulted in smoother morphology and the disappearance of the granular structure of starch [25]. In addition, no phase separation is observed in TPSB, which exhibits starch granules with and without benzoyl peroxide that are uniformly dispersed in the matrix. Further, it indicates a good compatibility and plasticization process of starch with glycerol in the twin-screw extruder [15,25].

Figure 3. Surface morphology of APPS and cross-section morphology of TPS and TPSB.

3.4. Crystallinity

The X-ray diffraction pattern of APPS, TPS and TPSB shows the semi-crystalline (presence of amorphous and crystalline) characteristic, as can be observed in Figure 4. In Figure 4, APPS shows diffraction peaks with high intensity at 2θ of 15.1°, 17.2°, 18.0° and 23.3°, which indicates that the characteristic palm starch has a C-type pattern crystalline structure. The XRD result of APPS was confirmed by previous studies [5,6,8]. The change in the crystalline structure of APPS after processing with glycerol and benzoyl peroxide is clearly visible in the X-ray diffraction pattern. The TPS and TPSB have similar diffraction patterns. The TPS and TPSB diffraction patterns show diffraction peaks at 2θ of 11.9°; 13.3° and 18.0° and did not show any diffraction peak crystallinity type C of APPS, proving that the initial APPS granules were gelatinized during the thermoplasticization process in the twin-screw extruder. In addition, TPS and TPSB show a broad hump diffraction peak pattern at 2θ 19°. The broad hump diffraction peak pattern at 19° is a characteristic of completely amorphous material [13]. This indicates that TPS and TPSB were not completely amorphous. Amorphous regions are caused by the disruption of the double-helix conformations of the starch due to the starch gelatinization, while the crystalline regions were formed by the recrystallization, favored by the formation of microcrystalline connections due to the presence of glycerol [16].

In addition, at TPS and TPSB, new diffraction peaks also appeared at 2θ 13.39°, 18.13° and 20.68°. These peaks are characteristic crystallinities type VH, which is formed during thermo-mechanical processing in the twin-screw extruder [13]. This proves that there was a change in the crystallinity structure of APPS to TPS and TPSB, from type C to VH. The diffraction peak at 2θ 20.68° in TPSB has a higher intensity compared to TPS. This kind of feature is caused by the presence of benzoyl peroxide during the extrusion process at TPSB. After the extrusion process, types of crystallinnity can be distinguished in TPSs: (i) residual crystallinity, native A, B or C crystallinity, which incompletely destroy and melt starch granules during the process and (ii) processing-induced crystallinity, V_H, V_A or E_H crystallinity, which is formed during the thermo-mechanical process [13]. TPS shows process-induced crystallinity due to the hot-processing process, caused by the crystallinity in the starch chain, compounded with plasticizer and water into a single-helix structure. This degree of crystallinity was induced by hot processing, attributed to a strong interaction among the hydroxyl groups of the starch molecular chain, which was replaced by a hydrogen bond formed between the starch and the plasticizer [18]. The presence of glycerol as a plasticizer increases the mobility of the molecular chains and causes crystallization [32]. The extrusion process resulted in damage to the crystal structure of the native starch, as shown by the difference in diffraction patterns between APPS, TPS and TPSB. However, the glycerol induces plasticization of the starch chains during extrusion and will recrystallize the starch [1].

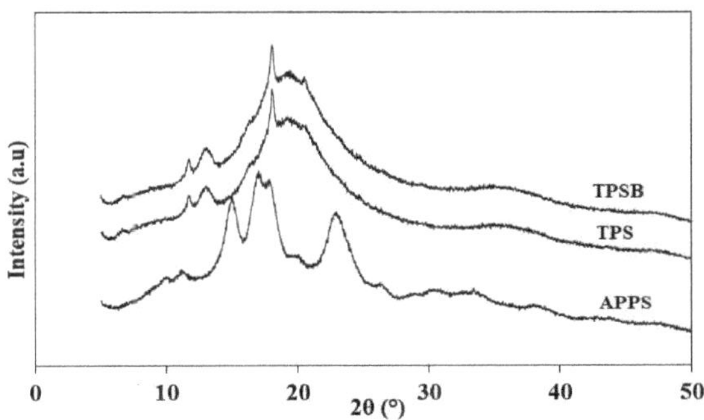

Figure 4. XRD pattern of APPS, TPS and TPSB.

3.5. Mechanical Properties

The mechanical properties in the form of tensile strength, elongation and elastic modulus of TPS and TPSB are shown in Figure 5. In Figure 5, the tensile strength, elongation at break and elastic modulus of TPS were 7.19 MPa, 33.95% and 0.56 Gpa and TPSB were 8.61 MPa, 30.16% and 0.54 Gpa, respectively. However, previous research on TPS preparation from APPS by solvent casting reported the tensile strength and elongation at break of the TPS films from APPS of 4.8 MPa and 38.10% [5] and 2.42 MPa and 8.03% [10], respectively. The tensile strength value of TPS obtained in this study was higher, which proved that the use of the extrusion method in the preparation of TPS could further increase the interaction between the plasticizer and APPS compared to the solvent-casting method. A strong correlation between the processing method applied and the mechanical properties of TPS was reported in the literature [5]. Further, Zhang et al., 2013 reported tensile strength of TPS from oxidized corn starch of 1.0–2.1 MPa and elongation at break of 131.7–170.2% [25]. These results are also different from the mechanical properties produced in this study. This might have occured due to the different types of starch used. In addition, the increase in tensile strength proves that the addition of benzoyl peroxide can increase the tensile strength of TPS. During the extrusion process, the benzoyl peroxide as an oxidizing agent resulted in the oxidation of the starch molecule, which was expected to cause the formation of a carbonyl or carboxyl group. Either carbonyl or carboxyl group are able to form strong hydrogen bonds with the hydroxyl groups of starch, resulting in a stiffer film and increasing the tensile strength of TPSB [16]. However, the addition of an oxidizing agent will also result in a decrease in the elongation at break of TPSB, from 33.95% to 30.16% and elastic modulus from 0.56 Gpa to 0.54 GPa.

The stiffer TPSB results in limited mobility of the molecular chain in TPSB, thereby reducing flexibility, which results in a decrease in elongation at break. In addition, in the extrusion process, depolymerization also occurs, which facilitates the tendency of molecular reassociation, with a greater potential for interaction [16]. Therefore, the new polymer matrix formed has different interactions between starch molecules and glycerol, so as to produce stronger bioplastics. In addition, the extrusion process can also support the alignment of the chain in the direction of flow, which results in a more flexible material. Therefore, extrusion is expected to produce thermoplastic materials whose molecules are more neatly and orderly arranged so as to increase the tensile strength of thermoplastic materials [16].

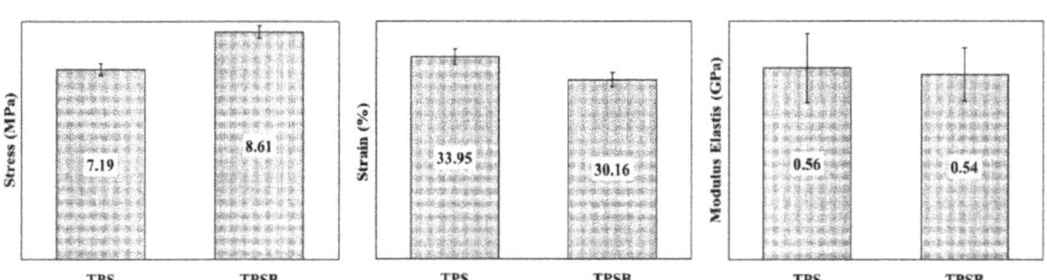

Figure 5. Mechanical properties of TPS and TPSB.

3.6. Rheological Properties

The melt flow rate (MFR), shear rate and viscosity of TPS and TPSB are shown in Figure 6. The presence of benzoyl peroxide in TPS showed a decrease in the MFR value from 7.13 gr/10 min to 5.73 gr/10 min. The benzoyl peroxide, as an oxidizing agent, resulted in the oxidation of the starch molecule, which was expected to cause the formation of a carbonyl or carboxyl group. Either carbonyl or carboxyl group are able to form strong hydrogen bonds with the hydroxyl groups of starch [16]. These strong hydrogen bonds with the hydroxyl groups of starch reduce the molecular mobility of polymers, resulting in a stiffer film of TPSB film and inhibiting the flow rate of the TPSB, thereby reducing the amount of TPSB material that comes out of the MFI instrument barrel, resulting in a decrease in the MFR value of TPSB. The presence of benzoyl peroxide, which makes TPSB stiffer and harder to flow, also causes an increase in viscosity from 2482.19 Pa.s to 2604.60 Pa.s and a decrease in shear rate from 9.61 s^{-1} to 7.63 s^{-1}.

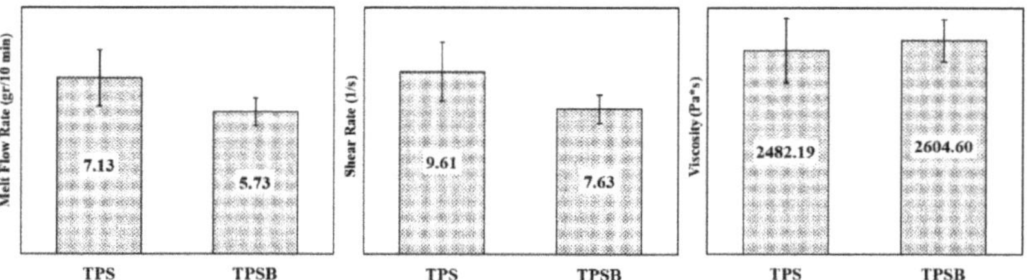

Figure 6. MFR, shear rate and viscosity TPS and TPSB.

The strong hydrogen bonds between oxidized starch with hydroxyl groups of starch will result in a decrease in chain mobility of TPSB, an increase in stiffness, an increase in viscosity and a decrease in shear rate, thereby reducing the melt flow rate of TPSB. The relationship between viscosity and shear rate has also been revealed [25]: when the flow resistance is reduced, the shear rate will increase and the viscosity will decrease, indicating that the melted starch mixture extruded behaves like a pseudoplastic liquid. Furthermore, as the shear rate increases, the chain entanglement in starch decreases, which leads to a weakening of the inter- and intra-molecular interactions between starches, thereby reducing the flow resistance. Therefore, a compatible and well-dispersed mixture can be characterized by increasing the shear rate and decreasing the viscosity [25].

3.7. Thermal Properties

The DSC curve and the glass transition temperature (Tg) APPS, TPS and TPSB are shown in Figure 7 and Table 1. It can be seen that the peak gelatinization temperature of APPS was 70°C, close to a previously reported peak temperature of gelatinization of APPS of 67 °C [6]. However, it was lower than the results in the literature, which reported that the gelatinization temperature of APPS was around 98 °C [8]. In addition, Tg values of TPS and TPSB were 65 °C and 52 °C, respectively. The Tg value of TPS was lower than the gelatinization temperature of APPS. The decrease in the Tg value was related to the structure of the APPS granules being destroyed by glycerol during the extrusion process at high temperatures [18]. The plasticization process by glycerol reduces and exchanges the inter- and intra-molecular bonds between starch with a starch–glycerol hydrogen bond, increases free volume and increases chain mobility and intermolecular spacing, thereby improving the flexibility of TPS, thus, leading to a reduction in Tg [3,10]. The use of benzoyl peroxide in TPS preparation further reduces the Tg value. These show that the chain mobilities and flexibilities of TPSB are enhanced due to the fact that the addition of oxidized starch interrupts the hydrogen bonds between starch chains. Similar results were also reported by a previous study, where the oxidized starch had a stronger interaction between oxidized starch and starch chains, compared to interaction between starch, which leads to an increase in the mobility of starch chains [25].

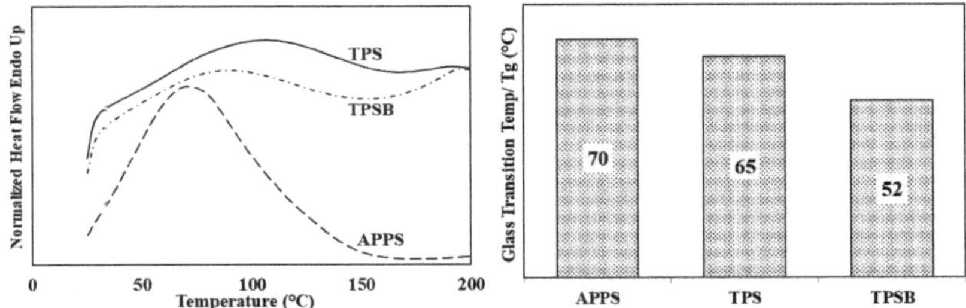

Figure 7. DSC curve and glass trasnsition temperature of APPS, TPS and TPSB.

Table 1. Thermal properties of APPS, TPS and TPSB.

Sample	APPS	TPS	TPSB
Gelatinization temperature (°C)	70	-	-
Glass transition temperature (°C)	-	65	52
Residual mass at 600 °C (%)	9.36	5.47	5.51

Thermal stability of APPS, TPS and TPSB was studied by TGA. Figure 8 shows the thermal stability curve of APPS, TPS and TPSB. The APPS showed two stages of thermal degradation. The initial stage of thermal degradation occurred at temperatures up to 150 °C, with the peak of thermal degradation occurring at 70 °C. This initial thermal degradation was correlated to the evaporation of water. A similar TGA pattern of APPS was reported in the literature [10]. Thermal degradation in the second stage, as the main thermal degradation, occurred in a temperature range of 260–400 °C, with the peak of main thermal degradation occurring at 300 °C. This main thermal degradation was also correlated to the dehydration of starch molecules to form glucose. Figure 8 and Table 1 also show that the final APPS residue at 600 °C was about 9.36%, associated with the partial carbonization of starch. These results are similar to the previous study's findings [18].

The thermal degradation curve of TPS also has two stages of thermal degradation, which are similar to those of APPS. The initial thermal degradation is also related to the evaporation of water. However, the initial thermal degradation of TPS has a different pattern with APPS and the thermal degradation temperature range in the early stages is higher, up to 180 °C. In addition to that, the peak of thermal degradation in the early stages of TPS occurs at a temperature of 150 °C. The main thermal degradation of TPS occurs in a temperature range of 260–400 °C and shows two main stages of thermal degradation. The first major thermal degradation is at a temperature of 260–310 °C, with a peak of thermal degradation at 290 °C, associated with the decomposition of the glycerol-rich phase, while the second major thermal degradation occurs in a temperature range of 310–400 °C, with a peak thermal degradation at a temperature of 330 °C, associated with the degradation of amylose and amylopectin starch. This result has good accordance with a previous study [15,16,26]. The presence of glycerol causes the peak temperature of TPS thermal degradation to shift towards higher temperatures. As a result, the glycerol in starch provides an advantage in thermal stability by increasing the mobility of the molecular chains due to the plasticization process, thereby increasing the fluidity of the material and delaying the decomposition of the material caused by the process at high temperatures [15]. The thermal degradation of TPSB shows a pattern of thermal degradation that is similar to the pattern of thermal degradation of TPS. However, the use of benzoyl peroxide in the preparation of TPSB increased the thermal resistance and extended the thermal degradation temperature range of TPSB when compared to TPS. This is evidenced by the mass remaining at a temperature of 600 °C, i.e., TPSB is 5.51%, while TPS is 5.47% (Table 1). This is presumably because oxidized starch increases the interaction of glycerol to starch by hydrogen bonding, thus, making them more difficult to evaporate during processing. This proves that oxidized starch can improve the thermal stability of TPS, which means that the addition of oxidized starch prevents the degradation of starch-based materials at processing temperatures [25].

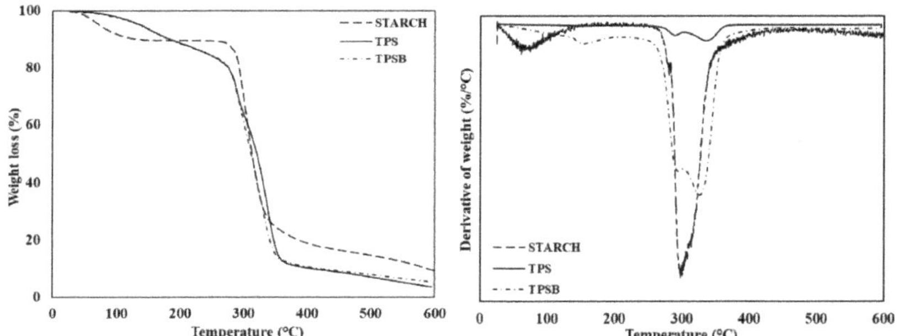

Figure 8. Thermal stability of APPS, TPS and TPSB.

4. Conclusions

In this research, in situ modification for TPS preparation based on *Arenga pinnata* palm starch was successfully carried out. Modified TPS was prepared by adding palm starch, glycerol and benzoyl peroxide simultaneously with the twin-screw extruder. Morphology analysis of TPS showed that the starch granules were damaged and gelatinized in the extrusion process. No phase separation is observed in TPSB, which exhibits that starch granules with and without benzoyl peroxide are uniformly dispersed in the matrix. The use of benzoyl peroxide in the preparation of TPS increases the density, tensile strength, viscosity and thermal stability as well as extending the thermal degradation temperature range of TPS. However, it reduces elongation at break, elastic modulus, melt flow rate, shear rate and glass transition temperature. The results of this study are expected to provide insight into the potential of *A. pinnata* as a new starch source for the development of biodegradable materials. In addition, in situ modification for the preparation of modified TPS with the extrusion process in the twin-screw extruder can reduce TPS preparation time because it only requires one preparation stage, meaning there is no need to modify the starch first. The characteristics of modified TPS produced in this study are expected to contribute to the further development of biodegradable materials.

Author Contributions: Conceptualization, M.G. and M.C.; methodology, M.G.; resources, M.G. and Y.M.; data curation, M.G.; writing—original draft preparation, M.G.; writing—review and editing, M.C. and Y.M.; visualization, M.G.; supervision, M.C. and Y.M. All authors have read and agreed to the published version of the manuscript.

Funding: The authors received no specific funding for this study.

Institutional Review Board Statement: Not applicable.

Data Availability Statement: The data presented in this study are available on request from the corresponding author.

Acknowledgments: This research was supported by LPDP project No. PRJ-97/LPDP/2019. The authors thank the facilities, scientific and technical support from Advanced Characterization Laboratories Serpong and Cibinong, National Research and Innovation Institute through E-Layanan Sains, Badan Riset dan Inovasi Nasional (BRIN).

Conflicts of Interest: The authors declare no conflict of interest.

References

1. Nessi, V.; Falourd, X.; Maigret, J.-H.; Cahier, K.; D'Orlando, A.; Descamps, N.; Gaucher, V.; Chevigny, C.; Lourdin, D. Cellulose nanocrystals-starch nanocomposites produced by extrusion: Structure and behavior in physiological conditions. *Carbohydr. Polym.* **2019**, *225*, 115123. [CrossRef] [PubMed]
2. Lendvai, L.; Karger-Kocsis, J.; Kmetty, Á.; Drakopoulos, S.X. Production and characterization of microfibrillated cellulose-reinforced thermoplastic starch composites. *J. Appl. Polym. Sci.* **2016**, *133*, 1–8. [CrossRef]
3. Sanyang, M.L.; Sapuan, S.M.; Jawaid, M.; Ishak, M.R.; Sahari, J. Recent developments in sugar palm (*Arenga pinnata*) based biocomposites and their potential industrial applications: A review. *Renew. Sustain. Energy Rev.* **2016**, *54*, 533–549. [CrossRef]
4. Kringel, D.H.; Dias, A.R.G.; Zavareze, E.R.; Gandra, E.A. Fruit Wastes as Promising Sources of Starch: Extraction, Properties, and Applications. *Starch/Staerke* **2020**, *72*, 3–4. [CrossRef]
5. Ilyas, R.A.; Sapuan, S.M.; Ibrahim, R.; Abral, H.; Ishak, M.R.; Zainudin, E.S.; Atikah, M.S.N.; Nurazzi, N.M.; Atiqah, A.; Ansari, M.N.M.; et al. Effect of sugar palm nanofibrillated cellulose concentrations on morphological, mechanical and physical properties of biodegradable films based on agro-waste sugar palm (*Arenga pinnata* (*Wurmb.*) *Merr*) starch. *J. Mater. Res. Technol.* **2019**, *8*, 4819–4830. [CrossRef]
6. Adawiyah, D.R.; Sasaki, T.; Kohyama, K. Characterization of arenga starch in comparison with sago starch. *Carbohydr. Polym.* **2013**, *92*, 2306–2313. [CrossRef]
7. Asyraf, M.R.M.; Rafidah, M.; Ebadi, S.; Azrina, A.; Razman, M.R. Mechanical properties of sugar palm lignocellulosic fibre reinforced polymer composites: A review. *Cellulose* **2022**, *29*, 6493–6516. [CrossRef]
8. Zhang, L.; Mei, J.Y.; Ren, M.H.; Fu, Z. Optimization of enzyme-assisted preparation and characterization of *Arenga pinnata* resistant starch. *Food Struct.* **2020**, *25*, 100149. [CrossRef]
9. Ishak, M.R.; Sapuan, S.M.; Leman, Z.; Rahman, M.Z.A.; Anwar, U.M.K.; Siregar, J.P. Sugar palm (*Arenga pinnata*): Its fibres, polymers and composites. *Carbohydr. Polym.* **2013**, *91*, 699–710. [CrossRef]
10. Sahari, J.; Sapuan, S.M.; Zainudin, E.S.; Maleque, M.A. Thermo-mechanical behaviors of thermoplastic starch derived from sugar palm tree (*Arenga pinnata*). *Carbohydr. Polym.* **2013**, *92*, 1711–1716. [CrossRef]
11. Sahari, J.; Sapuan, S.M.; Zainudin, E.S.; Maleque, M.A. A New Approach to Use *Arenga pinnata* as Sustainable Biopolymer: Effects of Plasticizers on Physical Properties. *Procedia Chem.* **2012**, *4*, 254–259. [CrossRef]
12. Kampangkaew, S.; Thongpin, C.; Santawtee, O. The synthesis of cellulose nanofibers from Sesbania Javanica for filler in thermoplastic starch. *Energy Procedia* **2014**, *56*, 318–325. [CrossRef]
13. Ghanbari, A.; Tabarsa, T.; Ashori, A.; Shakeri, A.; Mashkour, M. Preparation and characterization of thermoplastic starch and cellulose nanofibers as green nanocomposites: Extrusion processing. *Int. J. Biol. Macromol.* **2018**, *112*, 442–447. [CrossRef] [PubMed]
14. Teixeira, E.M.; Curvelo, A.A.S.; Corrêa, A.C.; Marconcini, J.M.; Glenn, G.M.; Mattoso, L.H.C. Properties of thermoplastic starch from cassava bagasse and cassava starch and their blends with poly (lactic acid). *Ind. Crops Prod.* **2012**, *37*, 61–68. [CrossRef]
15. Liu, W.; Wang, Z.; Liu, J.; Dai, B.; Hu, S.; Hong, R.; Xie, H.; Li, Z.; Chen, Y.; Zeng, G. Preparation, reinforcement and properties of thermoplastic starch film by film blowing. *Food Hydrocoll.* **2020**, *108*, 106006. [CrossRef]
16. La Fuente, C.I.A.; Siqueira, L.V.; Augusto, P.E.D.; Tadini, C.C. Casting and extrusion processes to produce bio-based plastics using cassava starch modified by the dry heat treatment (DHT). *Innov. Food Sci. Emerg. Technol.* **2022**, *75*, 102906. [CrossRef]
17. Gilfillan, N.W.; Moghaddam, L.; Bartley, J.; Doherty, W.O.S. Thermal extrusion of starch film with alcohol. *J. Food Eng.* **2016**, *170*, 92–99. [CrossRef]
18. Chen, J.; Wang, X.; Long, Z.; Wang, S.; Zhang, J.; Wang, L. Preparation and performance of thermoplastic starch and microcrystalline cellulose for packaging composites: Extrusion and hot pressing. *Int. J. Biol. Macromol.* **2020**, *165*, 2295–2302. [CrossRef]

19. Hietala, M.; Mathew, A.P.; Kristiina, O. Bionanocomposites of thermoplastic starch and cellulose nanofibers manufactured using twin-screw extrusion. *Eur. Polym. J.* **2013**, *49*, 950–956. [CrossRef]
20. Fourati, Y.; Magnin, A.; Putaux, J.L.; Boufi, S. One-step processing of plasticized starch/cellulose nanofibrils nanocomposites via twin-screw extrusion of starch and cellulose fibers. *Carbohydr. Polym.* **2020**, *229*, 115554. [CrossRef]
21. Bangar, S.P.; Whiteside, W.S.; Ashogbon, A.O.; Kumar, M. Recent advances in thermoplastic starches for food packaging: A review. *Food Packag. Shelf Life* **2021**, *30*, 100743. [CrossRef]
22. Ekielski, A.; Żelaziński, T.; Mishra, P.K.; Skudlarski, J. Properties of biocomposites produced with thermoplastic starch and digestate: Physicochemical and mechanical characteristics. *Materials* **2021**, *14*, 6092. [CrossRef] [PubMed]
23. Maulana, M.I.; Lubis, M.A.R.; Febrianto, F.; Hua, L.S.; Iswanto, A.H.; Antov, P.; Kristak, L.; Mardawati, E.; Sari, R.K.; Zaini, L.H.; et al. Environmentally Friendly Starch-Based Adhesives for Bonding High-Performance Wood Composites: A Review. *Forests* **2022**, *13*, 1614. [CrossRef]
24. Vanier, N.L.; El Halal, S.L.M.; Dias, A.R.G.; Zavareze, E.R. Molecular structure, functionality and applications of oxidized starches: A review. *Food Chem.* **2017**, *221*, 1546–1559. [CrossRef] [PubMed]
25. Zhang, Y.; Wang, X.; Zhao, G.; Wang, Y. Influence of oxidized starch on the properties of thermoplastic starch. *Carbohydr. Polym.* **2013**, *96*, 358–364. [CrossRef] [PubMed]
26. Herniou-Julien, C.; Mendieta, J.R.; Gutiérrez, T.J. Characterization of biodegradable/non-compostable films made from cellulose acetate/corn starch blends processed under reactive extrusion conditions. *Food Hydrocoll.* **2018**, *89*, 67–79. [CrossRef]
27. Jariyasakoolroj, P.; Supthanyakul, R.; Laobuthee, A.; Lertworasirikul, A.; Yoksan, R.; Phongtamrug, S.; Chirachanchai, S. Structure and properties of in situ reactive blend of polylactide and thermoplastic starch. *Int. J. Biol. Macromol.* **2021**, *182*, 1238–1247. [CrossRef]
28. Raabe, J.; Fonseca, A.S.; Bufalino, L.; Ribeiro, C.; Martins, M.A.; Marconcini, J.M.; Mendes, L.M.; Tonoli, G.H.D. Biocomposite of cassava starch reinforced with cellulose pulp fibers modified with deposition of silica (SiO_2) nanoparticles. *J. Nanomater.* **2015**, *2015*, 6. [CrossRef]
29. Teixeira, E.M.; Lotti, C.; Correa, A.C.; Teodoro, K.B.R.; Marconcini, J.M.; Mattoso, L.H.C.M. Thermoplastic Corn Starch Reinforced with Cotton Cellulose Nanofibers. *J. Appl. Polym. Sci.* **2011**, *120*, 2428–2433. [CrossRef]
30. Yan, Q.; Hou, H.; Guo, P.; Dong, H. Effects of extrusion and glycerol content on properties of oxidized and acetylated corn starch-based films. *Carbohydr. Polym.* **2012**, *87*, 707–712. [CrossRef]
31. Yusoff, N.H.; Pal, K.; Narayanan, T.; de Souza, F.G. Recent trends on bioplastics synthesis and characterizations: Polylactic acid (PLA) incorporated with tapioca starch for packaging applications. *J. Mol. Struct.* **2021**, *1232*, 129954. [CrossRef]
32. Özeren, H.D.; Olsson, R.T.; Nilsson, F.; Hedenqvist, M.S. Prediction of plasticization in a real biopolymer system (starch) using molecular dynamics simulations. *Mater. Des.* **2020**, *187*, 108387. [CrossRef]

Article

Chemical, Physical, and Mechanical Properties of Belangke Bamboo (*Gigantochloa pruriens*) and Its Application as a Reinforcing Material in Particleboard Manufacturing

Apri Heri Iswanto [1,2,*], Elvara Windra Madyaratri [3,4], Nicko Septuari Hutabarat [1], Eka Rahman Zunaedi [1], Atmawi Darwis [5], Wahyu Hidayat [6], Arida Susilowati [2,7], Danang Sudarwoko Adi [4], Muhammad Adly Rahandi Lubis [4,8], Tito Sucipto [1,2], Widya Fatriasari [4], Petar Antov [9,*], Viktor Savov [9] and Lee Seng Hua [10]

1 Department of Forest Product, Faculty of Forestry, Universitas Sumatera Utara, Kampus USU Padang Bulan, Medan 20155, Indonesia; nickohutabarat0709@gmail.com (N.S.H.); ekazunaidi99@gmail.com (E.R.Z.); titomedan@yahoo.com (T.S.)
2 JATI-Sumatran Forestry Analysis Study Center, Faculty of Forestry, Universitas Sumatera Utara, Kampus USU Padang Bulan, Medan 20155, Indonesia
3 Department of Forest Products, Faculty of Forestry and Environment, IPB University, Bogor 16680, Indonesia; elvarawindra@yahoo.com
4 Research Center for Biomass and Bioproducts, National Research and Innovation Agency, Jl Raya Bogor KM 46, Cibinong 16911, Indonesia; danang@biomaterial.lipi.go.id (D.S.A.); marl@biomaterial.lipi.go.id (M.A.R.L.); widya.fatriasari@biomaterial.lipi.go.id (W.F.)
5 School of Life Sciences and Technology, Institut Teknologi Bandung, Gedung Labtex XI, Jalan Ganesha 10, Bandung 40132, Indonesia; atmawidarwis@gmail.com
6 Department of Forestry, Faculty of Agriculture, University of Lampung, Jl Sumantri Brojonegoro, Gedung Meneng, Bandar Lampung 35145, Indonesia; wahyu.hidayat@fp.unila.ac.id
7 Department of Silviculture, Faculty of Forestry, Universitas Sumatera Utara, Kampus USU Padang Bulan, Medan 20155, Indonesia; arida.susilowati@usu.ac.id
8 Research Collaboration Center for Biomass and Biorefinery between BRIN, Universitas Padjadjaran, National Research and Innovation Agency, Jl Raya Bogor KM 46, Cibinong 16911, Indonesia
9 Faculty of Forest Industry, University of Forestry, 1797 Sofia, Bulgaria; victor_savov@ltu.bg
10 Laboratory of Biopolymer and Derivatives, Institute of Tropical Forestry and Forest Product, Universiti Putra Malaysia, Kuala Lumpur 43400, Malaysia; lee_seng@upm.edu.my
* Correspondence: apri@usu.ac.id (A.H.I.); p.antov@ltu.bg (P.A.)

Abstract: This study aimed to analyze the basic properties (chemical composition and physical and mechanical properties) of belangke bamboo (*Gigantochloa pruriens*) and its potential as a particleboard reinforcement material, aimed at increasing the mechanical properties of the boards. The chemical composition was determined by Fourier transform near infrared (NIR) analysis and X-ray diffraction (XRD) analysis. The physical and mechanical properties of bamboo were evaluated following the Japanese standard JIS A 5908 (2003) and the ISO 22157:2004 standard, respectively. The results showed that this bamboo had average lignin, holocellulose, and alpha-cellulose content of 29.78%, 65.13%, and 41.48%, respectively, with a degree of crystallinity of 33.54%. The physical properties of bamboo, including specific gravity, inner and outer diameter shrinkage, and linear shrinkage, were 0.59%, 2.18%, 2.26%, and 0.18%, respectively. Meanwhile, bamboo's mechanical properties, including compressive strength, shear strength, and tensile strength, were 42.19 MPa, 7.63 MPa, and 163.8 MPa, respectively. Markedly, the addition of belangke bamboo strands as a reinforcing material (surface coating) in particleboards significantly improved the mechanical properties of the boards, increasing the modulus of elasticity (MOE) and bending strength (MOR) values of the fabricated composites by 16 and 3 times.

Keywords: *Gigantochloa pruriens*; chemical properties; physical and mechanical properties; wood-based composites; particleboard

Citation: Iswanto, A.H.; Madyaratri, E.W.; Hutabarat, N.S.; Zunaedi, E.R.; Darwis, A.; Hidayat, W.; Susilowati, A.; Adi, D.S.; Lubis, M.A.R.; Sucipto, T.; et al. Chemical, Physical, and Mechanical Properties of Belangke Bamboo (*Gigantochloa pruriens*) and Its Application as a Reinforcing Material in Particleboard Manufacturing. *Polymers* **2022**, *14*, 3111. https://doi.org/10.3390/polym14153111

Academic Editors: Gianluca Tondi and Luis Alves

Received: 5 July 2022
Accepted: 29 July 2022
Published: 30 July 2022

1. Introduction

As a country known for its high biodiversity, Indonesia has hundreds of bamboo species. It has the third-largest bamboo population in the world after China and India. Bamboo grows in dry to wet, tropical climates [1–4]. In addition to being used as a nonpermanent material in construction and furniture making, it has various industrial applications, such as fiber reinforcement, paper, textiles, oriented strand board, particleboard, food, and bioenergy [5–10]. Furthermore, it is a fast-growing plant with a short harvest time (3–5 years old) and high productivity of 20–40 tons/ha/year, 7–30% higher than woody plants [5,11–13].

One of the most common bamboo species of the *Gigantochloa* genus, found in northern Sumatra, is the *Gigantochloa pruriens*, an endemic species spreading across the Karo and Gayo districts in Indonesia. The culms of *Gigantochloa pruriens* are regarded as valuable feedstocks widely used in construction applications for manufacturing walls, pillars, roofs, etc. Similar to wood, bamboo also has various properties. The characterization of chemical, physical, and mechanical properties of belangke bamboo raw materials is a very important factor in determining its suitability for various commercial applications, one of which is as a raw material for manufacturing particleboard. Currently, research data related to belangke bamboo is still rather limited. Based on the literature review results, only one study related to the anatomical properties of belangke bamboo was found, which was reported by Darwis et al. [14]. Several studies have shown that, in general, bamboo has good mechanical properties so that it can be used as a raw material for structural composites and as a surface layer modification material to improve the strength of particleboard [7,9,10,15–17].

In particleboard manufacturing, mixing particles of two different species is not a common practice. Wood particles with stronger mechanical strength could be used to compensate for the inferiority of the other weaker wood particles in the particleboard [18,19]. Lee et al. [18,19] used rubberwood (*Hevea brasiliensis*) particles as a surface layer and oil palm trunk particles as a core layer to produce three-layer particleboard. The mechanical properties of the particleboard were found to increase along with the increased proportion of rubberwood particles in the surface layer. The studies demonstrated that the employment of stronger wood species as surface layer could result in improved mechanical strength of the composites. Bamboo was found to be a good reinforcing material for particleboard. De Almeida et al. [20] reported that the incorporation of 25% and 50% of *Dendrocalamus asper* bamboo into particleboard made of *Eucalyptus urophylla* × *grandis* wood had improved the bending strength of the particleboard produced. Mixing a more compressible wood with other noncompressible wood could enhance the compression and consolidation of the particleboard [21]. A study by Zaia et al. [22] reported the application of bamboo laminas of *Dendrocalamus giganteus* as a reinforcement for particleboard. The board produced showed a great potential to be used as construction materials. However, relevant studies are scarce, particularly those involving bamboo strands. Therefore, the goal of this research work was to investigate the basic properties (chemical composition, physical, and mechanical properties) of belangke bamboo (*Gigantochloa pruriens*) and evaluate its potential as a particleboard reinforcing material, aimed at improving the boards' mechanical properties.

2. Materials and Methods

2.1. Materials

The *Gigantochloa pruriens* bamboo (Figure 1) was obtained from the Binjai region, North Sumatra (3°35′55″ N, 98°28′49″ E). The length and diameter of belangke bamboo used in this work were approximately 14 m and 6 cm, respectively.

Furthermore, the samples used for chemical and crystallinity composition observation were divided into three parts, namely bottom (B), middle (M), and top (T) as presented in Figure 2.

Figure 1. Bamboo clumps and bamboo middle culm cross-sections.

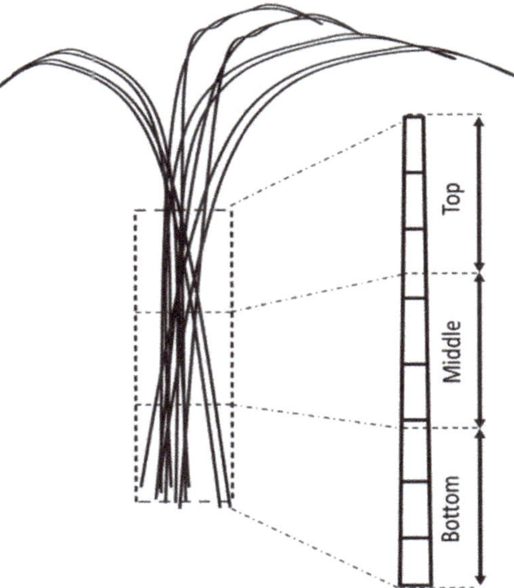

Figure 2. Bamboo culm parts (bottom, middle, and top) for analysis of chemical components and physical–mechanical properties of bamboo.

2.2. *Methods*

2.2.1. Chemical Component Analysis

Material Preparation

Sample preparation was performed according to the standards of the Technical Association of the Pulp and Paper Industry (TAPPI) T 257 cm-02 [23] and T 264 cm-97 [24]. Furthermore, samples were mashed using a ring flaker/hammer mill/disk mill until all filtered through a sieve, number 40 mesh.

Determination of Chemical Components

a. Acid Insoluble Lignin Contents

This test was performed based on National Renewable Energy Laboratory (NREL) Laboratory Analytical Procedure (LAP) 003 standard [25]. Empty filter funnel 1G3 was dried in an oven at 105 °C for at least 4 h before testing, cooled in a desiccator for 30 min, and weighed in the dry weight of the oven. Furthermore, a total of 0.3 g of the extractive free sample was weighed and put in a small vial with a wide mouth of ±20 mL, and the water content was measured based on TAPPI T 264 cm-97. In addition, it was added to 3 mL of 72% sulfuric acid (Merck, Darmstadt, Germany), stirred using a magnetic stirrer for 2 h with 320 rpm at room temperature (conditioned using a Petri dish filled with water), transferred to a 100 mL Duran bottle, and diluted using 84 mL of distilled water to a final concentration of 4% sulfuric acid (Merck). The sample was tightly closed and heated by autoclaving at 121 °C for 1 h. Furthermore, a total of ±10 mL was filtered using an IG3 glass filter with the help of a vacuum and stored for measurement of ASL. The samples in the IG3 filter funnel were washed with a minimum of 50 mL of hot water and dried in an oven at 105 °C for 24 h. Consequently, the sample was removed from the oven and cooled in a desiccator for 30 min, and weighed.

b. Acid Soluble Lignin Contents

This test was performed based on the NREL LAP 003 standard [25]. The required volume of hydrolyzed filtrate sample and sulfuric acid (Merck) were diluted in a centrifuge tube or other container and was used to blank the spectrophotometer (Shimadzu UV-1800, Kyoto, Japan), which was set at a wavelength of 205 nm. The absorbance was measured within the range of 0.2–0.8 AU (absorption units). An absorbance outside this range indicates that the dilution should be performed with a blank of sulfuric acid (Merck).

c. Holocellulose Contents

A 1G3 filter funnel was dried in an oven at 105 °C for at least 4 h before testing. Furthermore, it was cooled in a desiccator for 30 min, and the oven-dry weight was weighed. A total of 1.0 g of the extractive-free sample was weighed and put into a 100 mL Erlenmeyer. Simultaneously, the sample's water content was measured, and 40 mL of aquadest was added. A total of 1.5 mL of 25% sodium chlorite (Merck) was added to 0.125 mL of 100% glacial acetic acid (Merck), tightly closed using heat-resistant plastic, and tightly tied using a rubber band. It was heated in a water bath for 1 h at 80 °C and repeated three times, cooled in an ice bath, and filtered using a previously weighed IG3 filter funnel In addition, it was washed with 100 mL of cold water and 25 mL of acetone and dried in an oven at 105 °C for 24 h. Consequently, it was removed from the oven and cooled in a desiccator for 30 min, and weighed.

d. Alpha Cellulose Contents

The test was performed using an empty 1G3 filter funnel dried in an oven at 105 °C for at least 4 h before testing. The IG3 was cooled in a desiccator for 30 min, and the oven-dry weight was weighed. Furthermore, a total of 0.5 g of the holocellulose sample was placed into a ±20 mL wide-mouth vial. Subsequently, the water content of the sample was measured, and 6.25 mL of 17% sodium hydroxide (Merck) was added, ensuring that all samples had been moistened with the solvent. It was stirred using a magnetic stirrer for 15 min at 320 rpm and left without stirring for 30 min. Afterward, 8.25 mL was added stirred for 5 min with 320 rpm, and left without stirring for 1 h. The sample was filtered using an IG3 filter glass, rinsed using 25 mL of 8.3% sodium hydroxide (Merck), and washed with 100 mL of aquadest. Furthermore, it was added to 10 mL of 10% acetone (technical) and soaked for 3 min, rinsed with aquadest until the pH became neutral, and dried in an oven at 105 °C for 24 h. Afterward, it was removed from the oven and cooled in a desiccator for 30 min, and weighed.

e. Extractive Contents in Ethanol Benzene (1:2)

This test was performed based on the standard TAPPI T 204 cm-97 [24] by drying the boiling flask in an oven at 105 °C for at least 4 h and cooling it in a desiccator for 30 min. A total of 3 g oven-dry weight (record the weight) of the empty boiling flask was weighed, inserted, and wrapped in filter paper with cotton at both ends and tied using mattress thread. The sample's moisture content was calculated simultaneously at the time of weighing the sample for extractives. Meanwhile, the extractive device was arranged and turned on, adding 150 mL of 150 mL ethanol–benzene (1:2) into a boiling flask. It was extracted for a period of 24 cycles for 4–5 h until all were dissolved in the extracting solution (marked by the extracting solution in a colorless soxhlet) and removed from the soxhlet. Furthermore, the extract solution in the flask was boiled in an oven at 40 °C and stored for further testing (holocellulose and lignin), evaporated until about 5 mL was left, and dried in an oven at 105 °C for 24 h. Afterward, it was cooled in a desiccator for 30 min and weighed.

f. Ash Contents

This test was performed based on TAPPI T 211 cm-02 standard [26]. The crucible porcelain was dried in a furnace (Blue M, New Columbia, United States) for 30–60 min at 525 ± 25 °C and cooled for 30–60 min in a desiccator, and then the weight was recorded. A total of 1 g of dried bamboo powder was weighed in crucible porcelain of known weight and put into a furnace at 525 ± 25 °C for 6 h. Afterward, it was removed from the furnace, cooled in a 30–60 min desiccator, and weighed.

2.2.2. Degree of Crystallinity

The degree of crystallinity was estimated using X-ray diffraction (XRD) analysis diffraction intensity data. At 40 kV and 30 mA electrical current. An XRD (MaximaX-XRD 700, Shimadzu, Kyoto, Japan) was employed with a Cu K X-ray source (0.15406 nm). A small amount of sample (40–60 mesh particle size) was placed in a holder glass and evaluated at room temperature with a 2°/min scanning speed and a 10–40° scanning angle. As shown below, the degree of crystallinity was estimated using Equation (1) [27].

$$Xc\ (\%) = \frac{Fc}{Fc + Fa} \times 100\% \tag{1}$$

where *Fc* denoted crystalline domains and *Fa* denoted amorphous domains.

With the assumption that the sharp and broad peaks related to crystalline and amorphous domains, the peaks (crystalline and amorphous) were fitted using the Lorentz function applied to the diffractograms.

2.2.3. Near-Infrared Spectroscopy Acquisition

Powdered bamboo samples were divided into three parts i.e., bottom (B), middle (M), and top (T), and then they were analyzed using a Fourier transform near-infrared (NIR) spectrometer (Perkin Elmer Frontier, Waltham, MA, USA) to obtain chemical information based on NIR specific bands. Scanning for each part was conducted at a wavenumber of 10,000–4000 cm^{-1}. The spectral resolution and scanning parameter was set to 16 cm^{-1} with 32 scans. The absorbance spectrum was recorded after the normalizing with the spectral as the background for the NIR machine [28]. After obtaining the original spectra, the second derivatives were made using a Savitzky–Golay method [29], smoothed at nine points, and in fifth polynomial order [30]. The script for second derivative spectra was written on Python programming 3.7 by the authors. The spectral region above 8000 cm^{-1} was excluded from this analysis due to the possibility of noises, and no significant information can be seen in this area [31]. The important chemical contents of belangke bamboos, such as lignin, extractives, cellulose, and hemicellulose, were analyzed based on the specific NIR band spectra by Schwanninger et al. [32].

2.2.4. Physical and Mechanical Properties Tests of Bamboo

Test samples and procedures for physical (specific gravity and shrinkage) and mechanical (compression and shear strength) properties were made according to the ISO 22157-1:2019 standard [33].

a. Specific gravity (SG)

Bamboo was cut into sizes of 2.5 cm × 2.5 cm and as thick as the wall thickness. Specific gravity of bamboo was calculated using Equation (2).

$$SG = \frac{ODW}{V} \times [1/\rho] \quad (2)$$

where *ODW* is oven-dry conditions (g), *V* is volume (cm^3), and ρ is density of water at 4 °C (1000 kg/m^3).

b. Shrinkage

The test sample was made from a bamboo column with a height or length of 10 cm (Figure 3). In the shrinkage test, the parameters determined are outer diameter, inner diameter, and linear shrinkage. Samples were measured as initial conditions before drying (*I*). The sample was put into an oven with a temperature of 103 ± 2 °C and then measured again (*F*). The value of shrinkage was calculated according to the Equation (3):

$$Shrinkage = \frac{I - F}{I} \times 100\% \quad (3)$$

where *I* is the initial reading and *F* is the final reading.

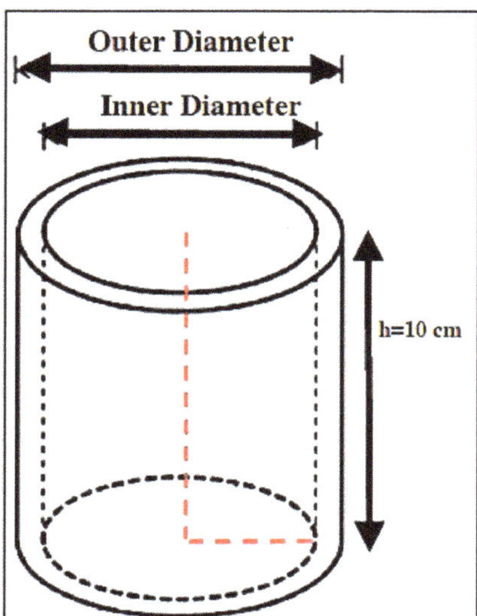

Figure 3. The sample used for the shrinkage test.

c. Compression strength test

The sample size of the compression test adjusts to the outer diameter of the bamboo. The height/length of the bamboo is the same as the outer diameter of the bamboo. For a bamboo diameter of less than 2 cm, the height of the bamboo sample is twice the outer diameter (Figure 4). The compressive strength was calculated using the Equation (4):

$$\sigma_{ult} = \frac{F_{ult}}{A} \tag{4}$$

where σ_{ult} is the ultimate compressive stress, in MPa, rounded off to the nearest 0.5 MPa; F_{ult} is the maximum load at which the specimen fails, N; and A is the cross-sectional area, mm^2.

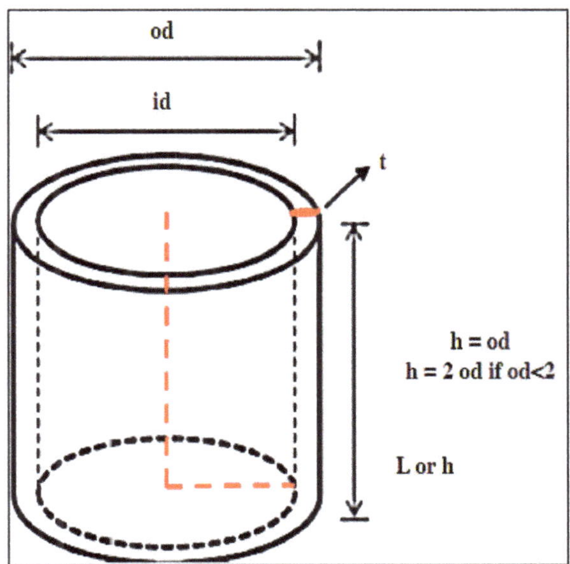

Figure 4. Sample used for the compression test.

d. Shear strength test

The sample size for the shear strength test adjusts to the outside diameter. Tests were performed parallel to the fibers. The height of the bamboo is equal to the outer diameter (Figure 4). The shear strength of bamboo can be calculated using Equation (5):

$$\tau_{ult} = \frac{F_{ult}}{\sum(t \times L)} \tag{5}$$

where τ_{ult} is the ultimate shear strength, MPa, rounded off to the nearest 0.1 MPa; F_{ult} is the maximum load at which the specimen fails, N; and $\sum(t \times L)$ is the sum of the four products of t (wall thickness) and L (height).

e. Tension strength test

The tensile strength test sample was prepared according to the ISO/TR 22157-2: 2004 standard [34]. The test sample is shown in Figure 5. The tensile strength was calculated using Equation (6):

$$\sigma_{ult} = \frac{F_{ult}}{A} \tag{6}$$

where σ_{ult} is the ultimate tension strength, MPa; F_{ult} is the maximum load at which the specimen fails, N; and A is the volume of the sample (mm^3) $A = \frac{\pi}{4} \times \left[D^2 - (D - 2t)^2\right]$.

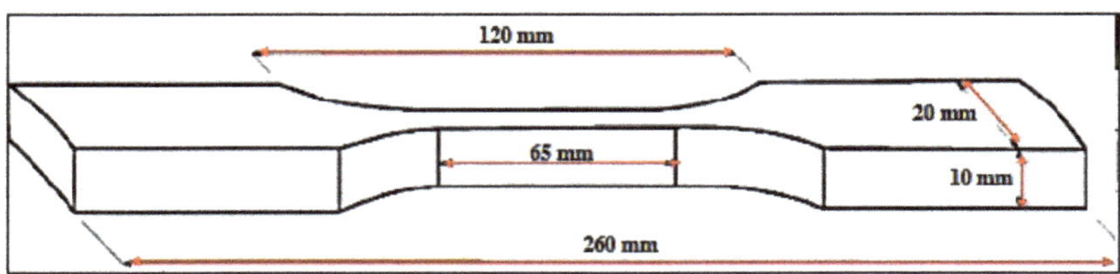

Figure 5. The sample used for the tension test.

2.2.5. Particleboard

Material Preparation

Kemenyan wood (*Styrax sumatrana*) was converted into two types of particles: sawdust and wood shavings. Then, the belangke bamboo (*Gigantochloa pruriens*) was converted into a barkless strand. The particle sizes are presented in Table 1. The commercial adhesive used in this research was isocyanate type H3M with the specifications shown in Table 2. The isocyanate adhesive was produced by PT. Polychemie Asia Pacific (Jakarta, Indonesia).

Table 1. The particle size of *Styrax sumatrana* wood and *Gigantochloa pruriens.*

Particle Type	Length (cm) *	Width (cm) *	Thickness (cm) *	Mesh	Slenderness Ratio
Sawdust	-	-	-	20	-
Shaving	35.26 (11.44)	18.66 (5.76)	0.14 (1.08)	N/A	252.02 (35.09)
Strand 7.5 cm	7.50 (0.03)	2.51 (0.03)	0.13 (0.02)	N/A	62.52 (8.25)
Strand 25 cm	25.34 (0.19)	2.79 (0.40)	0.09 (0.25)	N/A	297.69 (91.73)

Remark: (*) the number of measurement samples is 100 units. Values in parentheses represent the standard deviation.

Table 2. Properties of isocyanate adhesive used in this work.

Properties	Information
Solids content	98%
Viscosity	150–250 cps/23 °C
pH	6.5–8.5
Pot life mixture	Within 60 min
Spread volume	200~250 g/m^2
Assembly time	Within 10 min
Cold press for a low-density wood	0.69–0.98 MPa
Cold press for a high-density wood	0.98–1.47 MPa
Press time Cold Press	30–45 min depends on wood species, size of lamella, temperature, and spread volume

Source: PT. Polychemie Asia Pasific (Jakarta, Indonesia).

The raw materials used, i.e., wood shavings, sawdust, and belangke bamboo, were placed in the oven until they reached a moisture content of 8%. In this work, three-layered particleboard (face, core, and back, at the ratio of 2:1:2) with dimensions of 250 mm × 250 mm a thickness of 10 mm, and a target density of 750 kg/m^3 were fabricated in the laboratory.

Particleboard Manufacturing

The dried particles were sprayed with 7% isocyanate adhesive. Meanwhile, for the surface layer (face and back), belangke bamboo strands were glued together using a brush in one of the surface areas. The bamboo strand as the back layer was arranged first into a mold measuring 250 mm × 250 mm, followed by the particles as the core layer. The next stage was laying the bamboo strand as a surface layer. The mattress was prepressed manually, then subjected to hot pressing in an automatically controlled laboratory hot press (custom by home industry in Bandung, Indonesia) at a temperature of 160 °C and pressure of 2.94 MPa for 5 min. The specifications of the laboratory hot press are presented in Table 3.

Table 3. Specification of the hot press used.

No	Specification	
1	Height	160 cm
2	Length	90 cm
3	Width	50 cm
4	Pressure plate area	35×35 cm^2
5	Maximum hydraulic pressure	20.59 MPa
6	Hydraulic lifting	100 ton
7	Maximum heating power	2×3000 watt
8	Maximum temperature	250 °C

Physical and Mechanical Properties of Particleboard

After hot pressing, the laboratory-fabricated particleboard panels (Figure 6) were conditioned for seven days in a conditioning room at a relative humidity of 65 ± 5% and 20 ± 2 °C prior to properties evaluation. The cutting of the test sample and evaluation of the boards' physical and mechanical properties, i.e., density, moisture content, water absorption, thickness swelling, modulus of elasticity (MOE), modulus of rupture (MOR), and internal bond (IB), were performed according to the standard JIS A 5908 [35]. The sizes of the test samples used are presented in Table 4.

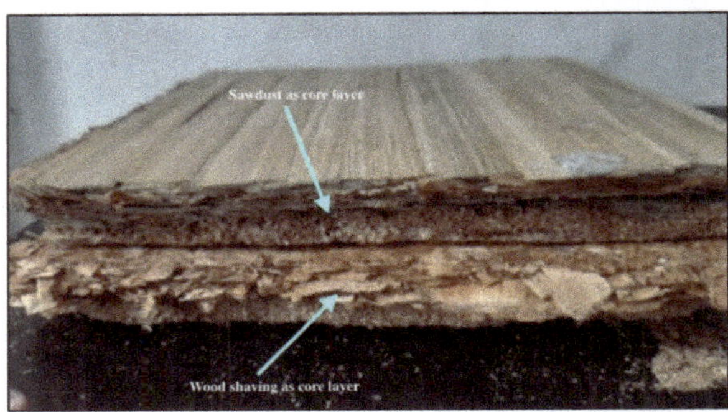

Figure 6. The appearance of coated and uncoated particleboard with belangke bamboo strands.

Table 4. Particleboard test sample sizes.

Parameter	Size
Density	10 cm (length) × 10 cm (width)
Moisture content (MC)	10 cm (length) × 10 cm (width)
Water absorption (WA)	5 cm (length) × 5 cm (width)
Thickness swelling (TS)	5 cm (length) × 5 cm (width)
Modulus of elasticity (MOE)	20 cm (length) × 5 cm (width)
Modulus of rupture (MOR)	20 cm (length) × 5 cm (width)
Internal bond (IB)	5 cm (length) × 5 cm (width)

2.3. Data Analysis

The chemical components and physical–mechanical properties of bamboo were performed using a nonfactorial, completely randomized design. Chemical analysis was performed in duplicate, making a total of six-unit test samples. Meanwhile, for the physical and mechanical properties of bamboo, repeated testing of each test parameter was performed four times. The total test samples in this sub-research are twenty test sample units. The particleboard study used a factorial completely randomized design (CRD) model. The first factor is the type of particle (sawdust and shaving), and the second factor is the length of the bamboo strand as a coating material (75 mm and 250 mm). The number of replications for each treatment was three units. The total test samples in particleboard research are eighteen test sample units.

3. Results and Discussions

3.1. Chemical Components of Bamboo

3.1.1. Structural Chemical Components of *Gigantochloa pruriens*

Acid Insoluble Lignin (AIL) and Acid Soluble Lignin (ASL)

Lignin content is the total of AIL and ASL with the average value presented in Figure 7.

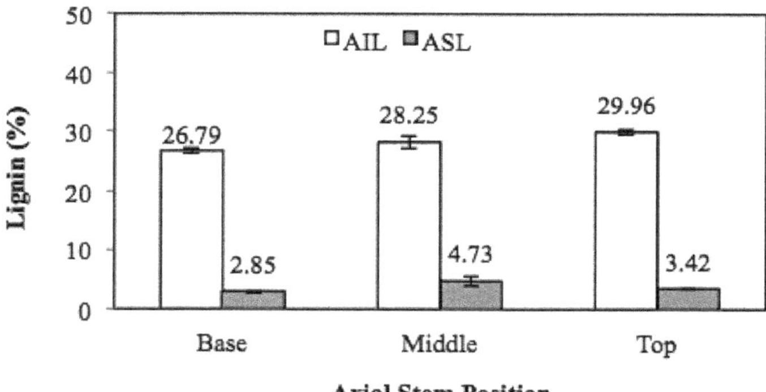

Figure 7. Lignin contents of *Gigantochloa pruriens*.

The AIL and ASL values ranged from 25.25–27.56% and 2.73–4.47%, respectively. The total lignin (AIL + ASL) content of *Gigantochloa pruriens* ranged from 27.97–30.79%, with the average value of the lignin content in all parts of the culm being 29.79%. Lignin in bamboo plants consists of p-coumaryl units with guaiacyl, syringyl, and p-hydroxyphenyl groups, which is similar to grasses or lignin in *Poaceae* plants [36].

The study related to the chemical components of bamboo performed by Li [37] reported that the lignin content of *Phyllostachys pubescens* ranged from 20–26%, while that of *Bambusa vulgaris* ranged from 11.6–20.3% [38]. Furthermore, Sharma et al. [39] reported that *P. pubescens* bamboo had a lignin content of around 22%. Loiwatu and Manuhuwa [40] reported that the lignin content of three types of bamboo, namely *Dendrocalamus asper*,

Schizostachyium brachycladum, and *Schizostachyium lima*, was each 27, 37, 26, 18, and 26.05%. The lignin content of the species *Bambusa vulgaris* Schard var. vitata, *Gigantochloa nigrocillata*, *Gigantochloa apus*, *Gigantochloa verticillata*, and *Dendrocalamus asper*, obtained from Cibinong, West Java-Indonesia, varied between 30.01–36.88% [6].

Meanwhile, several researchers have reported the lignin content in other bamboo species from the *Gigantochloa* genus, including *Gigantochloa apus* with 22–25% [41] and *Gigantochloa brang*, *Gigantochloa levis*, *Gigantochloa Scortechinii*, and *Gigantochloa wrayi*—24%, 26%, 32%, and 26%, respectively [42]. In addition, Yusoff et al. [43] reported that the lignin content of *Gigantochloa albociliata* bamboo ranged from 22–24%. Based on the content of several *Gigantochloa* species, the average value of *Gigantochloa pruriens* lignin content is 32%, which is relatively higher compared to the other bamboo species. The high lignin content contributes to the high calorific value and structural rigidity, making it ideal for building components [5]. Statistical analysis showed that the axial position of the culm had no significant effect at the 95% confidence interval on the AIL (sig. 0.071) and ASL (sig. 0.054) values.

Holocellulose Content

The holocellulose content of *Gigantochloa pruriens* bamboo ranged from 63.56% to 66.66% (Figure 8), with the lowest and highest content being at the top and bottom of the culm, respectively. The difference in each part of the culm was also reported by Li [37], who determined that the culm's position affects the amount of holocellulose content. *Gigantochloa pruriens* bamboo has holocellulose content equivalent to *Gigantochloa pseudoarundinacea* [44], which is slightly lower compared to *Dendrocalamus asper*, *Schizostachyium brachycladum*, *Schizostachyium lima* [40], and *Bambusa vulgaris* Var vulgaris [44].

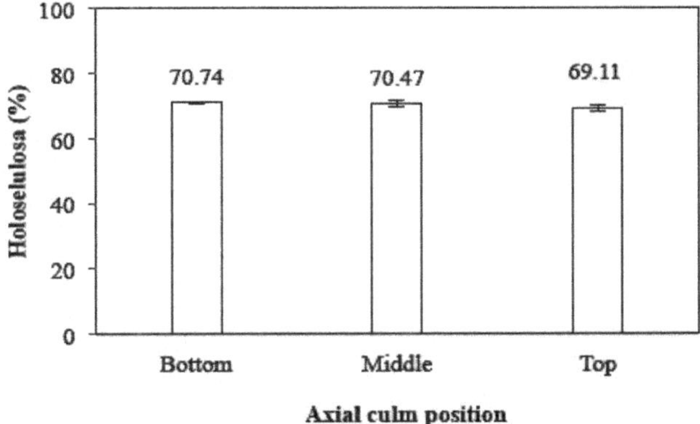

Figure 8. Holocellulose content of *Gigantochloa pruriens*.

Wahab et al. [42] obtained a holocellulose content value of 79%, 84%, 74%, and 84% for the bamboo species *Gigantochloa brang*, *Gigantochloa levis*, *Gigantochloa scortechinii*, and *Gigantochloa wrayi*, respectively. Yusoff et al. [43] reported that the holocellulose value of *Gigantochloa albociliata* ranged from 53–57%, higher than *Gigantochloa albociliata*. Statistical analysis showed that the axial position of the culm has no significantly different effect on the 95% confidence interval for holocellulose (sig. 0.239).

Alpha-Cellulose Content

A graphical representation of the alpha-cellulose content of *Gigantochloa pruriens* is given in Figure 9. The values ranged from 39.70–44.40%, with the lowest and highest holocellulose being in the top and at the bottom of the culm, respectively. The average value of alpha-cellulose in all parts of the culm was 41.48%, which falls within the normal range of alpha-cellulose (40–55%) as reported by Li [37]. This indicated that it was equivalent to *Schizostachyium brachycladum* but relatively lower compared to *Dendrocalamus asper* and *Schizostachyium lima*, as reported by Loiwatu and Manuhuwa [40]. However, it was still far below *Bambusa vulgaris*, as reported by Nahar and Hasan [45].

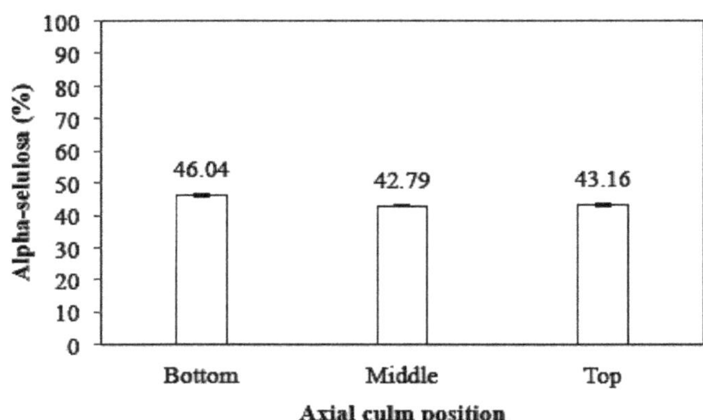

Figure 9. The alpha-cellulose contents of *Gigantochloa pruriens*.

Li [37] stated that the range of cellulose content of bamboo was very suitable for its application as a raw material for the pulp and paper industry. Meanwhile, Nugroho et al. [46] reported that the parallel tensile strength of the bamboo fiber is quite highly correlated with the content of alpha-cellulose. Furthermore, it was stated that the mechanical properties and cellulose fraction in the cell wall play a greater role compared to the total polysaccharide content. Cellulose which has a linear polymer structure may affect the mechanical properties of wood, while more amorphous hemicellulose may affect the hygroscopic properties of bamboo.

Indriatie et al. [41] reported that *Gigantochloa apus* has an alpha-cellulose content of 69–72%, while Wahab et al. [42] reported that the α-cellulose content of *Gigantochloa brang*, *Gigantochloa levis*, *Gigantochloa scortechinii*, and *Gigantochloa wrayi* were 51, 33, 46, and 37%, respectively. Yusoff et al. [43] reported that the alpha-cellulose content of *Gigantochloa albociliata* ranged from 42–43%, which was relatively lower compared to *Gigantochloa apus*, *Gigantochloa brang*, and *Gigantochloa scortechinii*. The axial position of the culm had a significant effect on the 95% confidence interval for alpha-cellulose (sig. 0.005).

3.1.2. Nonstructural Chemical Components of *Gigantochloa pruriens* Bamboo

Extractive Solubility in Ethanol Benzene (1:2)

Extractives are nonpolymer organic materials soluble in nonpolar neutral organic solvents, such as alcohol-benzene, including hydrophobic oils, waxes, and resins [47] The average value of the extractive solubility in benzene ethanol (1:2) of *Gigantochloa pruriens* bamboo is presented in Figure 10, ranging from 2.18–4.01%. The lowest and highest values were determined in the middle and at the bottom of the culm, respectively. This value is similar to that stated by Tolessa et al. [48], who state that the extractive content of *Oxytenanthera abyssinica* bamboo ranges from 3–5%.

The low extractive content is favorable for use in the pulp and paper industry [49]. The determined extractive content of *Gigantochloa pruriens* bamboo was relatively lower compared to *Gigantochloa apus* [41], *Gigantochloa brang*, *Gigantochloa levis*, *Gigantochloa scortechinii* and *Gigantochloa wrayi* [42]. However, it was slightly higher compared to *Gigantochloa albociliata* bamboo [43].

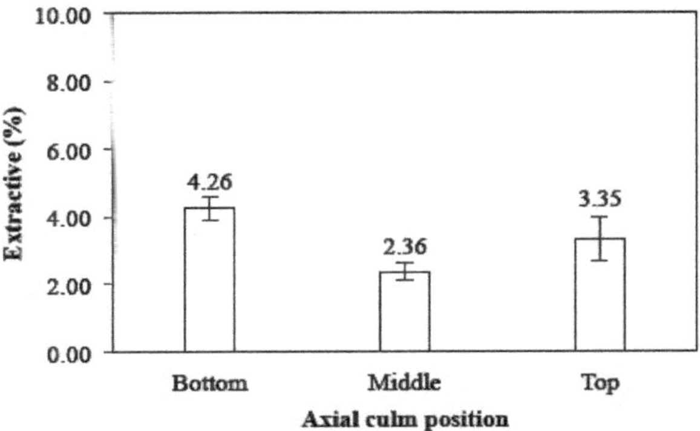

Figure 10. Extractive solubility in ethanol benzene (1:2) of *Gigantochloa pruriens*.

The extractive content of *Gigantochloa pruriens* bamboo was different in each part of the culm. The extractive content of ethanol–benzene was less than 3–5%, contributing to several properties, such as color, odor, decay resistance, density, and flammability [48]. Extractive substances are a determining factor in the shrinkage properties of wood through physical and chemical mechanisms. These extractive substances play a vital role in wood stability as a bulking agent. Furthermore, they play a chemical role through the hydrophobic nature of the compounds contained in extractive substances. It is generally nontoxic, therefore enabling the preservation of all bamboo species before use because of their susceptibility to be attacked by powdery mildew, fungi, and termites compared to wood [40]. The higher alcohol–toluene extractive content in bamboo may be useful as an anti-rot agent and provides good strength due to its higher specific gravity. Generally, the higher solubility of tannins, gums, sugars, starches, and dyes indicates easier access and penetration of chemicals into cell walls [48]. Statistical analysis showed that the axial position of the culm has a significantly different effect on the 95% confidence interval on the solubility of the extractive in benzene–ethanol (sig. 0.055).

Ash Contents

The average value of the ash content of *Gigantochloa pruriens* bamboo ranged from 1.36 to 2.57%, with an increase in ash content from the position of the bottom of the culm to the top, as seen in Figure 11. Ashes may be identified due to unburned compounds containing calcium, potassium, magnesium, manganese, and silicon.

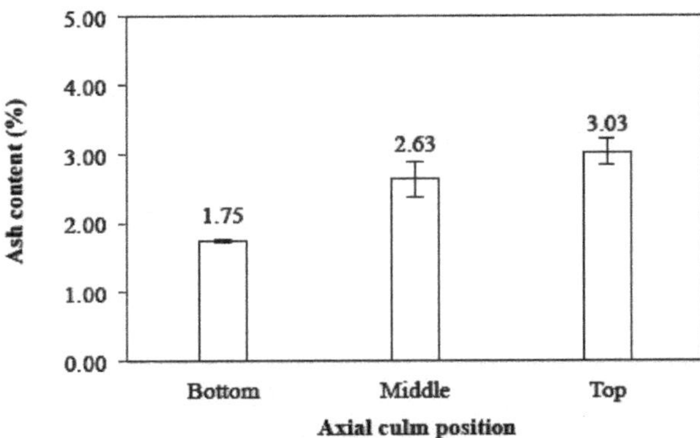

Figure 11. Ashes content based on *Gigantochloa pruriens*.

The average value of ash content in all culms was 2%, while the ash content of *Gigantochloa brang*, *Gigantochloa levis*, *Gigantochloa scortechinii*, and *Gigantochloa wrayi* are 1.2; 1.3; 2.8; and 0.8% [42], respectively. Yusoff et al. [43] reported that the ash content of *Gigantochloa albociliata* ranged from 1.5–1.8%. Meanwhile, that of bangkele bamboo is equivalent to *Gigantochloa apus* (2.45%) and *Gigantochloa levis* (2.45%) but lower compared to that of *Gigantochloa atroviolacea* (3.29%) [50]. However, it is relatively higher than *Gigantochloa brang*, *Gigantochloa levis*, *Gigantochloa scortechinii*, and *Gigantochloa albociliata*. The axial position of the culm had a significantly different effect on the 95% confidence interval on the ash content of *Gigantochloa pruriens* bamboo (sig. 0.013).

3.2. Degree of Crystallinity

The crystallinity of woody plants is the crucial factor for responding to the properties of tree growth, anatomical structure, and chemical characteristics. Furthermore, it also has a great effect on Young's modulus, dimensional stability, density, and wood hardness [51]. The degree of crystallinity of *Gigantochloa pruriens* bamboo is based on axial position and was calculated based on the relative amounts of crystalline and amorphous parts in the bamboo (Table 5). The X-ray diffractogram of bamboo in the top, middle, and bottom had a similar pattern as shown in Figure 12, with the highest peak in the middle position as the present Cellulose I structure with a monoclinic crystal structure. This structure is commonly found in plants, including bamboo. Although the alpha-cellulose content of bamboo in the middle position was the lowest, the degree of crystallinity in the middle position was the highest. It can be influenced by the difference in chemical composition in each axial position. There is no direct correlation between the cellulose content of hemicellulose and the lignin content of bamboo with the degree of crystallinity. A similar report was also found to have no significant variation in crystallinity in the longitudinal direction within each internode of bamboo. Further observation is needed to ensure the reason for the relationship between bamboo growth and lignification and chemical composition [52]. Lignin and hemicellulose can be included in the amorphous part (2θ of 15–22°), while the crystalline part is derived from cellulose with 2θ of around 22° [51,53,54]. Compared to a previous report on betung bamboo [55], which found that the degree of crystallinity was 30.43%, the degree of crystallinity In the bottom was higher than the value. This degree of crystallinity is lower than that of acacia and pine wood by as much as 44.10% and 35.73%, respectively [56]. The variation in the degree of crystallinity based on the axial position in *Gigantochloa pruriens* bamboo had a similar tendency to the variation in chemical composition in the axial position of raru wood [57]. In addition to affecting the chemical properties, the degree of crystallinity also affects the mechanical properties in this study. The high degree of crystallinity in the middle of the bamboo belangke culm is shown to cause the tensile strength value to be higher than the other culm positions. Barrios et al. [58] and Takeuchi et al. [59] state that the degree of crystallinity positively correlates with dimensional stability and tensile strength. Furthermore, Li et al. [60] also reported that a high degree of crystallinity in the outer wood of *Pinus radiata* caused an increase in the value of MOR, MOE, and compression parallel to the fiber.

Table 5. Degree of crystallinity of *Gigantochloa pruriens* bamboo based on the axial position.

Sample	*Fc* (kcps.deg)	*Fa* (kcps.deg)	*Xc*%
Bottom	22.8276	46.5733	32.89
Middle	41.598	68.0017	37.95
Top	26.567	62.6533	29.78

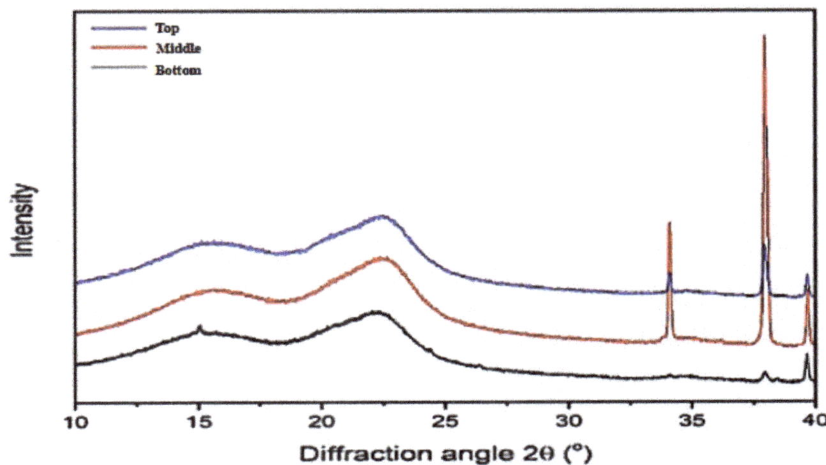

Figure 12. X-ray diffractograms of *Gigantochloa pruriens* bamboo based on axial position.

3.3. Near-Infrared Analysis

A graphical representation of the original (A) and second derivative spectra (B) of belangke bamboo in the axial direction is given in Figure 13. At first glance, the original spectra of belangke bamboo in the axial direction (Figure 13A) were almost similar, thus the important bands associated with major chemical components of bamboo were not well resolved. These spectra have only visualized the discrepancies in baseline level [61], where the top part of the bamboo was slightly higher than other regions. Via et al. [62] reported that density was one of the factors that affected the absorbance of NIR spectra. The higher density corresponded to a higher absorbance level. Typically, the bamboo's top is higher than its other portions [56].

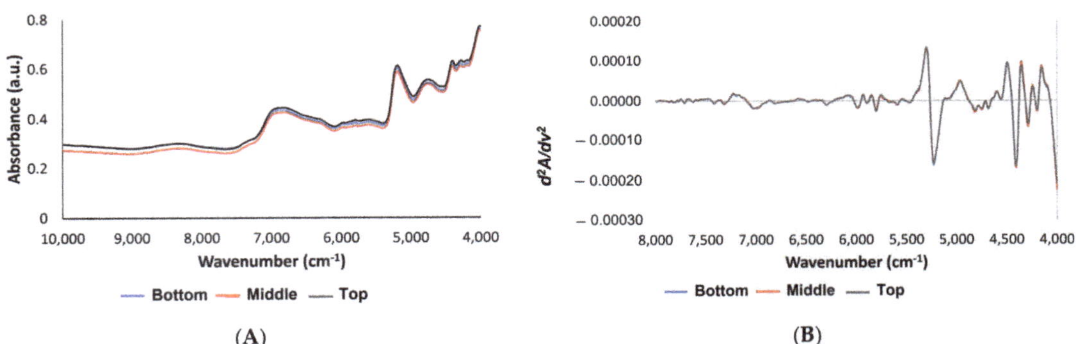

Figure 13. The original spectra of NIR at the wavenumber 10,000–4000 cm^{-1} (**A**); second derivative spectra at the wavenumber 8000–4000 cm^{-1} (**B**).

On the contrary, second derivative spectra (Figure 13B) visualized more different bands than the original ones [63]. In addition, these spectra revealed some important bands of the chemical components [28], thus analyzing the chemical contents, such as lignin, cellulose, holocellulose, and extractives, which were based on these spectra. The second derivative spectra were divided into four regions based on the characteristics of the chemical functional groups: the first region 10,000–7300 cm^{-1} were second or third overtones, but there was less information on wood components; the second region 7300–6050 cm^{-1} were OH overtone vibrations; the third range 6050–5500 cm^{-1} were CH vibrations and the vibrations from aromatic structure; and the fourth region 5500–4000 cm^{-1} were several combinations of vibrations and coupling vibrations [28,32]. Although the NIR second derivative spectra of belangke bamboo were almost identical, some peak regions had different characteristics in the axial direction of the stem.

Therefore, this study analyzed the peak characteristics of the NIR spectra of belangke bamboo, which are assigned to the main chemical components, such as cellulose, holocellulose, lignin, and extractives, through enlarging the band regions on the second derivative spectra (Figure 14). These bands were mainly located in the third range for lignin and the fourth range for cellulose. According to the NIR second derivative spectra, the band assigned to holocellulose, such as 4404 cm^{-1} (cellulose and hemicellulose), was identical. This result corresponded to the wet chemical analysis for holocellulose in terms that the values were not significantly different based on the axial direction. Markedly, bands 4283 cm^{-1} assigned to cellulose, hemicellulose, and xylan seem slightly different with the bottom part having the lowest peak. In the band at 4808 cm^{-1} at the semi-crystalline and crystalline regions as well as 4739 cm^{-1} (cellulose) at the cellulose group region, the middle part had the lowest value. It was assumed that this region had a higher degree of crystallinity, as demonstrated by the positive correlation with the results of the XRD analysis. The extractive content with the bands at 4686 cm^{-1} showed that the bottom part was supposed to have high extractive content while the band at 5980 cm^{-1} showed that the middle part tended to have more lignin content than another part. However, the wet chemical analysis revealed that there were nonsignificant values of lignin content at the different high levels of the bamboo stem.

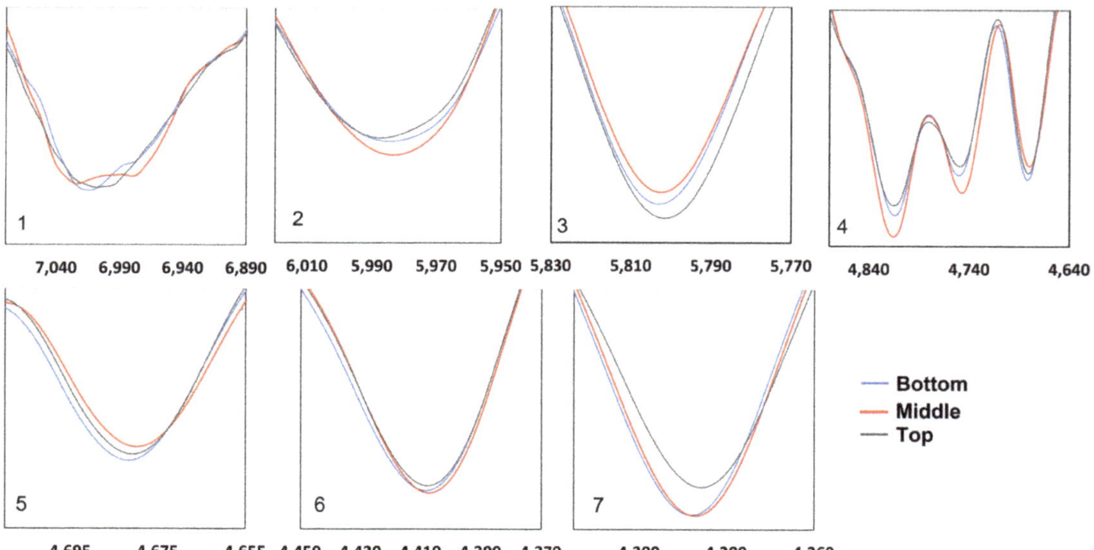

Figure 14. Enlarged bands at specific regions of second derivative spectra: (**1**) bands 7000 cm^{-1} assigned to an amorphous region of cellulose; (**2**) bands 5980 cm^{-1} assigned to aromatic skeletal due to lignin; (**3**) bands 5800 cm^{-1} assigned to furanose or pyranose due to hemicellulose; (**4**) bands 4890–4,620 cm^{-1} assigned to cellulose region; (**5**) bands 4686 cm^{-1} assigned to lignin or extractives; (**6**) bands 4404 cm^{-1} assigned cellulose and hemicellulose; and (**7**) bands 4283 cm^{-1} assigned to cellulose, hemicellulose, and xylan.

3.4. Physical and Mechanical Properties of Bamboo

3.4.1. Physical Properties of Bamboo

A summary of the physical properties of belangke bamboo is presented in Table 6.

Table 6. Physical properties of *Gigantochloa pruriens*.

Physical Properties	Bottom	Middle	Top
Specific gravity	0.60 (0.08)	0.58 (0.03)	0.60 (0.03)
Outer Diameter Shrinkage (%)	2.33 (0.45)	1.29 (0.45)	2.33 (0.10)
Inner Diameter Shrinkage (%)	1.94 (0.62)	0.94 (0.62)	3.67 (0.14)
Linear Shrinkage (%)	0.13 (0.07)	0.2 (0.04)	0.21 (0.09)

Values in parentheses are standard deviation.

Specific Gravity (SG)

The SG value of belangke bamboo ranged from 0.58 to 0.60. The bottom had a higher SG value than the middle, which was confirmed by its chemical components where this bamboo had a higher alpha-cellulose content value at the bottom. High SG indicates a higher proportion of cell walls and cellulose content, which will affect the mechanical properties, especially the MOR value, as stated by Baharoğlu et al. [64]. The SG value of belangke bamboo was equivalent to that of apus bamboo (*Gigantochloa apus*), as reported by Abdullah et al. [65]. Analysis of variance on the SG value showed that the axial position of the bamboo culm was not significantly different at the 95% confidence level (sig. 0.696).

Shrinkage

The inner diameter, outer diameter, and linear shrinkage ranged from 0.94–3.67%, 1.29–3.17%, and 0.13–0.21%, respectively. The shrinkage of the outer diameter was greater than the inner diameter due to the denser vascular bundle diameter on the outside of the bamboo [14]. The effect of density is evident in this shrinkage parameter. The greater the density, the greater the shrinkage value of the bamboo. Furthermore, the linear shrinkage tends to increase from the bottom to the top of the bamboo stem. Darwis et al. [14] reported that the percentage of belangke bamboo vascular bundles at the bottom of the culm was lower than at the middle and top. The high proportion of vascular bundles impacts the amount of bamboo shrinkage. Analysis of variance on the inner and outer shrinkage values showed that the axial position of the bamboo culm was not significantly different at the 95% confidence level. At the same time, linear shrinkage was not significantly different.

3.4.2. Mechanical Properties of Bamboo

A summary of the results obtained for the mechanical properties of belangke bamboo is presented in Table 7.

Table 7. Mechanical properties of *Gigantochloa pruriens*.

Mechanical Properties	Bottom	Middle	Top
Compression Strength (MPa)	45.44 (0.66)	42.15 (6.63)	38.96 (8.84)
Tensile Strength (MPa)	116.77 (20.90)	278.74 (15.20)	95.88 (5.46)
Shear Strength (MPa)	7.39 (0.61)	7.69 (1.59)	7.79 (0.94)

Values in parentheses are standard deviation.

Compression Strength

The compressive strength values ranged from 38.96–45.44 MPa. The highest value was determined at the bottom of the bamboo culm, and the lowest value was obtained at the top. The compressive strength value of belangke bamboo decreased from the bottom to the top. Darwis et al. [14] reported that the number of vascular bundles in belangke bamboo at the position of the bottom of the stem was the lowest. Chowdhury et al. [66] stated that the fiber's compression value was negatively correlated with fiber diameter and vessel cells but positively correlated with vessel wall thickness. Analysis of variance on compression strength value showed that the axial position of the bamboo culm was not significantly different at the 95% confidence level (sig. 0.396).

Tensile Strength

The tensile strength values ranged from 95.88 to 278.74 MPa, the highest value was in the middle of the bamboo culm, and the lowest was at the top. According to Epsiloy [67], fiber length affects the mechanical properties of bamboo. Darwis et al. [14] reported that the middle part of the belangke bamboo culm has longer fibers than the bottom and ends of the bamboo, so it has the highest tensile strength value. In addition to differences in anatomical structure, the chemical components that comprise bamboo can also cause these differences. Lignin and alpha-cellulose significantly affect the tensile strength of bamboo. Lignin contributes to the stiffness properties, while cellulose has a linear polymer structure with a high crystalline fraction. The results showed that the center of the bamboo culm had the highest crystallinity value. Analysis of variance on the tensile strength value showed that the axial position of the bamboo culm was significantly different at the 95% confidence level (sig. 0.000).

Shear Strength

The value of shear strength ranged from 7.39 to 7.79 MPa; the highest value for shear strength was at the top of the bamboo culm, and the lowest was at the bottom. It was attributed to the presence of belangke bamboo parenchyma. The proportion of vascular bundles and parenchyma as the matrix affects the shear strength of bamboo. When the number of proportions of the matrix gets bigger, the tendency to decrease the shear strength gets bigger. Darwis et al. [14] reported that the bottom of the belangke bamboo culm has the largest parenchyma. The cause of the low shear strength at the bottom is also due to the low lignin content. Longui et al. [68] stated that lignin is proven to be able to increase the stiffness properties of the secondary walls and cohesion between wood tissues and has a strong effect on the mechanical properties of wood, especially in the transverse direction. Analysis of variance on the shear strength value showed that the axial position of the bamboo culm was not significantly different at the 95% confidence level (sig. 0.874).

3.5. Physical and Mechanical Properties of Particleboard

A summary of the physical and mechanical properties of the laboratory-fabricated three-layered particleboards, manufactured with belangke bamboo as a reinforcing material, is given in Table 8.

Table 8. Physical and mechanical properties value of particleboard produced in this work. Particleboard made with 100% wood sawdust were denoted as A0 while those reinforced with 75 mm and 250 mm bamboo strands were denoted as A1 and A2, respectively. Meanwhile, particleboard made with 100% wood shavings were denoted as B0 while those reinforced with 75 mm and 250 mm bamboo strand were denoted as B1 and B2.

Parameter	20 Mesh Sawdust			Wood Shaving		
	Control (A0)	**Strand 7.5 (A1)**	**Strand 25 (A2)**	**Control (B0)**	**Strand 7.5 (B1)**	**Strand 25 (B2)**
Density (kg/m^3)	690 (30)	640 (30)	680 (40)	670 (10)	670 (30)	620 (10)
MC (%)	7.96 (0.25)	7.10 (0.18)	6.89 (0.37)	7.10 (0.46)	6.66 (0.17)	6.56 (0.30)
WA (%)	47.70 (5.99)	45.94 (8.83)	68.41 (6.73)	21.72 (5.39)	31.46 (6.72)	46.12 (3.60)
TS (%)	4.38 (0.77)	6.50 (1.56)	7.37 (1.38)	5.29 (0.86)	7.01 (1.05)	9.57 (2.42)
MOE (GPa)	0.32 (0.12)	5.28 (1.08)	3.34 (0.45)	2.41 (0.19)	13.63 (0.47)	6.91 (0.27)
MOR (MPa)	6.39 (1.99)	22.38 (2.40)	11.38 (3.88)	28.52 (0.38)	43.17 (0.89)	48.37 (8.19)
IB (MPa)	0.39 (0.09)	0.19 (0.04)	0.29 (0.08)	0.29 (0.04)	0.15 (0.04)	0.18 (0.01)

Values in parentheses are standard deviation.

3.5.1. Density

As seen from the results in Table 8, the average values of particleboard density ranged from 620–690 kg/m^3. The lowest value was found on the B2 type board, while the highest value was obtained for the A0 type board. The density of the laboratory-produced boards was below the targeted value of 750 kg/m^3. This might be attributed to: (1) the loss in particles during the mat-forming and compression processes following the statement of Bufalino et al. [69] that the low value of particleboard density is due to the presence of some particles that are wasted during the manufacturing process; (2) spring back, in which the average spring back value in this study was almost 30%, meaning that the board after conditioning experienced a thickness swelling of nearly 1/3 of the target thickness of 1 cm. Several other factors that affect the density value of particleboard include the density of wood, compression pressure, and the amount of adhesive used [70]. Bowyer et al. [71] stated that the density value is highly dependent on the wood density and compression pressure.

Analysis of variance on the density value showed that the interactions between particle types and strand length are not significantly different at the 95% confidence level (sig. 0.070). Overall, the density value of this particleboard has met the JIS A 5908-2003 standard, which requires the board density value to be between 400–900 kg/m^3.

3.5.2. Moisture Content (MC)

According to Table 7, the moisture content values of the particleboard ranged from 6.56–7.96%. The lowest value was found for the B2 type board, while the highest value was determined for the A0 type board. Particleboard with sawdust core layer had a higher moisture content than the wood shavings core layer. Farrokhpayam et al. [72] reported that the fine particle size would absorb water vapor more significantly than the coarse particle size. Using bamboo strand as a coating on the board's surface tends to reduce the board's MC. Analysis of variance on the MC value showed that the interactions between particle types and strand length are not significantly different at the 95% confidence level (sig. 0.107). Overall, the MC of this particleboard met the JIS A 5908-2003 standard requirements, i.e., MC values between 5–13%.

3.5.3. Water Absorption (WA)

The WA value of the board ranges from 21.72–68.41%. The lowest value was found for the B0 type board, while the highest one was determined for the A2 type board. A similar trend with the MC parameter also occurs in the WA parameter. A board with a smaller particle size results in a relatively higher WA value than a board with larger particles. Sawdust has a greater high-surface area per unit weight compared to that of wood shaving; therefore, it tends to absorb more water [19]. In this study, using bamboo strands as a surface layer resulted in increased WA value compared to the uncoated boards—density is a factor that affects the board's WA value. A higher board density causes a lower board WA value. Bufalino et al. [69] stated that the density value and adhesive content affect water absorption. Meanwhile, Vital et al. [73] noted that the board density value is the opposite of the WA value. Analysis of variance on the WA value showed that the interactions between particle types and strand length are significantly different at the 95% confidence level (sig. 0.000).

3.5.4. Thickness Swelling (TS)

Thickness swelling values ranged from 4.38–9.57%. The lowest value was determined for the A0 type board, while the highest value was obtained for the B2 type board. Using bamboo strands as a surface layer increases the board's TS value. It is related to the density of the panels produced in this study. Strand-coated boards on the surface tend to have lower density values. The board with belangke bamboo strands on the surface exhibited a spring back of 60% higher than the control. Iswanto et al. [74] reported the effect of relatively high spring back as an indicator of the cause of the high TS value of the sorghum stem particleboard coated with rope bamboo strand (*Gigantochloa apus*). The high spring back of the board is thought to be due to the weak bond between sawdust or wood shavings and bamboo. The inhomogeneity of these two types of particles is a weak point, resulting in the board's high spring back value being coated with bamboo strands on the surface.

Overall, boards with a core layer of sawdust had a lower TS value than wood shavings. Farrokhpayam et al. [72] reported that the finer particle size resulted in a lower TS value of the board than the coarser or larger particle size. Analysis of variance on the TS showed that the interaction between particle types and strand length are not significantly different at a 95% confidence interval for the resulting particleboard thickness swelling value (sig. 0.531). Markedly, the particleboard panels fabricated in this work met the JIS A 5908-2003 standard requirements of a maximum thickness swelling of 12%.

3.5.5. Modulus of Elasticity (MOE) and Modulus of Rupture (MOR)

The average MOE value of the laboratory-fabricated particleboards ranged from 0.32–13.63 GPa, and the MOR value ranged from 6.39–48.37 MPa. The lowest MOE value of 0.3 GPa was determined for the A0 board, while the highest value of 13.6 GPa was obtained for the B1 board. Regarding the MOR, the lowest value of 6.4 MPa was determined for the A0 particleboard, and the highest value of 48.4 MPa was obtained for the B2 board. The appearance of damage in the test specimen in bending tests is presented in Figure 15.

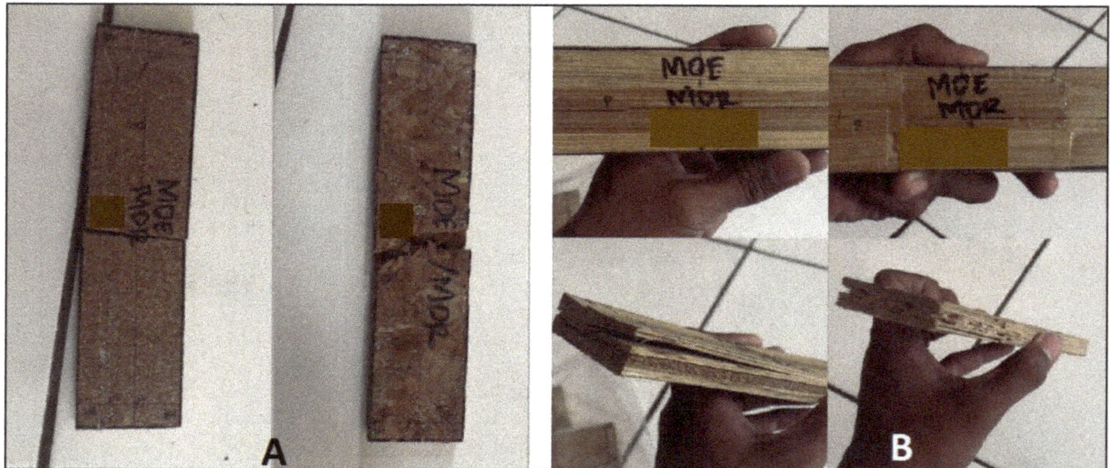

Figure 15. The appearance of damage in samples in MOE and MOR testing: nonstrand-coated board (**A**), and strand-coated board (**B**).

Overall, using belangke bamboo strands as a surface layer resulted in higher MOE and MOR values compared to uncoated boards. The use of belangke bamboo as a coating can increase the MOE and MOR values by 16 and 3 times, respectively (Figure 16). This is due to the good density and flexural strength of bamboo. Rofii et al. [75] stated that high-density materials in the surface layer obtained better curvature than materials with low specific gravity. Iswanto et al. [9] reported that using the belangke bamboo strand as a surface layer on particleboard resulted in higher MOE and MOR values than Tali (*Gigantochloa apus*) bamboo strand because the specific gravity of belangke bamboo was higher than that of *Gigantochloa apus*. According to Iswanto et al. [74], boards coated with bamboo strands have higher flexural strength than boards coated with thin plywood. Using bamboo as a coating material on the surface layer has a positive effect on increasing mechanical properties especially flexural strength and fracture strength, as reported by Iswanto et al. [9,74].

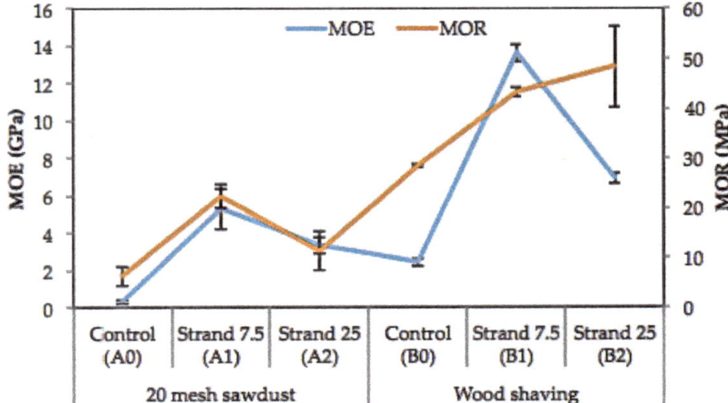

Figure 16. Modulus of rupture (MOR) and modulus of elasticity (MOE) of particleboard fabricated in this work.

Overall, the use of larger particles, in this case, shavings, showed higher MOE and MOR values when compared to powders [76–79]. This condition is in line with the research conducted by Farrokhpayam et al. [72], who reported that fine-sized particles have lower MOE and MOR values than medium- and coarse-sized particles. Wood shavings have a high slenderness ratio value when compared to powder. Badejo [80] and Rokiah et al. [81] stated that strands or particles that are longer and thinner would produce higher MOR values compared to short and thick sizes.

Analysis of variance on the MOE and MOR values showed that the interaction between particle type and strand length are significantly different at a 95% confidence level (sig. 0.000). Overall, the MOE and MOR values of the particleboard met the JIS A 5908-2003 standard requirements of a minimum of 2.05 GPa and 7.85 MPa, respectively.

3.5.6. Internal Bond (IB)

The IB values of the boards ranged from 0.15–0.39 MPa. The lowest value was found for the B1 type board, while the highest value was obtained for the A0 type board. The laboratory boards with surfaces coated with bamboo strands exhibited lower IB values. This was attributed to the weak particle bonds between the surface layer and the core layer due to the use of two different types of particles.

Overall, the core layer's finer particle size (powder) resulted in higher IB values than the shavings particles. Cosereanu et al. [82] stated that fine particles with a flat surface and diffuse appearance could achieve better bond contact so that the structure obtained was more compact and homogeneous. Meanwhile, coarse particles with localized concave geometry produce adhesive agglomeration and reduce compactness, thereby reducing internal bonding strength.

The density of the boards also affects the IB values. In this study, the core layer in the form of sawdust resulted in a higher board density than the type of shavings particles. Wong et al. [83] stated a linear relationship between the board density and IB value. Warmbier et al. [84] reported that IB and screw holding strength increased with increasing board density values.

Analysis of variance on the IB value showed that the interaction between particle type and strand length are significantly different at a 95% confidence level (sig. 0.027). Overall, the resulting IB value met the JIS A 5908-2003 standard, which requires a minimum value of 0.15 MPa.

4. Conclusions

From this study, it was observed that the lignin content of *Gigantochloa pruriens* bamboo includes AIL and ASL, with the former having the lowest and highest values at the bottom and top of the culm, respectively (25.25–27.56%). Meanwhile, ASL had the lowest and highest values at the bottom and middle of the culm (2.73–4.47%), respectively, with the lowest holocellulose content being at the top of the culm and the highest at the bottom (63.56–66.66%). Furthermore, it has the lowest alpha-cellulose in the top and the highest at the bottom (39.70–44.40%). The lowest extractive solubility in ethanol–benzene (1:2) was at the middle culm position and the highest at the bottom (2.18–4.01%). Meanwhile, the lowest ash content was determined in the bottom of the culm and the highest at the top, respectively (1.36–2.57%). The crystallinity degree of *Gigantochloa pruriens* bamboo in the axial position varied between 29.78–37.95%. In this work, the degree of crystallinity also influences mechanical and chemical properties of bamboo.

The physical properties of bamboo, including density, inner and outer diameter shrinkage, and linear shrinkage, were 0.59, 2.18%, 2.26%, and 0.18%, respectively. Meanwhile, bamboo's mechanical properties, including compressive strength, shear strength, and tensile strength, were 42.19 MPa, 7.63 MPa, and 163.8 MPa, respectively. Based on its density, belangke bamboo was assigned to strength class III.

The study demonstrated the potential of using belangke bamboo strands as a surface coating material in particleboard manufacturing. Despite having higher WA and TS values compared to the uncoated counterparts, particleboard panels reinforced with belangke bamboo strands demonstrated satisfactory mechanical strength. Markedly, the addition of belangke bamboo strands as a reinforcing material in particleboards significantly improved the mechanical properties of the boards, resulting in increased MOE and MOR values of particleboards by 16 and 3 times, respectively.

Author Contributions: Conceptualization, A.H.I. and W.F.; methodology, E.W.M., A.H.I., W.F., N.S.H. and D.S.A.; software, A.H.I., W.F. and D.S.A.; validation, W.F. and A.H.I.; formal analysis, A.H.I. and W.F.; investigation, A.H.I. and W.F.; resources, A.H.I.; data curation, A.H.I. and W.F.; writing—original draft preparation, W.F., A.H.I., M.A.R.L., E.W.M., N.S.H., E.R.Z., A.D., W.H., A.S., D.S.A., T.S., P.A., V.S. and L.S.H.; writing—review and editing, A.H.I., W.F., P.A. and L.S.H.; visualization, W.F. and A.H.I.; supervision, A.H.I.; project administration, A.H.I.; funding acquisition, P.A. and A.H.I. All authors have read and agreed to the published version of the manuscript.

Funding: This research was funded by Deputy of Strengthening Research and Development, Ministry of Research and Technology/National Research and Innovation Agency for funding this study for the Penelitian Dasar Unggulan Perguruan Tinggi Scheme for the FY 2021 (Number: 12/E1/KP.PTNBH/2021, 8 March 2021).

Institutional Review Board Statement: Not applicable.

Informed Consent Statement: Not applicable.

Data Availability Statement: The data presented in this study are available on request from the corresponding author.

Acknowledgments: The authors express gratitude to the Deputy of Strengthening Research and Development, Ministry of Research and Technology/National Research and Innovation Agency for funding this study for the Penelitian Dasar Unggulan Perguruan Tinggi Scheme for the FY 2021 (Number: 12/E1/KP.PTNBH/2021, 8 March 2021). Furthermore, the facilities and technical support from the Advanced Characterization Laboratories Cibinong—Integrated Laboratory of Bioproduct, Indonesian Institute of Sciences, through the E-Services of the Indonesian Institute of Sciences. This research was also supported by project No. НИС-Б-1145/04.2021 "Development, Properties, and Application of Eco-Friendly Wood-Based Composites" performed at the University of Forestry, Sofia, Bulgaria.

Conflicts of Interest: The authors declare no conflict of interest.

References

1. Tripathi, S.; Prakash Mishra, O.; Bhardwaj, N.; Varadhan, R. Pulp and papermaking properties of bamboo species *Melocanna baccifera*. *Cellul. Chem. Technol.* **2018**, *52*, 81–88.
2. Buziquia, S.T.; Lopes, P.V.F.; Almeida, A.K.; de Almeida, I.K. Impacts of bamboo spreading: A review. *Biodivers. Conserv.* **2019**, *28*, 3695–3711. [CrossRef]
3. Sanchez-Echeverri, L.; Aita, G.; Robert, D.; Rodriguez-Garcia, M. Correlation between chemical compounds and mechanical response in culms of two different ages of *Guadua angustifolia* Kunth. *Madera Y Bosques* **2014**, *20*, 87–94. [CrossRef]
4. Shangguan, W.; Gong, Y.; Zhao, R.; Ren, H. Effects of heat treatment on the properties of bamboo scrimber. *J. Wood Sci.* **2016**, *62*, 383–391. [CrossRef]
5. Scurlock, J.M.O.; Dayton, D.; Hames, B. Bamboo: An Overlooked Biomass Resource? *Biomass Bioenergy* **2000**, *19*, 229–244. [CrossRef]
6. Fatriasari, W.; Hermiati, E. Analissi morfologi serat dan sifat fisis-kimia pada enam jenis bambu sebagai bahan pulp dan kertas. *J. Ilmu Teknol. Has. Hutan* **2008**, *1*, 67–72.
7. Febrianto, F.; Jang, J.; Lee, S.-H.; Santosa, I.; Hidayat, W.; Kwon, J.H.; Kim, N. Effect of Bamboo Species and Resin Content on Properties of Oriented Strand Board Prepared from Steam-treated Bamboo Strands. *BioResources* **2015**, *10*, 2642–2655. [CrossRef]
8. Park, S.H.; Jang, J.-H.; Wistara, N.; Hidayat, W.; Lee, M.; Febrianto, F. Anatomical and physical properties of Indonesian bamboos carbonized at different temperatures. *J. Korean Wood Sci. Technol.* **2018**, *46*, 656–669. [CrossRef]
9. Iswanto, A.; Susilowati, A.; Putra, R.; Nopriandi, D.; Windra, E. Natural durability of raru wood (*Cotylelobium melanoxylon*) against subterranean termite attack. *J. Phys. Conf. Ser.* **2020**, *1542*, 012051. [CrossRef]
10. Febrianto, F.; Sahroni; Hidayat, W.; Bakar, E.; Kwon, G.; Kwon, J.H.; Hong, S.-I.; Kim, N. Properties of oriented strand board made from Betung bamboo (*Dendrocalamus asper* (Schultes.f) Backer ex Heyne). *Wood Sci. Technol.* **2010**, *46*, 53–62. [CrossRef]
11. Chaowana, K.; Wisadsatorn, S.; Chaowana, P. Bamboo as a Sustainable Building Material—Culm Characteristics and Properties. *Sustainability* **2021**, *13*, 7376. [CrossRef]
12. Engler, B.; Schoenherr, S.; Zhong, Z.; Becker, G. Suitability of Bamboo as an Energy Resource: Analysis of Bamboo Combustion Values Dependent on the Culm's Age. *Int. J. For. Eng.* **2012**, *23*, 114–121. [CrossRef]
13. Dewi, O. Overview of Bamboo Preservation Methods for Construction Use in Hot Humid Climate. *Int. J. Built Environ. Sci. Res.* **2020**, *4*, 1–10. [CrossRef]
14. Darwis, A.; Apri Heri, I.; Jeon, W.-S.; Kim, N.; Wirjosentono, B.; Susilowati, A.; Rudi, H. Variation of quantitative anatomical characteristics in the culm of belangke bamboo (*Gigantochloa pruriens*). *BioResources* **2020**, *15*, 6617–6626. [CrossRef]
15. Sun, Y.; Yahui, Z.; Huang, Y.; Wei, X.; Yu, W. Influence of Board Density on the Physical and Mechanical Properties of Bamboo Oriented Strand Lumber. *Forests* **2020**, *11*, 567. [CrossRef]
16. Chen, G.; Wu, J.; Jiang, H.; Zhou, T.; Li, X.; Yu, Y. Evaluation of OSB webbed laminated bamboo lumber box-shaped joists with a circular web hole. *J. Build. Eng.* **2020**, *29*, 101129. [CrossRef]
17. Iswanto, A.H.; Ompusunggu, P.L. Sandwich Particleboard (SPb): Effect of particle length on the quality of board. *IOP Conf. Ser. Earth Environ. Sci.* **2019**, *374*, 012002. [CrossRef]
18. Lee, S.H.; Chin, K.L.; Lum, W.; Ashaari, Z.; Bakar, E.; Nurliyana, M.; Chai, E.; H'ng, P. Mechanical and physical properties of oil palm trunk core particleboard bonded with different UF resins. *J. Oil Palm Res.* **2014**, *26*, 163–169.
19. Lee, S.H.; Lum, W.C.; Zaidon, A.; Maminski, M. Microstructural, mechanical and physical properties of post heat-treated melamine-fortified urea formaldehyde-bonded particleboard. *Eur. J. Wood Wood Prod.* **2015**, *73*, 607–616. [CrossRef]
20. De Almeida, A.C.; De Araujo, V.A.; Morales, E.A.M.; Gava, M.; Munis, R.A.; Garcia, J.N.; Barbosa, J.C. Wood-bamboo particleboard: Mechanical properties. *BioResources* **2017**, *12*, 7784–7792. [CrossRef]
21. Sutiawan, J.; Hadi, Y.S.; Nawawi, D.S.; Abdillah, I.B.; Zulfiana, D.; Lubis, M.A.R.; Nugroho, S.; Astuti, D.; Zhao, Z.; Handayani, M.; et al. The properties of particleboard composites made from three sorghum (*Sorghum bicolor*) accessions using maleic acid adhesive. *Chemosphere* **2022**, *290*, 133163. [CrossRef] [PubMed]
22. Zaia, U.; Barbosa, J.; Morales, E.; Rocco Lahr, F.; Nascimento, M.; De Araujo, V. Production of Particleboards with Bamboo (*Dendrocalamus giganteus*) Reinforcement. *Bioresources* **2015**, *10*, 1424–1433. [CrossRef]
23. Technical Association of the Pulp and Paper Industry (TAPPI). *T 257 cm-02 Sampling and Preparing Wood for Analysis*; Technical Association of the Pulp and Paper Industry (TAPPI): Peachtree Corners, GA, USA, 2002.
24. Technical Association of the Pulp and Paper Industry (TAPPI). *T 264 cm-97 Preparation of Wood for Chemical Analysis*; Technical Association of the Pulp and Paper Industry (TAPPI): Peachtree Corners, GA, USA, 1997.
25. Sluiter, A.; Hames, B.; Ruiz, R.; Scarlata, C.; Sluiter, J.; Templeton, D.; Crocker, D.L.A.P. Determination of structural carbohydrates and lignin in biomass. *Lab. Anal. Proced.* **2008**, *1617*, 1–16.
26. TAPPI. Technical Association of the Pulp and Paper Industry (TAPPI). T 211 om-02. In *Ash in Wood, Pulp, Paper and Paperboard: Combustion at 525 °C*; Technical Association of the Pulp and Paper Industry (TAPPI): Peachtree Corners, GA, USA, 2007.
27. Segal, L.; Creely, J.J.; Martin, A.E.; Conrad, C.M. An empirical method for estimating the degree of crystallinity of native cellulose using the x-ray diffractometer. *Text. Res. J.* **1959**, *29*, 786–794. [CrossRef]
28. Horikawa, Y.; Mizuno-Tazuru, S.; Sugiyama, J. Near-infrared spectroscopy as a potential method for identification of anatomically similar Japanese diploxylons. *J. Wood Sci.* **2015**, *61*, 251–261. [CrossRef]

29. Savitzky, A.; Golay, M.J.E. Smoothing and Differentiation of Data by Simplified Least Squares Procedures. *Anal. Chem.* **1964**, *36*, 1627–1639. [CrossRef]
30. Hwang, S.-W.; Horikawa, Y.; Lee, W.-H.; Sugiyama, J. Identification of Pinus species related to historic architecture in Korea using NIR chemometric approaches. *J. Wood Sci.* **2016**, *62*, 156–167. [CrossRef]
31. Sudarwoko Adi, D.; Hwang, S.-W.; Pramasari, D.; Amin, Y.; Cipta, H.; Damayanti, R.; Dwianto, W.; Sugiyama, J. Anatomical Properties and Near Infrared Spectra Characteristics of Four Shorea Species from Indonesia. *HAYATI J. Biosci.* **2020**, *27*, 247. [CrossRef]
32. Schwanninger, M.; Rodrigues, J.; Fackler, K. A Review of Band Assignments in near Infrared Spectra of Wood and Wood Components. *J. Near Infrared Spectrosc.* **2011**, *19*, 287–308. [CrossRef]
33. *ISO 22157-1:2019*; Bamboo-Determination of Physical and Mechanical Properties-Part 1: Requirements. International Organization for Standardization: Geneva, Switzerland, 2019.
34. *ISO/TR 22157-2:2004*; Bamboo-Determination of Physical and Mechanical Properties-Part 2: Laboratory Manual. International Organization for Standardization: Geneva, Switzerland, 2004.
35. *JIS A 5908*; Japanese Industrial Standard for Particleboard. Japanese Standard Association Japan: Tokyo, Japan, 2003.
36. Banik, N.; Dey, V.; Sastry, G.R.K. An Overview of Lignin & Hemicellulose Effect Upon Biodegradable Bamboo Fiber Composites Due to Moisture. *Mater. Today Proc.* **2017**, *4*, 3222–3232. [CrossRef]
37. Li, X. Physical, Chemical and Mechanical Properties of Bamboo and Its Utilization Potential for Fibreboard Manufacturing. Master's Thesis, Chinese Academy of Forestry, Beijing, China, 2004.
38. Jara, A.; Migo, V.; Acda, M.; Calderon, M.; Florece, L.; Razal, R. Chemical Composition of *Bambusa vulgaris* Shoots as Influenced by Harvesting Time and Height. *Philipp. J. Crop Sci. (PJCS)* **2018**, *43*, 1–9.
39. Sharma, B.; Shah, D.U.; Beaugrand, J.; Janeček, E.-R.; Scherman, O.A.; Ramage, M.H. Chemical composition of processed bamboo for structural applications. *Cellulose* **2018**, *25*, 3255–3266. [CrossRef] [PubMed]
40. Loiwatu, M.; Manuhuwa, E. Komponen kimia dan aatomi tiga jenis bambu dari seram, Maluku. *AGRITECH* **2008**, *28*, 76–83.
41. Indriatie, R.; Mudaliana, S.; Masruri, M. Microbial resistant of building plants of Gigantochloa apus. *IOP Conf. Ser. Mater. Sci. Eng.* **2019**, *546*, 042013. [CrossRef]
42. Wahab, R.; Mustafa, M.; Salam, M.; Sudin, M.; Samsi, W.H.; Sukhairi, M.; Mat Rasat, M.S. Chemical Composition of Four Cultivated Tropical Bamboo in Genus Gigantochloa. *J. Agric. Sci.* **2013**, *5*, 66–75. [CrossRef]
43. Yusoff, M.; Bahari, S.; Haliffuddin, R.; Zakaria, M.N.; Jamaluddin, M.; Rashid, N. Chemical contents and thermal stability of Madu bamboo (*Gigantochloa albociliata*) for natural-bonded fiber composites. *IOP Conf. Ser. Earth Environ. Sci.* **2021**, *644*, 012009. [CrossRef]
44. Maulana, M.; Marwanto, M.; Nawawi, D.; Nikmatin, S.; Febrianto, F.; Kim, N. Chemical components content of seven Indonesian bamboo species. *IOP Conf. Ser. Mater. Sci. Eng.* **2020**, *935*, 012028. [CrossRef]
45. Nahar, S.; Hasan, M. Effect of Chemical Composition, Anatomy and Cell Wall Structure on Tensile Properties of Bamboo Fiber *Eng. J.* **2013**, *17*, 61–68. [CrossRef]
46. Nugroho, N.; Bahtiar, E.T.; Lestari, D.; Nawari, D. Variasi Kekuatan Tarik dan Komponen Kimia Dinding Sel pada Empat Jenis Bambu (Variation of Tensile Strength and Cell Wall Component of Four Bamboos Species). *J. Ilmu Dan Teknol. Kayu Trop.* **2013**, *11*, 153–160.
47. Selvan, R.T.; Parthiban, K.T.; Khanna, S. Physio-Chemical Properties of Bamboo Genetic Resources at Various Age Gradations *Int. J. Curr. Microbiol. Appl. Sci.* **2017**, *6*, 1671–1681. [CrossRef]
48. Tolessa, A.; Woldeyes, B.; Feleke, S. Chemical Composition of Lowland Bamboo (*Oxytenanthera abyssinica*) Grown around Asossa Town, Ethiopia. *World Sci. News* **2017**, *74*, 141–151.
49. Fatriasari, W.; Supriyanto, S.; Apri Heri, I. The Kraft Pulp And Paper Properties of Sweet Sorghum Bagasse (*Sorghum bicolor* L Moench). *J. Eng. Technol. Sci.* **2015**, *47*, 149–159. [CrossRef]
50. Purbasari, A.; Samadhi, T.; Bindar, Y. Thermal and Ash Characterization of Indonesian Bamboo and Its Potential for Solid Fuel and Waste Valorization. *Int. J. Renew. Energy Dev.* **2016**, *5*, 95–100. [CrossRef]
51. Fatriasari, W.; Masruchin, N.; Hermiati, E. *Selulosa: Karakteristik dan Pemanfaatannya*; LIPI Press: Jakarta, Indonesia, 2019.
52. Toba, K.; Nakai, T.; Shirai, T.; Yamamoto, H. Changes in the cellulose crystallinity of moso bamboo cell walls during the growth process by X-ray diffraction techniques. *J. Wood Sci.* **2015**, *61*, 517–524. [CrossRef]
53. Elanthikkal, S.; Unnikrishnan, G.; Varghese, S.; Guthrie, J. Cellulose microfibers produced from banana plant wastes: Isolation and characterisation. *Carbohydr. Polym.* **2010**, *80*, 582–859. [CrossRef]
54. Fatriasari, W.; Ridho, M.R.; Karimah, A.; Sudarmanto; Ismadi; Amin, Y.; Ismayati, M.; Lubis, M.A.R.; Solihat, N.N.; Sari, F.P.; et al. Characterization of Indonesian Banana Species as an Alternative Cellulose Fibers. *J. Nat. Fibers* **2022**, 1–18. [CrossRef]
55. Fatriasari, W.; Syafii, W.; Wistara, N.; Syamsu, K.; Prasetya, B. The Characteristic Changes of Betung Bamboo (*Dendrocalamus asper* Pretreated by Fungal Pretreatment. *Int. J. Renew. Energy Dev.* **2014**, *3*, 133–143. [CrossRef]
56. Darmawan, T.; Bahanawan, A.; Sudarwoko Adi, D.; Dwianto, W.; Nugroho, N. Fixation process of laminated bamboo compression from curved cross-section slats Indones. *J. For. Res.* **2021**, *8*, 159–171. [CrossRef]
57. Iswanto, A.H.; Tarigan, F.O.; Susilowati, A.; Darwis, A.; Fatriasari, W. Wood Chemical Compositions of Raru Species Originating from Central Tapanuli, North Sumatra, Indonesia: Effect of Differences in Wood Species and Log Positions. *J. Korean Wood Sci. Technol.* **2021**, *49*, 416–429. [CrossRef]

58. Barrios, A.; Trincado, G.; Watt, M. Wood Properties of Juvenile and Mature Wood of *Pinus radiata* D. Don Trees Growing on Contrasting Sites in Chile. *For. Sci.* **2017**, *63*, 184–191. [CrossRef]
59. Takeuchi, R.; Wahyudi, I.; Aiso, H.; Ishiguri, F.; Istikowati, W.T.; Ohkubo, T.; Ohshima, J.; Iizuka, K.; Yokota, S. Anatomical characteristics and wood properties of unutilized *Artocarpus* species found in secondary forests regenerated after shifting cultivation in Central Kalimantan, Indonesia. *Agrofor. Syst.* **2019**, *93*, 745–753. [CrossRef]
60. Li, M.-Y.; Ren, H.-Q.; Wang, Y.-R.; Gong, Y.-C.; Zhou, Y.-D. Comparative studies on the mechanical properties and microstructures of outerwood and corewood in *Pinus radiata* D. Don. *J. Wood Sci.* **2021**, *67*, 60. [CrossRef]
61. Gierlinger, N.; Schwanninger, M.; Wimmer, R. Characteristic and classification of Fourier-transform near infrared spectra of heartwood of different larch species (*Larix* sp). *J. Near Infrared Spectrosc.* **2004**, *12*, 113–119. [CrossRef]
62. Via, B.; McDonald, T.; Fulton, J. Nonlinear multivariate modeling of strand density from near-infrared spectra. *Wood Sci. Technol.* **2012**, *46*, 1073–1084. [CrossRef]
63. Sandak, J.; Sandak, A.; Meder, R. Assessing trees, wood and derived products with NIR spectroscopy: Hints and tips. *J. Near Infrared Spectrosc.* **2016**, *24*, 485–505. [CrossRef]
64. Baharoğlu, M.; Nemli, G.; Sarı, B.; Birtürk, T.; Bardak, S. Effects of anatomical and chemical properties of wood on the quality of particleboard. *Compos. Part B Eng.* **2013**, *52*, 282–285. [CrossRef]
65. Abdullah, A.H.; Karlina, N.; Rahmatiya, W.; Patimah; Fajrin, A. Physical and mechanical properties of five Indonesian bamboos. *IOP Conf. Ser. Earth Environ. Sci.* **2017**, *60*, 012014. [CrossRef]
66. Chowdhury, M.Q.; Ishiguri, F.; Hiraiwa, T.; Matsumoto, K.; Takashima, Y.; Iizuka, K.; Yokota, S.; Yoshizawa, N. Variation in anatomical properties and correlations with wood density and compressive strength in *Casuarina equisetifolia* growing in Bangladesh. *Aust. For.* **2012**, *75*, 95–99. [CrossRef]
67. Espiloy, Z.B. Physico-mechanical properties and anatomical relationships of some Philippine bamboos. In Proceedings of the International Bamboo Workshop, Hangzhou, China, 6–14 October 1985; pp. 257–264.
68. Longui, E.; Brémaud, I.; da Silva Júnior, F.; Lombardi, D.; Alves, E. Relationship among extractives, lignin and holocellulose contents with performance index of seven wood species used for bows of string instruments. *IAWA J.* **2012**, *33*, 141–149. [CrossRef]
69. Bufalino, L.; Albino, V.C.S.; Sá, V.; Corrêa, A.A.R.; Mendes, L.; Almeida, N.A. Particleboards made from Australian red cedar: Processing variables and evaluation of mixed-species. *J. Trop. For. Sci.* **2012**, *24*, 162–172.
70. Kelly, M. *Critical Literature Review of Relationship between Processing Parameter and Physical Properties of Particleboard;* General Technical Report; Wisconsin University: Madison, WI, USA, 1977.
71. Bowyer, J.; Shmulsky, R. *Forest Products and Wood Science: An Introduction*, 4th ed.; Wiley-Blackwell: Hoboken, NJ, USA, 2003.
72. Farrokhpayam, S.R.; Valadbeygi, T.; Sanei, E. Thin particleboard quality: Effect of particle size on the properties of the panel. *J. Indian Acad. Wood Sci.* **2016**, *13*, 38–43. [CrossRef]
73. Vital, B.; Lehmann, W.; Boone, R. How Species and Board Density Affect Properties of Exotic Hardwood Particleboard. *For. Prod. J.* **1974**, *24*, 37–45.
74. Iswanto, A.; Aritonang, W.; Irawati, A.; Supriyanto; Widya, F. The physical, mechanical and durability properties of sorghum bagasse particleboard by layering surface treatment. *J. Indian Acid Wood Sci. Technol* **2017**, *14*, 1–8. [CrossRef]
75. Rofii, M.N.; Yumigeta, S.; Kojima, Y.; Suzuki, S. Effect of furnish type and high-density raw material from mill residues on properties of particleboard panels. *J. Wood Sci.* **2013**, *59*, 402–409. [CrossRef]
76. Hidayat, W.; Aprilliana, N.; Asmara, S.; Bakri, S.; Hidayati, S.; Banuwa, I.S.; Lubis, M.A.R.; Iswanto, A.H. Performance of eco-friendly particleboard from agroindustrial residues bonded with formaldehyde-free natural rubber latex adhesive for interior applications. *Polym. Compos.* **2022**, *43*, 2222–2233. [CrossRef]
77. Antov, P.; Savov, V.; Neykov, N. Influence of the composition on the exploitation properties of combined medium density fibreboards manufactured with coniferous wood residues. *Eur. Mech. Sci. J.* **2018**, *2*, 140–145.
78. Faria, D.; Lopes, T.; Mendes, L.; Guimarães, J. Valorization of wood shavings waste for the production of wood particulate composites. *Matéria* **2020**, *25*, 1–11. [CrossRef]
79. Gößwald, J.; Barbu, M.C.; Petutschnigg, A.; Krišťák, L'.; Tudor, E.M. Oversized Planer Shavings for the Core Layer of Lightweight Particleboard. *Polymers* **2021**, *13*, 1125. [CrossRef]
80. Badejo, S.O.O. Effect of flake geometry on properties of cement-bonded particleboard from mixed tropical hardwoods. *Wood Sci. Technol.* **1988**, *22*, 357–369. [CrossRef]
81. Hashim, R.; Saari, N.; Sulaiman, O.; Sugimoto, T.; Hiziroglu, S.; Sato, M.; Tanaka, R. Effect of particle geometry on the properties of binderless particleboard manufactured from oil palm trunk. *Mater. Des.* **2010**, *31*, 4251–4257. [CrossRef]
82. Cosereanu, C.; Brenci, L.; Zeleniuc, O.; Fotin, A.N. Effect of particle size and geometry on the performance of single-layer and three-layer particleboard made from sunflower seed husks. *BioResources* **2015**, *10*, 1127–1136. [CrossRef]
83. Wong, E.D.; Zhang, M.; Wang, Q.; Kawai, S. Formation of the density profile and its effects on the properties of particleboard. *Wood Sci. Technol.* **1999**, *33*, 327–340. [CrossRef]
84. Warmbier, K.; Wilczyński, M. Resin Content and Board Density Dependent Mechanical Properties of One-Layer Particleboard Made from Willow (*Salix viminalis*). *Drv. Ind.* **2016**, *67*, 127–131. [CrossRef]

Article

Utilization of Suberinic Acids Containing Residue as an Adhesive for Particle Boards

Raimonds Makars [1,2,*], Janis Rizikovs [1], Daniela Godina [1], Aigars Paze [1] and Remo Merijs-Meri [3]

[1] Latvian State Institute of Wood Chemistry, Dzerbenes iela 27, LV-1006 Riga, Latvia; janis.rizikovs@kki.lv (J.R.); daniela.godina@kki.lv (D.G.); aigars.paze@kki.lv (A.P.)
[2] PolyLabs SIA, Mukusalas iela 46, LV-1004 Riga, Latvia
[3] Institute of Polymer Materials, Riga Technical University, Paula Valdena iela 3/7, LV-1048 Riga, Latvia; remo.merijs-meri@rtu.lv
* Correspondence: raimonds.makars@kki.lv; Tel.: +371-29-71-73-80

Abstract: The birch (*Betula* spp.) outer bark is a valuable product rich in betulin. After removal of betulin extractives, suberin containing tissues are left. Suberin is a biopolyester built from α,ω-bifunctional fatty acids (suberinic acids), which after depolymerization together with ligno-carbohydrate complex is a potential adhesive as a side-stream product (residue) from obtaining suberinic acids for polyol synthesis. In this work, we studied the utilization possibilities in particleboards of the said residue obtained by depolymerization in four different solvents (methanol, ethanol, isopropanol and 1-butanol). The adhesives were characterised by chemical (acid number, solubility in tetrahydrofuran, epoxy and ash content) and instrumental analytical methods (SEC-RID, DSC, TGA and FTIR). Based on the results of mechanical characteristics, ethanol was chosen as the most suitable depolymerization medium. The optimal hot-pressing parameters for particleboards were determined using the design of experiments approach: adhesive content 20 wt%; hot-pressing temperature 248 °C, and hot-pressing time 6.55 min.

Keywords: birch outer bark; suberinic acids; particle boards

Citation: Makars, R.; Rizikovs, J.; Godina, D.; Paze, A.; Merijs-Meri, R. Utilization of Suberinic Acids Containing Residue as an Adhesive for Particle Boards. *Polymers* **2022**, *14*, 2304. https://doi.org/10.3390/polym14112304

Academic Editors: Petar Antov, Pavlo Bekhta, Yonghui Zhou and Viktor Savov

Received: 13 May 2022
Accepted: 3 June 2022
Published: 6 June 2022

1. Introduction

Birch (*Betula* spp.) trees are very common in the Northern hemisphere. Veneer production is considered the most efficient way of processing birch wood, and birch bark is accumulated as the by-product in this process and mostly used for the production of thermal energy by incineration. Birch bark is composed of two layers: the outer bark and the inner bark [1]. The outer bark proportion of the total birch biomass is roughly 3 to 5 wt% [2]. Although the amount of bark is relatively small, the large production volumes of veneer result in a noticeable accumulation of this material.

The birch outer bark (BOB) contains a high proportion of lupane-type triterpene extractives (betulin, lupeol, betulinic acid) which can be isolated by various methods [3–5]. After the removal of extractives from BOB, mainly suberin-containing external protective tissues are left [6].

Suberin is a biopolymer of plant origin occurring in specific parts of tissue, where protection from the surrounding environment is required. Chemically, suberin comprises two chemically distinct parts: (a) an aliphatic region built from α,ω-bifunctional fatty acids (suberinic acids (SA)) and glycerol through ester linkages; (b) polyaromatic lignin-like structures [7]. These regions are cross-linked and various depolymerization methods are proposed to obtain SA monomers and oligomers. The most common method is alkaline hydrolysis in water [8,9] and other solvents, such as isopropanol [2,6]. Our previous research showed that SA obtained by hydrolytical depolymerization can be further used as a bio-based adhesive in the preparation of environmentally friendly wood composites, such as plywood and particle boards. The particle boards made from SA-based adhesive in hot-pressing process exhibited satisfactory water-resistant properties as determined by low thickness swelling values. In addition, the high internal bonding values showed that the SA-based adhesive has as good compatibility with wood particles as a filler [9,10]. The satisfactory water resistance is explained by the lipophilic properties of suberin [11].

In addition, in our previous work [12] we have demonstrated that SA can be obtained from extractive-free BOB by alkaline depolymerization in ethanol-water solution for further utilization in polyol synthesis. During this process, residue containing SA and lignocellulosic components is accumulated. One way to utilize this residue after suberin depolymerization is as solid fuel or fuel filler [13]. However, these products have low added-value, therefore the aim of this study was to investigate the utilization of the residue obtained after depolymerization in four different solvents (methanol, ethanol, isopropanol and 1-butanol) in particle board preparation. For the most promising adhesive, the design of experiments (DOE) approach was used to determine the optimal hot-pressing parameters.

2. Materials and Methods

2.1. Raw Material—Birch Outer Bark

Isolated and fractionated BOB was kindly supplied by AS Latvijas Finieris (Riga, Latvia). BOB samples were dried at room temperature (moisture content 4–5 wt%) and milled in an SM 100 cutting mill (Retsch GmbH, Haan, Germany) to pass through a sieve with holes measuring 4 mm in diameter. Milled BOB was fractionated by sieving using an AS 200 Basic vibratory sieve shaker (Retsch GmbH, Haan, Germany). Fraction of >4 mm was collected and repeatedly milled in the same cutting mill to pass through a sieve with 2 mm aperture. BOB was further extracted with ethanol twice, as described by Godiņa et al. [14]. Extracted BOB was dried at room temperature and further used as a feedstock for depolymerization.

2.2. Other Materials and Chemicals

Ethanol (96,3% v/v) was supplied by SIA Kalsnavas elevators (Jaunkalsnava, Latvia). Methanol (≥99.8%), hydrochloric acid (HCl) (≥37%), and tetrahydrofuran (THF) (anhydrous, ≥99.9%) were obtained from Sigma-Aldrich (Steinheim, Germany). Isopropanol (Reag. Ph Eur, ≥99.8%) was received from Merck (Darmstadt, Germany). 1-Butanol (99%) was purchased from Acros Organics (Geel, Belgium). Potassium hydroxide (KOH) (Reag. Ph Eur, 85.0–100.5%) was provided by VWR International (Leuven, Belgium). Nitric acid (HNO_3) (≥65%) and acetone (≥99%) were obtained from Honeywell (Seelze, Germany).

2.3. Obtaining SA-Based Adhesive

BOB depolymerization was carried out in KOH solution for 60 min at 66 to 80 °C (depending on the solvent). Four different solvents were used: methanol (MeOH), ethanol (EtOH), isopropanol (i-PrOH) and 1-butanol (BuOH). The depolymerization conditions are given in Table 1.

Table 1. Depolymerization conditions.

Solvent/Sample	KOH, g L^{-1}	BOB: Solvent, g L^{-1}	Temperature, °C
MeOH	41.5	100	66
EtOH [1]	29.2	100	80
i-PrOH	41.5	100	80
BuOH	41.5	100	80

[1] ethanol-water solution 1:10 (m/m).

After depolymerization, the solution was cooled down and filtered. The filtrate can be further used for preparing polyols as described by Rizikovs et al. [12]. The precipitate (residue) was further used to prepare the adhesive by suspending it in water to the ratio of 100 g per litre. The suspension was acidified with HNO_3 to pH 2.0, then it was filtered and rinsed with deionized water. As a result, four different SA-based adhesive samples were obtained.

2.4. Characterization of SA-Based Adhesive

For SA characterization, adhesive samples were dried at room temperature and milled with a CryoMill cryogenic mill (Retsch GmbH, Haan, Germany) under LN_2 purge at −196 °C.

2.4.1. Acid Number

To about 0.2 g of the sample, 5 mL of DMSO was added and stirred for 1 h. Afterwards, 20 mL of i-PrOH and 5 mL of water were added, and the solution was titrated with 0.1 M KOH solution. Two replicate experiments were performed for each sample.

2.4.2. Epoxy Groups

To about 0.2 g of the sample, 10 mL of 0.2 M HCl in acetone was added. The solution was stirred for 1 h and titrated with a known concentration of 0.1 M KOH solution. Two replicate experiments were performed for each sample.

2.4.3. Soluble Substance in THF

To about 0.1 g sample 5 mL of THF was added. The mixture was heated at 50 °C for 0.5 h, cooled and stirred for another hour. The soluble substance in THF was determined by filtration through a filter crucible (porosity 2). Two replicate experiments were performed for each sample. The filtrate was further used for size exclusion chromatography analysis.

2.4.4. Ash Content

The ash content was determined at 715 ± 10 °C according to ISO standard 1171:2010 [15].

2.4.5. Size-Exclusion Chromatography

GPC analysis of sample (100 µL) was performed using an Agilent Infinity 1260 HPLC system with degasser, auto sampler, RI detector and MALS (miniDAWN) detector. Two GPC analytical columns connected in line were used for the analysis: PLgel Mixed-E (3 uL, 300 × 7.5 mm). The flow rate was 1 mL/min and the RI detector temperature was 35 °C.

The polystyrene calibration graph was obtained by preparing polystyrene standard solutions in THF at a mass concentration of 2 mg/mL and analysing them with GPC instrument. Molecular weights of polystyrene standard substances were 500, 850, 1000, 2500, 5000, 9000, 17,500, 20,000, and 30,000 Da.

2.4.6. Differential Scanning Calorimetry (DSC)

DSC822 differential scanning calorimeter (Mettler-Toledo, Greifensee, Switzerland) was used to analyse the thermal behaviour of adhesive samples. The analysis was performed in pierced aluminium pans under N_2 purging at a rate of 10 °C min^{-1} in two stages: (1) heating from 20 °C to 260 °C; (2) cooling from 260 °C to 20 °C.

2.4.7. Thermogravimetric Analysis (TGA)

TGA analysis was performed using a TA Instruments Discovery TGA 5500 thermogravimetric analyzer. Mass loss was determined in Pt sample pans under an N_2 purge at 50 mL min^{-1} by isothermally treating the sample at 30 °C, followed by heating to 700 °C at a rate of 10 °C min^{-1}.

2.4.8. Fourier Transform Infrared Spectroscopy (FTIR) Analysis

FTIR spectrometry data was collected with a Nicolet iS50 spectrometer (Thermo Fisher Scientific, Waltham, MA, USA) at a resolution of 4 cm^{-1}, 32 scans. The FTIR data were collected using the attenuated total reflectance (ATR) technique with a diamond crystal prism.

2.5. Particle Board Preparation

The SA-based adhesive with moisture content ~80 wt% was mixed with fractionated (0.4–2.0 mm) birch wood particles generated from veneer shorts. Afterwards, the mixture was oven-dried at 100 °C to moisture content <1 wt% and hot-pressed with thickness bars by a LAP 40 single-stage press (Gottfried Joos Maschinenfabrik GmbH & Co. KG, Pfalzgrafenweiler, Germany). The dimensions and density of the boards were designed to be 180 × 150 × 7 mm and 0.8 g cm^{-3}, respectively.

2.5.1. Comparison of Particle Board Properties Depending on SA-Based Adhesive Sample

To choose the most suitable adhesive, MeOH, EtOH, i-PrOH, and BuOH adhesive-based particle boards were made, as described above. Based on previous experience [9] particle board hot-pressing temperature (T) was set to 225 °C and the pressing time (t) was 5 min. Pressure (p) was varied in two cycles: 3.5 MPa in the first cycle (t = 2 min) followed by pressure release to 0.1 Mpa per 30 s; during the second cycle, the pressure was increased to 1.7 Mpa (t = 1 min 50 s) followed by another pressure release to 0.7 Mpa for 40 s. The adhesive content (c) in the particle boards varied from 20 to 40 wt% (dry basis).

2.5.2. Experimental Design to Determine Optimal Hot-Pressing Parameters

After choosing the most suitable adhesive, the design of experiments (DOE) approach was used to determine the optimal hot-pressing parameters. Three variable factors were defined: c, T, and t. At first, full factorial design methodology (2^3) using Design Expert 13 software (Stat-Ease Inc., Minneapolis, MN, USA) was used, which consisted of 8 runs. The variable factor levels are given in Table 2. Depending on t, the hot-pressing was performed in 2 or 3 cycles (at t = 2 and t = 8 min, respectively). The duration of the hot-pressing cycles is given in Table 3.

Table 2. Variable factor levels for DOE.

Variable	Factor Level	
	Low	**High**
c, wt%	20	40
T, °C	200	250
t, min	2	8

Table 3. Duration of the hot-pressing cycles.

Cycle	t = 2 min	t = 8 min
1	3.5 MPa (1 min 20 s) 0.1 MPa (10 s)	3.5 MPa (2 min 30 s) 0.1 MPa (30 s)
2	1.7 MPa (20 s) 0.7 MPa (10 s)	1.7 MPa (2 min) 0.1 MPa (30 s)
3	- -	1.7 MPa (2 min) 0.7 MPa (30 s)

The effects of the hot-pressing parameters on response values (modulus of elasticity (MOE), bending strength (MOR), and thickness swelling after a 24 h immersion test (TS 24 h), density) were evaluated using the software. To improve the resolution, the full factorial experimental plan was augmented by 8 additional runs using D-optimality criteria.

2.6. Evaluation of the Particle Board Properties

The obtained PBs were conditioned (RH = 65 ± 5%, T = 20 ± 2 °C) and characterized according to relevant standards by density [16], MOE and MOR in the 3-point bending test [17], and TS 24 h [18]. The particle boards obtained using the optimal hot-pressing parameters were characterized by internal bonding (IB), determining the tensile strength perpendicular to the plane of the board [19]. Mechanical tests (MOE, MOR, IB) were performed on a Z010 (Zwick Roell AG, Ulm, Germany) universal machine for testing the resistance of materials. The obtained particle boards were compared with the EN 312 standard requirements [20].

3. Results and Discussion

3.1. Characterization of SA-Based Adhesive Samples

3.1.1. Acid Number, Epoxy Group Content, Soluble Substance in THF, Ash Content

Chemical analysis (acid number, epoxy group content, soluble substance in THF, ash content) was carried out to determine the properties of adhesive samples. The results are summarized in Table 4. Acid number values show that the highest acid functionality was for i-PrOH samples (122 mg KOH g^{-1}), which could be a desirable characteristic for an adhesive, because acid groups are involved in adhesive cross-linking reactions. However, the i-PrOH adhesive had the lowest epoxy group content (0.11 mmol g^{-1}). A high epoxy group content is a desirable property since epoxy groups play a role in the formation of cross-linked networks [21]. From that point of view, the epoxy group value for the EtOH adhesive was almost 6 times higher—0.61 mmol g^{-1}. i-PrOH and BuOH adhesives showed higher values for the soluble substance in THF (58.1 and 57.5 wt%, respectively), compared to the MeOH sample (44.0 wt%), which suggests that a higher proportion of suberin-based monomers and oligomers were present in those samples. On the contrary, the MeOH sample had a relatively high ash content (12.9 wt%). This could be because of the higher SA particle aggregation in the acidification process. As a result, unacidified SA salts remained in the agglomerates, so the adhesive potentially could be less effective. The lowest ash content values were for SA obtained in the BuOH and EtOH (6.6 and 6.7 wt%, respectively).

Table 4. Chemical properties of SA-based adhesive samples.

Sample	Acid Number, mg KOH g^{-1}	Epoxy Groups, mmol g^{-1}	Soluble Substance in THF, wt%	Ash Content, wt%
MeOH	70.9	0.45	44.0	12.9
EtOH	95.8	0.61	51.8	6.7
i-PrOH	122.0	0.11	58.1	9.3
BuOH	91.1	0.25	57.5	6.6

Because of the high amount of acid number, epoxy groups, soluble substance in THF, and the low ash content adhesive obtained in ethanol showed the most promising properties for obtaining wood-based panels.

3.1.2. Size-Exclusion Chromatography

The adhesive molecular weight is also an important characteristic of the adhesive. During SEC-RID analysis of the soluble substance in THF, seven to eight (depending on the sample) different molecular weight fractions were separated, ranging from 94 to 69,909 Da. Since adhesives have limited solubility in THF, it is possible that larger molecular weight fractions (including lignocellulose) were left out. The chromatograms with the relevant molecular weight fractions are given in Figure 1. For a more convenient result comparison, lower (oligomeric + monomeric) molecular weight fraction (<3000 Da) percentages were estimated.

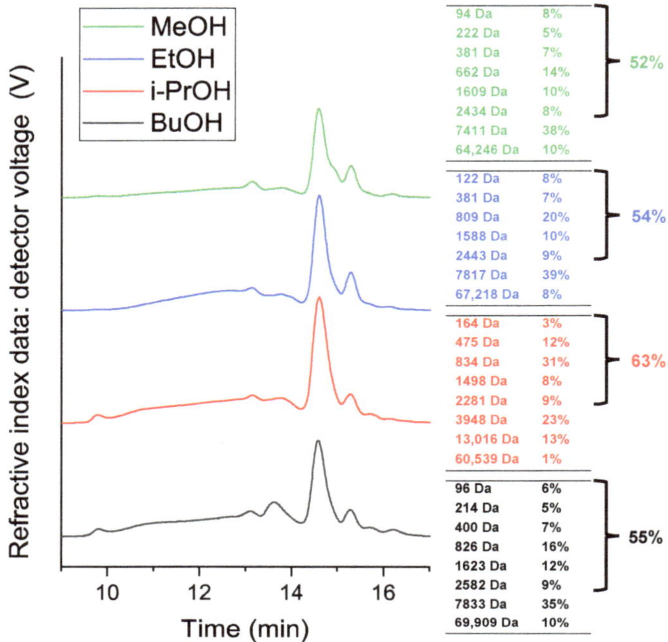

Figure 1. SA-based adhesive sample SEC-RID chromatograms.

Overall, the SEC-RID analysis shows similarities in the separation patterns. The analysis showed that the highest low molecular weight portion was for the i-PrOH sample, which corresponds to the higher acid number for the sample (Table 4). In addition, the solubility in THF was the highest and the M_n and M_w values were the lowest for this sample (Table 5), both indicating that this adhesive contained a higher proportion of low molecular weight fraction. Consequently, the polydispersity was higher for MeOH, EtOH and BuOH adhesives. The high polydispersity values show that a variety of monomers and oligomers were represented in all the samples. Suberin's aliphatic structure mostly comprises C_{16} to C_{24} α,ω-diacid and ω-hydroxyacid monomers [7] with molecular weights ranging from around 270 to 400 Da. Thus, the most pronounced peak at $t_R \approx 14.5$ min for molecular weights ranging from 662 to 834 Da is most likely attributed to SA dimers in the adhesives.

Table 5. Molecular weights of adhesive samples.

Sample	M_n, kDa	M_w, kDa	M_w/M_n
MeOH	9.886	7526	761
EtOH	9.036	6577	728
i-PrOH	3.926	1600	407
BuOH	10.502	7525	717

3.1.3. DSC Analysis

DSC thermograms in Figure 2 show that all four adhesives have two common crystalline phase melting zones (see red curve) at 84–87 °C and 133–139 °C. Heating curves for MeOH, EtOH and i-PrOH samples also had an additional shoulder at 65–70 °C, and for the MeOH adhesive, an additional peak at 78 °C was observed. Both EtOH and i-PrOH samples exhibited more pronounced crystalline phase melting areas, showing that more crystalline SA were present on these samples. The acid number values were also higher for these samples (Table 4) and this may suggest that acidic groups form more crystalline structures, most likely due to intermolecular hydrogen bonding. The cooling curves (in blue) showed crystallisation zones of the corresponding melting peaks. The areas of the corresponding crystallisation peaks at around 110 °C and 50 °C were smaller than the melting peaks, which could indicate to the thermal destruction (see TGA curves in Figure 3) and to intramolecular reactions in the heating process, and, thus, resulting in cross-linked structures. This suggests that all adhesives have thermosetting properties, which is necessary for obtaining wood composites with desirable properties.

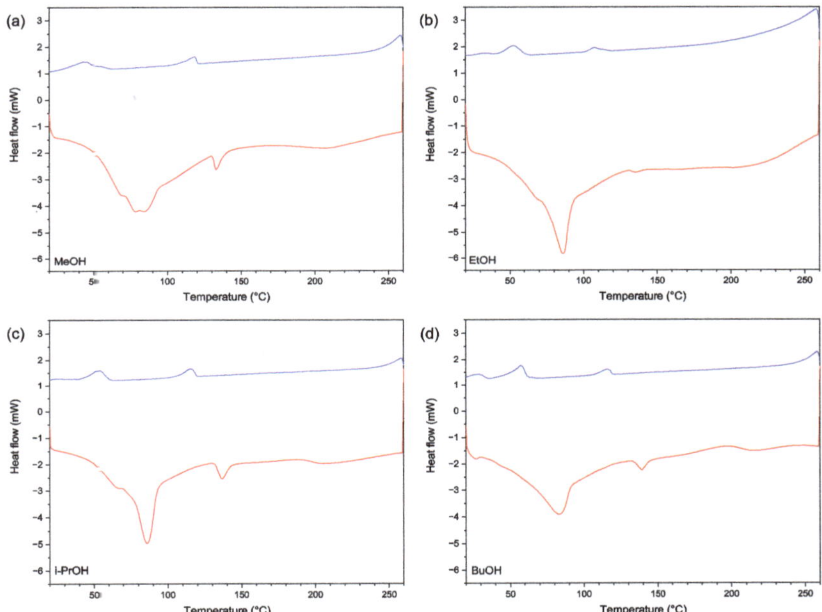

Figure 2. DSC thermograms for (**a**) MeOH, (**b**) EtOH, (**c**) i-PrOH, and (**d**) BuOH adhesive samples (the red curve represents heating from 20 °C to 260 °C and the blue curve represents cooling from 260 °C to 20 °C).

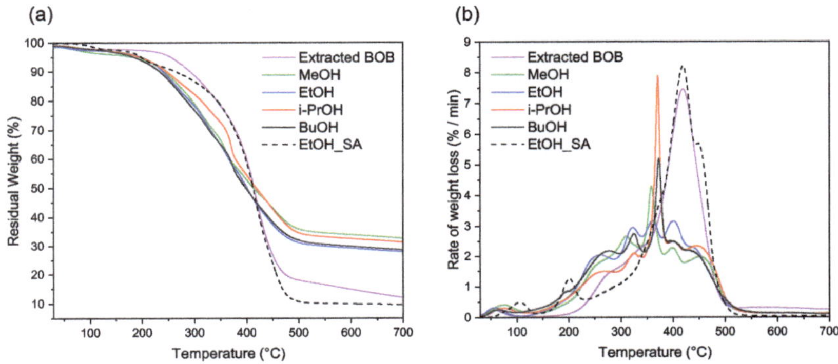

Figure 3. TGA (**a**) and DTGA (**b**) curves for adhesive and feedstock samples.

3.1.4. TGA Analysis

As seen in the TGA DTGA curves in Figure 3, all the adhesive samples showed similar weight loss patterns. The first weight loss was observed below 100 °C due to the moisture content in the sample. Onset temperature values in TGA curves for the i-PrOH and EtOH adhesives were the lowest at 207.6 °C and 207.8 °C, respectively, whereas MeOH was more heat resistant with an onset temperature of 232.2 °C. The second decomposition in DTGA starts at around 200 °C, which could be attributed to the decomposition of suberin-like structures. This was confirmed by acquiring an additional TGA curve from pure SA obtained from the filtrate, as mentioned in Section 2.3. (sample EtOH_SA). In addition, there is a possibility that suberin mass loss over 200 °C temperature overlaps with lignin decomposition [22], resulting in a second peak at 255–277 °C. According to Şen et al. [23] the hemicellulose decomposition of *Betula pendula* bark samples occurs at 309 °C, which coincides with the third peak at 310–325 °C. A sharp peak at 358–373 °C is characteristic because of the degradation of cellulose [23] followed by a decomposition of suberin components at ~400 °C and 443–450 °C, attributing to ω-hydroxy acids and α, ω-diacids respectively [22,23]. When comparing the DTGA curves of adhesive samples to feedstock (extracted BOB), there is a clear decrease in SA-related decomposition. This is due to the removal of some of the suberinic acids after depolymerization for further use in polyols.

Based on experimental findings from *Betula pendula* bark pyrolysis [23], suberin composition can be estimated using a multi-peak fitting technique of the DTGA signal. With the help of Origin Pro 2021b software deconvolution tool, we estimated the cellulose and suberinic acid content using Lorentzian multi-peak fitting. Aromatic suberin and lignin content were estimated from resulted char (at 700 °C) by subtracting ash content from the residue after TGA analysis. The results of chemical composition are summarized in Table 6. Cellulose content, as well as aromatic suberin plus lignin values, were close to all the samples. EtOH sample had the highest ω-hydroxy acid and total suberinic acid (ω-hydroxy + α, ω-diacids) content, which is important for the adhesive effectiveness.

Table 6. Estimations of chemical composition (dry-basis) of adhesive samples from TGA analysis.

Sample	Cellulose, wt%	Aromatic Suberin + Lignin, wt%	ω-Hydroxy Acids, wt%	α, ω-Diacids, wt%
MeOH	9.2	20.3	12.2	13.3
EtOH	9.0	21.4	17.5	11.9
i-PrOH	11.3	22.3	13.8	13.9
BuOH	8.3	22.1	14.2	11.2

3.1.5. FTIR Analysis

Infrared spectra (Figure 4) for adhesive samples and feedstock (extracted BOB) showed a difference for C=O absorption bands. The band at 1734 cm^{-1} was attributed to esters, whereas the band at 1702 cm^{-1} corresponds to C=O vibration in carboxylic acids [24,25]; thus confirming that adhesive samples consisted mostly of depolymerized SA. Another band at 1160 cm^{-1} was present just for the feedstock sample. The C–O–C vibrations at this wavenumber were most likely attributed to ester linkages [26]. Only for BuOH sample, the peak at 1734 cm^{-1} was present, as well as a band at 1160 cm^{-1} was more pronounced than for other adhesives. This may suggest that 1-butanol was the least suitable solvent for BOB depolymerization. Asymmetric (2922 cm^{-1}) and symmetric (2852 cm^{-1}) vibrations, as well as bands at 1464, 1455, and 1374 cm^{-1} were another characteristic of suberin, corresponding to C–H bonds in alkyl chains [24–27]. The vibrations at 721 cm^{-1} are inherent to suberin resulting from aliphatic $R_1CH=CHR_2$ structures [24,27]. Absorbance at the 1243 cm^{-1} region can be attributed to epoxy groups in suberin structures and to cellulose and lignin [26,27]. It appears that this band was slightly less pronounced for the MeOH sample, although MeOH epoxy group content (Table 4) was higher when compared to i-PrOH and BuOH samples, confirming that there is an overlap in the epoxy, cellulose, and lignin signals.

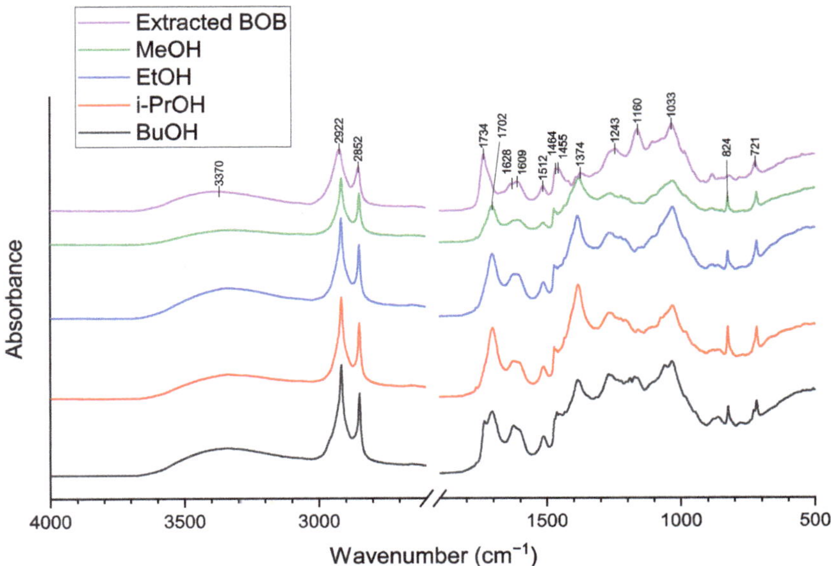

Figure 4. FT-IR spectra for adhesive samples and feedstock (extracted BOB).

Lignin and suberin aromatic components in the adhesive and feedstock samples were identified at 1628 and 1609 cm^{-1} resulting from C=C vibrations from the conjugated carbonyl groups of aromatic components. Absorbances at 1512 and 824 cm^{-1} are another characteristic band for lignin, corresponding to aromatic C=C stretching and ring vibrations [27]. The more pronounced band at 1628, 1609, and 824 cm^{-1} points to the fact that the lignin content has increased in the adhesive again, confirming the removal of some of the suberinic acids after depolymerization for further use in polyols. The peak at 1033 cm^{-1} is associated with C–O vibrations in polysaccharide components [24,27]. The stretching at around 3370 cm^{-1} can be attributed to O–H stretching in polysaccharides and due to the moisture in the sample.

3.2. Choosing the Most Suitable Adhesive for PB Hot-Pressing

To assess the properties of MeOH, EtOH, i-PrOH, and BuOH adhesive-based composites and compare their mechanical properties, particle board samples from the corresponding adhesives were made. The adhesive content was the only variable in these experiments (c = 20–40 wt%).

The data collected in Table 7 show that the density values for all the composites obtained were relatively similar and were just slightly higher than the designed density (0.830 g cm^{-3}). The differences between the other properties of the particle boards were more noticeable. Higher TS 24 h value was achieved with higher adhesive content. TS 24 h requirements were met by all samples made from i-PrOH, BuOH and by two samples from EtOH adhesive (c = 30 wt% and 40 wt%). Overall, MeOH adhesive-based particle boards had the poorest mechanical properties. Regarding the MOE values, very good mechanical properties were achieved by composites based on EtOH and i-PrOH adhesives. BuOH adhesive-based composites showed satisfactory MOE values at c = 20 and 30 wt% when compared to standard P2 requirements. Overall, with an increase in adhesive content, the MOE value decreases for the particleboards, and, therefore, higher adhesive content corresponds to less stiff composites. MOR values are amongst the lowest for BuOH adhesive-based boards. EtOH and i-PrOH adhesive-based particle boards showed the highest values of MOR. However, none of them met the requirements of standards P2 and P3.

Table 7. Mechanical properties of particle boards obtained from MeOH, EtOH, i-PrOH and BuOH adhesives.

Adhesive_c	MOE, N mm^{-2}	MOR, N mm^{-2}	TS 24 h, %	Density, g cm^{-3}
MeOH_20	1690	6.18	32.1	0.847
MeOH_30	1211	5.65	24.5	0.857
MeOH_40	1572	5.19	19.0	0.884
EtOH_20	2266	8.39	23.6	0.855
EtOH_30	2331	7.79	14.7	0.867
EtOH_40	2040	6.99	10.1	0.864
i-PrOH_20	2866	9.55	16.1	0.860
i-PrOH_30	2862	10.17	10.9	0.869
i-PrOH_40	2459	8.23	8.0	0.864
BuOH_20	2036	5.96	14.3	0.880
BuOH_30	1934	6.26	11.7	0.868
BuOH_40	1292	4.98	7.4	0.869
EN 312 P3 [1]	≥2050	≥15	≤17	–
EN 312 P2 [2]	≥1600	≥11	–	–

[1] Standard requirements for non-load-bearing boards (>6–13 mm thick) for use in humid conditions. [2] Standard requirements for boards (>6–13 mm thick) for interior fitments (including furniture) for use in dry conditions.

It seems that the higher acid number value (Table 4) corresponded to the overall better mechanical properties of the particle boards. The poor mechanical properties of MeOH adhesive-based boards could also be explained both by too little content of suberinic acids and their oligomers in the sample (low solubility in THF, Table 4) and because this adhesive had higher thermal stability, which requires a higher hot-pressing temperature. In addition unwanted condensation reactions during the drying stage of adhesive and wood particle mixture may have resulted in low particle board mechanical properties.

In summary, the most perspective depolymerization mediums for obtaining particle board adhesive from BOB seem to be EtOH and i-PrOH. From a technological point of view, obtaining the EtOH adhesive was less complicated. In addition, the cooling of i-PrOH depolymerizate resulted in rapid precipitation. It seems that this precipitation was a reason that i-PrOH adhesive exhibited highest acid number (Table 4). However, the yield of SA that can be obtained from filtrate after depolymerization in an i-PrOH was about 18% lower when compared to EtOH-obtained SA, which is important for SA-based polyol synthesis [12]. Therefore, it is important to add that the SA-based adhesives that are studied in this paper are obtained as a side-stream product. Based on this evidence the EtOH adhesive was considered the most suitable for further investigation of optimal hot-pressing parameters.

3.3. Experimental Design for Obtaining Particle Boards

As mentioned, the EtOH adhesive was chosen as the most suitable adhesive from the obtained data on the properties of the particle boards, for the full factorial experiment (FFE). The adhesive was further used to make particle board composite materials for investigation of the effects of the hot-pressing parameters on the properties of the particle boards through the FFE approach.

Depending on hot-pressing parameters, the properties (MOE, MOR, TS 24 h, density of particle boards were studied. The results in Table 8 show that lower c and higher T and τ values corresponded to better mechanical properties. It can be seen that temperature plays a very important role in the curing properties of SA-based adhesive, judging by MOE and MOR values. It also seems that the ability of the adhesive to penetrate the filler (wood particles) improves with an increase of the temperature, indicated by lower TS 24 h values. Overall, the statistical analysis showed that all hot-pressing variables had a significant effect on the particle board properties ($p < 0.05$).

Table 8. Properties of particle boards, depending on the hot-pressing parameters.

Variable Parameters			MOE, N mm^{-2}	MOR, N mm^{-2}	TS 24 h, %	Density, g cm^{-3}
c, wt%	T, °C	t, min				
20	210	2	659	2.08	41.8	0.688
20	210	8	1831	6.63	22.5	0.854
20	250	2	2820	7.64	17.3	0.845
20	250	8	3518	11.44	3.9	0.846
40	210	2	1403	5.45	20.9	0.844
40	210	8	2187	7.77	12.4	0.891
40	250	2	2612	7.92	8.3	0.880
40	250	8	2723	10.31	2.1	0.853

To improve the resolution, the experimental plan was augmented with eight additional runs using D-optimality criteria. The mechanical properties of the obtained particle boards are summarised in Table 9. The boards with t = 2 min were hot-pressed according to the parameters given in Table 3. The rest of the boards were hot-pressed in two cycles: during the first cycle, the pressure was 3.5 MPa (t = 120 s) followed by a reduction in the pressure to 0.1 MPa for 30 s; in the second cycle, the pressure was 1.7 MPa (t = 50 to 180 s) with a subsequent pressure release to 0.7 MPa for 15 to 28 s.

Table 9. Characteristics of particle boards, depending on the pressing parameters (additional points).

Variable Parameters			MOE, N mm^{-2}	MOR, N mm^{-2}	TS 24 h, %	Density, g cm^{-3}
c, wt%	T, °C	t, min				
29	232	5.63	2709	9.51	11.9	0.894
40	219	5.00	2149	7.07	12.3	0.876
29	224	5.90	2453	8.35	14.0	0.883
29	217	4.70	2073	7.34	18.6	0.863
33.5	230	2.00	1593	5.49	19.2	0.850
39	240	5.03	2679	8.74	6.9	0.896
20	233	5.30	2840	9.03	15.6	0.889
30	250	3.59	2951	8.76	7.0	0.870

Overall, the densities were slightly higher than the designed values. MOE values for most of the boards met the EN 312 P3 and P2 requirements, however, this was not the case with the MOR values. At this stage of the study, it seemed that the standard requirements P2 can only be met due to the comparatively low MOR value. In addition, for P2 requirements TS 24 h value was not relevant. However, most of the particle boards exhibited good water resistance (TS 24 h values below 15%) confirming the beneficial water-proofing properties of SA.

After summarizing the data, 4D mathematical models depending on the obtained results were developed and their visual representations are further discussed in the following subsubsections.

3.3.1. MOE of the Particle Boards

Changes in MOE of the obtained boards depending on the hot-pressing parameters are shown in Figure 5. A quadratic numerical model was chosen, as suggested by the software. To simplify the model, statistically insignificant values were excluded. According to ANOVA results, T, t, s . . . T, s . . . t, T . . . t and t^2 variables had a statistically significant effect ($p < 0.05$) on the MOE value. It can be seen that higher MOE values can be achieved at higher hot-pressing temperatures. These results corroborate that the formation of more cross-linked structures, and, thus, higher stiffness can be achieved at higher temperatures. In addition, as the hot-pressing time increases from 2 to 5 min, the value of MOE increases, but it decreases as it approaches t = 8 min, suggesting that at prolonged hot-pressing periods the degradation of both wood particles and adhesive occur.

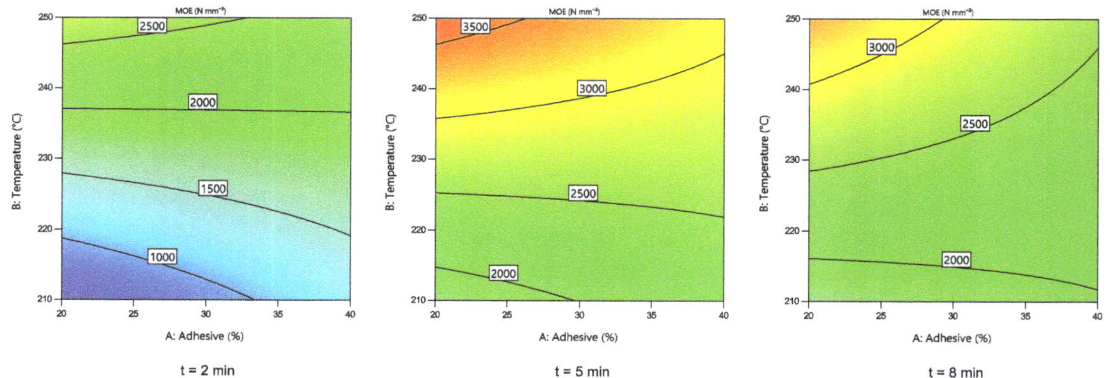

Figure 5. Changes in the MOE value of the boards depending on the hot-pressing parameters.

3.3.2. MOR of the Particle Boards

Changes in MOR of the obtained boards depending on the hot-pressing parameters are shown in Figure 6. A quadratic numerical model was chosen, as suggested by the software. To simplify the model, statistically insignificant values were excluded. According to ANOVA results, T, t, s . . . T, s . . . t and t^2 variables had a statistically significant effect ($p < 0.05$) on MOR value. The graphic interpretation of the model shows that at higher hot-pressing times and higher temperatures, it is possible to reach higher MOR values at relatively low adhesive contents.

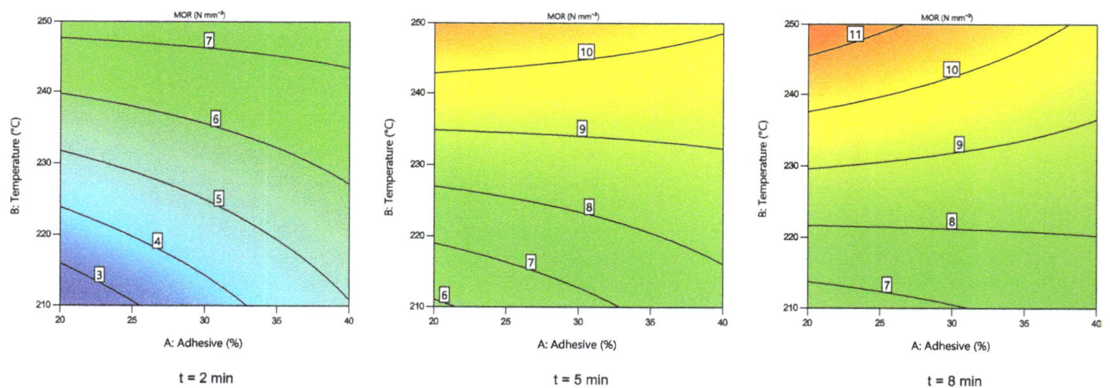

Figure 6. Changes in the MOR value of the boards depending on the hot-pressing parameters.

3.3.3. TS 24 h of the Particle Boards

Changes in TS 24 h of the obtained boards depending on the hot-pressing parameters are shown in Figure 7. A quadratic numerical model was chosen, as suggested by the software. To simplify the model, one statistically insignificant value (T^2) was excluded. All the rest of the values showed statistically significant effects ($p < 0.05$) on the water-resistance of the boards, which is also shown in the graphic interpretation of the model. It can be seen that the hot-pressing time had a noticeable effect as the moisture resistance (lower TS 24 h values) was significantly improved at t = 5 min (compared to t = 2 min). After a further increase of duration to t = 8 min, the improvement of the 24 h value is less pronounced.

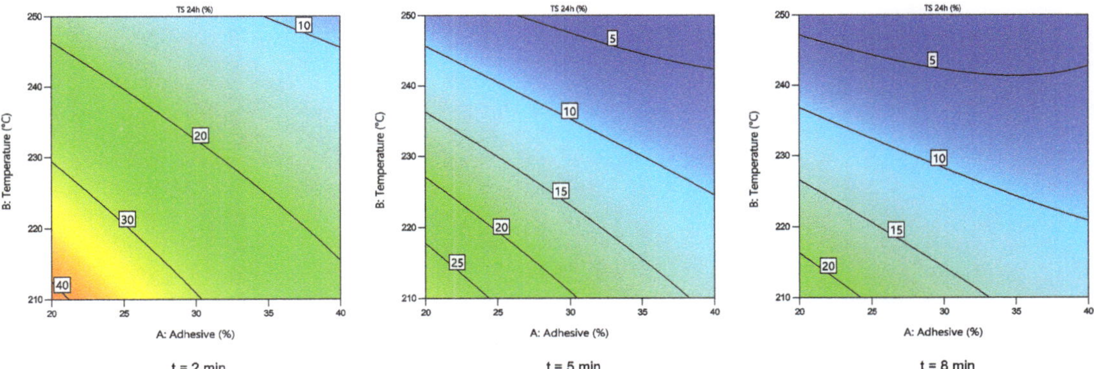

Figure 7. Changes in the TS 24 h value of the boards depending on the hot-pressing parameters.

3.3.4. Particle Board Density

Changes in the density of the obtained boards depending on the hot-pressing parameters are shown in Figure 8. The two-factor interaction numerical model was chosen, as suggested by the software. Statistical analysis showed that all hot-pressing parameters significantly ($p < 0.05$) affected the density of the boards. An interesting pattern can be seen for higher t values (8 min): at lower hot-pressing temperatures and higher c, the density is rather high. However, with an increase in T, the density further decreases. It could be explained by the rapid release of condensation and decomposition products at higher temperatures, resulting in lower density boards.

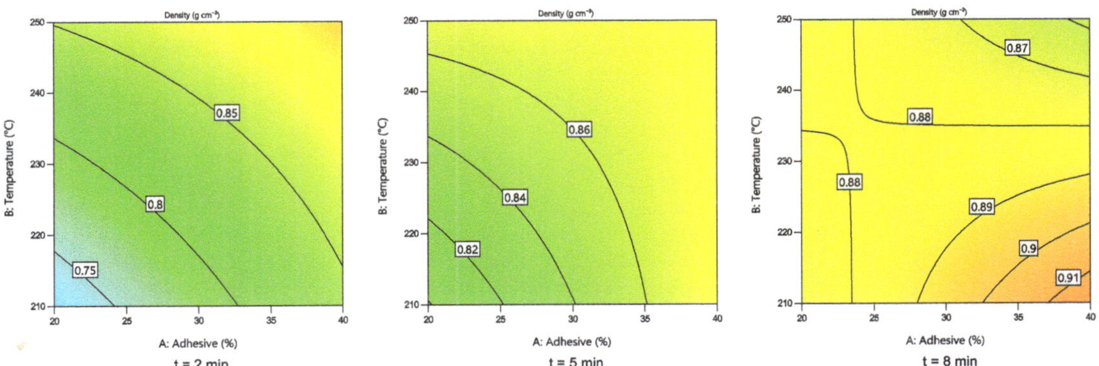

Figure 8. Changes in the density of the boards depending on the hot-pressing parameters.

3.3.5. Determination of Optimal Parameters for Particle Board Hot-Pressing

The constraints of the variables and response values given in Table 10 were used for the determination of optimal hot-pressing parameters.

Table 10. Constraints for hot-pressing parameters and response values.

Value	Goal	Lower Limit	Upper Limit
c, wt%	is in range	20	40
T, °C	minimize	210	250
t, min	minimize	2	8
MOE N, mm^{-2}	maximize	1800	3917
MOR N, mm^{-2}	maximize	11.00	12.59
TS 24 h, %	minimize	1.88	44.94
Density, g cm^{-3}	is target = 0.83	0.671	0.952

After setting the goals, the software suggested the following optimal hot-pressing parameters: c = 20 wt%; T = 248 °C and t = 6.55 min. To confirm the results, particle boards were hot-pressed according to the suggested parameters in two cycles. During the first cycle, the pressure was 3.5 MPa (t = 150 s) followed by a reduction in the pressure to 0.1 MPa for 30 s; while in the second cycle, the pressure was 1.7 MPa (t = 180 s) with a subsequent pressure release to 0.7 MPa for 34 s. All four response values as well as additionally IB were determined for the resulting boards and the results are given in Table 11. It can be seen that all values were within the limits of the model's 95% prediction interval and also met the EN 312 P2 requirements (Table 7).

Table 11. Mechanical properties of particle boards obtained under optimal conditions.

Response Value	Result	95% Prediction Interval Lower Limit	95% Prediction Interval Upper Limit
MOE, N mm^{-2}	3833	3249	3939
MOR, N mm^{-2}	11.27	10.17	12.55
TS 24 h, %	6.26	4.49	7.12
Density, g cm^{-3}	0.903	0.837	0.913
IB, N mm^{-2}	1.33 [1]	-	-

[1] EN 312 P2 requirement $\geq$ 0.40 N mm^{-2}.

4. Conclusions

The present work studied the possibilities of obtaining and utilising suberinic acid-containing residues after birch outer bark depolymerization in the alkanol environment. Four different adhesives obtained in four different solvents (methanol, ethanol, isopropanol, 1-butanol) were compared by chemical and thermal properties, as well as their performance in particle board bonding. The highest acid number (122.0 mg KOH g^{-1}) was reached when the adhesive was obtained in isopropanol, but this adhesive was characterised with the lowest epoxy content (0.11 mmol g^{-1}). For ethanol-based adhesive, the epoxy content (0.61 mmol g^{-1}) and ω-hydroxy suberinic acid (17.5 wt%) content as determined with DTGA was the highest. TGA analysis also showed that all obtained adhesives showed heat resistance from 207 °C for adhesives obtained in ethanol and isopropanol to 232 °C (methanol adhesive). Based on particle board mechanical tests, ethanol was chosen as the most appropriate depolymerization medium for obtaining adhesive. The optimal hot-pressing parameters were determined using the design of experiments approach: adhesive content 20 wt%; hot-pressing temperature 248 °C and hot-pressing time 6.55 min. The obtained optimal particle boards met the EN 312 P2 requirements and had a satisfactory water resistance. One limitation of this process is the relatively high hot-pressing temperature, and, therefore, the focus on the future research can be aimed at reducing it with polymerization catalysts or cross-linkers.

Author Contributions: Conceptualization, R.M. and J.R.; methodology, J.R., D.G. and A.P.; formal analysis, R.M.; data curation, R.M., J.R. and D.G.; writing—original draft preparation, R.M. and J.R.; writing—review and editing, R.M., J.R., A.P. and D.G. and R.M.-M.; visualization, R.M.; supervision

J.R. and R.M.-M.; project administration, J.R.; funding acquisition, J.R. All authors have read and agreed to the published version of the manuscript.

Funding: This research was funded by ERDF project no. 1.1.1.1/19/A/089 "Birch bark as a valuable renewable raw material for producing formaldehyde-free particle boards and suberinic acids polyols for the development of polyurethanes".

Institutional Review Board Statement: Not applicable.

Informed Consent Statement: Not applicable.

Data Availability Statement: Not applicable.

Conflicts of Interest: The authors declare no conflict of interest.

References

1. Vedernikov, D.N.; Shabanova, N.Y.; Roshchin, V.I. Change in the Chemical Composition of the Crust and Inner Bark of the Betula Pendula Roth. Birch (Betulaceae) with Tree Height. *Russ. J. Bioorganic Chem.* **2011**, *37*, 877–882. [CrossRef]
2. Heinämäki, J.; Pirttimaa, M.M.; Alakurtti, S.; Pitkänen, H.P.; Kanerva, H.; Hulkko, J.; Paaver, U.; Aruväli, J.; Yliruusi, J.; Kogermann, K. Suberin Fatty Acids from Outer Birch Bark: Isolation and Physical Material Characterization. *J. Nat. Prod.* **2017**, *80*, 916–924. [CrossRef] [PubMed]
3. Rizhikovs, J.; Zandersons, J.; Dobele, G.; Paze, A. Isolation of Triterpene-Rich Extracts from Outer Birch Bark by Hot Water and Alkaline Pre-Treatment or the Appropriate Choice of Solvents. *Ind. Crops Prod.* **2015**, *76*, 209–214. [CrossRef]
4. Krasutsky, P.A.; Kolomitsyn, I.V.; Krasutskyy, D.A. Depolymerization Extraction of Compounds from Birch Bark. 2012. Available online: https://patents.google.com/patent/US8197870B2/en (accessed on 20 April 2022).
5. Koptelova, E.N.; Kutakova, N.A.; Tret'yakov, S.I. Isolation of the Extractives and Betulin from Birch Bark Exposed to a Microwave Field. *Russ. J. Bioorganic Chem.* **2014**, *40*, 791–795. [CrossRef]
6. Rižikovs, J.; Zandersons, J.; Paže, A.; Tardenaka, A.; Spince, B. Isolation of Suberinic Acids from Extracted Outer Birch Bark Depending on the Application Purposes. *Balt. For.* **2014**, *20*, 98–105.
7. Graça, J. Suberin: The Biopolyester at the Frontier of Plants. *Front. Chem.* **2015**, *3*, 1–11. [CrossRef]
8. Paze, A.; Rizhikovs, J.; Godiņa, D.; Makars, R.; Berzins, R. Development of Plywood Binder by Partial Replacement of Phenol-Formaldehyde Resins with Birch Outer Bark Components. *Key Eng. Mater.* **2021**, *903*, 229–234. [CrossRef]
9. Tupciauskas, R.; Rizhikovs, J.; Grinins, J.; Paze, A.; Andzs, M.; Brazdausks, P.; Puke, M.; Plavniece, A. Investigation of Suberinic Acids-Bonded Particleboard. *Eur. Polym. J.* **2019**, *113*, 176–182. [CrossRef]
10. Paze, A.; Rizhikovs, J. Study of an Appropriate Suberinic Acids Binder for Manufacturing of Plywood. *Key Eng. Mater.* **2019**, *800*, 251–255. [CrossRef]
11. Pollard, M.; Beisson, F.; Li, Y.; Ohlrogge, J.B. Building Lipid Barriers: Biosynthesis of Cutin and Suberin. *Trends Plant Sci.* **2008**, *13*, 236–246. [CrossRef] [PubMed]
12. Rizikovs, J.; Godina, D.; Makars, R.; Paze, A.; Abolins, A.; Fridrihsone, A.; Meile, K.; Kirpluks, M. Suberinic Acids as a Potential Feedstock for Polyol Synthesis: Separation and Characterization. *Polymers* **2021**, *13*, 4380. [CrossRef] [PubMed]
13. Rizikovs, J.; Paze, A.; Makars, R.; Tupciauskas, R. A Method for Obtaining Thermoreactive Binders for a Production of Wood Composite Materials From Birch Outer Bark. 2021. Available online: https://patents.google.com/patent/EP3807341A1/ (accessed on 30 May 2022).
14. Godiņa, D.; Paze, A.; Rizhikovs, J.; Stankus, K.; Virsis, I.; Nakurte, I. Stability Studies of Bioactive Compounds from Birch Outer Bark Ethanolic Extracts. *Key Eng. Mater.* **2018**, *762*, 152–157. [CrossRef]
15. *ISO 1171:2010*; Solid Mineral Fuels—Determination of Ash. International Organization for Standardization: Geneva, Switzerland, 2010.
16. *EN 323:2000*; Wood-Based Panels—Determination of Density. European Committee for Standardisation: Brussels, Belgium, 2000.
17. *EN 310:2001*; Wood-Based Panels—Determination of Modulus of Elasticity in Bending and of Bending Strength. European Committee for Standardisation: Brussels, Belgium, 2001.
18. *EN 317:2000*; Particleboards and Fibreboards—Determination of Swelling in Thickness after Immersion in Water. European Committee for Standardisation: Brussels, Belgium, 2000.
19. *EN 319:2000*; Particleboards and Fibreboards—Determination of Tensile Strength Perpendicular to the Plane of the Board. European Committee for Standardisation: Brussels, Belgium, 2000.
20. *EN 312:2011*; Particleboards—Specifications. European Committee for Standardisation: Brussels, Belgium, 2011.
21. Sagnelli, D.; Vestri, A.; Curia, S.; Taresco, V.; Santagata, G.; Johansson, M.K.G.; Howdle, S.M. Green Enzymatic Synthesis and Processing of Poly (Cis-9,10-Epoxy-18-Hydroxyoctadecanoic Acid) in Supercritical Carbon Dioxide (ScCO2). *Eur. Polym. J.* **2021**, *161*, 110827. [CrossRef]
22. Shangguan, W.; Chen, Z.; Zhao, J.; Song, X. Thermogravimetric Analysis of Cork and Cork Components from Quercus Variabilis. *Wood Sci. Technol.* **2018**, *52*, 181–192. [CrossRef]
23. Şen, U.; Pereira, H. Pyrolysis Behavior of Alternative Cork Species. *J. Therm. Anal. Calorim.* **2022**, *147*, 4017–4025. [CrossRef]

24. Shiqian, W.; Xiaozhou, S.; Yafang, L.; Mingqiang, Z. Characterizations and Properties of Torrefied Quercus Variabilis Cork. *Wood Res.* **2018**, *63*, 947–958.
25. Rizhikovs, J.; Brazdausks, P.; Paze, A.; Tupciauskas, R.; Grinins, J.; Puke, M.; Plavniece, A.; Andzs, M.; Godina, D.; Makars, R. Characterization of Suberinic Acids from Birch Outer Bark as Bio-Based Adhesive in Wood Composites. *Int. J. Adhes. Adhes.* **2022**, *112*, 102989. [CrossRef]
26. Graça, J.; Pereira, H. Methanolysis of Bark Suberins: Analysis of Glycerol and Acid Monomers. *Phytochem. Anal.* **2000**, *11*, 45–51. [CrossRef]
27. Karnaouri, A.; Rova, U.; Christakopoulos, P. Effect of Different Pretreatment Methods on Birch Outer Bark: New Biorefinery Routes. *Molecules* **2016**, *21*, 427. [CrossRef] [PubMed]

Article

Effect of the Adhesive System on the Properties of Fiberboard Panels Bonded with Hydrolysis Lignin and Phenol-Formaldehyde Resin

Viktor Savov [1,*], Ivo Valchev [2], Petar Antov [1,*], Ivaylo Yordanov [2] and Zlatomir Popski [3]

1 Faculty of Forest Industry, University of Forestry, 1797 Sofia, Bulgaria
2 Faculty of Chemical Technologies, University of Chemical Technology and Metallurgy, 1757 Sofia, Bulgaria; ivoval@uctm.edu (I.V.); yordanov@uctm.edu (I.Y.)
3 Welde Bulgaria, 5600 Troyan, Bulgaria; popski@welde.bg
* Correspondence: victor_savov@ltu.bg (V.S.); p.antov@ltu.bg (P.A.)

Abstract: This study aimed to propose an alternative technological solution for manufacturing fiberboard panels using a modified hot-pressing regime and hydrolysis lignin as the main binder. The main novelty of the research is the optimized adhesive system composed of unmodified hydrolysis lignin and reduced phenol–formaldehyde (PF) resin content. The fiberboard panels were fabricated in the laboratory with a very low PF resin content, varying from 1% to 3.6%, and hydrolysis lignin addition levels varying from 7% to 10.8% (based on the dry wood fibers). A specific two-stage hot-pressing regime, including initial low pressure of 1.2 MPa and subsequent high pressure of 4 MPa, was applied. The effect of binder content and PF resin content in the adhesive system on the main properties of fiberboards (water absorption, thickness swelling, bending strength, modulus of elasticity, and internal bond strength) was investigated, and appropriate optimization was performed to define the optimal content of PF resin and hydrolysis lignin for complying with European standards. It was concluded that the proposed technology is suitable for manufacturing fiberboard panels fulfilling the strictest EN standard. Markedly, it was shown that for the production of this type of panels, the minimum total content of binders should be 10.6%, and the PF resin content should be at least 14% of the adhesive system.

Keywords: wood-based panels; fiberboards; adhesive system; hydrolysis lignin; phenol–formaldehyde resin; optimization; hot-pressing

Citation: Savov, V.; Valchev, I.; Antov, P.; Yordanov, I.; Popski, Z. Effect of the Adhesive System on the Properties of Fiberboard Panels Bonded with Hydrolysis Lignin and Phenol-Formaldehyde Resin. *Polymers* **2022**, *14*, 1768. https://doi.org/10.3390/polym14091768

Academic Editor: Antxon Santamaria

Received: 26 March 2022
Accepted: 25 April 2022
Published: 27 April 2022

1. Introduction

The growing demand for eco-friendly wood-based panels with lower environmental footprints and reduced hazardous emissions of volatile organic compounds, such as free formaldehyde from the finished wood composites, have imposed new stricter regulations and requirements on both researchers and industrial practice, related to the development of sustainable, "green" composites [1–5]. The production of fiberboards, with an estimated global production of more than 104 million m^3 in 2020, is the second largest worldwide, surpassed only by the production of plywood [6]. In the production of dry-process fiberboards, which accounts for about 74% of the total output, the problem with the hazardous formaldehyde emissions from the finished panels is also relevant [7,8]. A viable approach to solving this issue is using sustainable, bio-based, formaldehyde-free wood adhesives to partially or completely replace the conventional synthetic formaldehyde-based wood adhesives, such as urea–formaldehyde (UF), melamine–urea formaldehyde (MUF), and phenol–formaldehyde (PF) resins, commonly used in the panel industry [9–14]. Successful attempts for the development of bio-based adhesives, including modified condensed and hydrolyzed tannins, proteins, starch, lignin, carbohydrates, etc., have been reported [15–20]. The main drawbacks of using 100% bio-based adhesive formulations for bonding wood composites are related to the need for additional modification of natural raw materials to improve their chemical reactivity, the deteriorated dimensional stability and mechanical properties of the wood-based panels produced, and the need to modify the technological parameters, e.g., through the extension of pressing time. In terms of industrial utilization, significant positive results have been obtained in producing wood-based panels, mainly particleboards, with tannin-based bio-adhesives [21–23].

Lignin is an amorphous, three-dimensional complex biopolymer, composed of phenyl-propanoid units linked by intramolecular bonds, and the second most abundant natural material, surpassed only by cellulose. Lignin contains a large number of functional groups, e.g., aliphatics, phenolic, hydroxyl, and carbonyl groups, and acts as a natural binder in wood, being the main component of the middle lamella connecting wood cells [24]. In the production of wet process fiberboards, the properties of the panels are mainly due to the lignin bonds arising from the hot-pressing. This makes lignin a particularly promising bio-based adhesive for manufacturing dry-process fiberboards.

Significant quantities of lignin by-products, estimated to approximately 100 million tons per year, are generated worldwide, mostly as a waste and side streams of the pulp and paper industries, of which only about 2% is used for conversion into value-added products [25–27]. The enhanced valorization and commercial utilization of that lignin will support the transition to circular economy [28,29]. Lignin can be extracted from lignocellulosic biomass by applying physical, chemical, and biological treatment methods. Depending on the method by which they are obtained, the residual lignin products, i.e., technical lignins, are sulfate (Kraft) lignin, sulfite lignin (lignosulfonate), organosolv lignin and hydrolysis lignin. Although Kraft lignin is the most widespread globally, much of it is burned in the factories where it is obtained, which regenerates some of the chemical reagents used and produces heat and energy [30,31]. The main drawbacks for using lignosulfonates in wood adhesive formulations are related to the higher number of impurities, e.g., high sulfur and ash content, compared to the Kraft lignin, and the deteriorated hydrophobic properties of the wood-based panels, so it is recommended to be used in combination with synthetic binders with or without additional cross-linking [32–42]. Although there are some studies on the use of organosolv lignin in wood adhesives, mostly as a partial replacement of phenol in PF resins, its wider use is limited due to the significantly smaller quantities [43], compared to Kraft lignin and lignosulfonates. In the case of hydrolysis lignin (HL), these shortcomings are greatly avoided [44]. As HL is a by-product of bioethanol production, its amount is expected to increase worldwide [45,46].

Previous efforts to recover lignin have been focused mainly on its modification, its use as a substitute for phenol in lignin–phenol–formaldehyde (LPF) resins, used primarily in the production of plywood or the use of lignin for biofuels [47–53]. There are also successful attempts to produce fiberboards with lignin as a binder. Previous studies on the use of HL in wood adhesives showed that if it is introduced in the dry state in the pulp, lignin cannot be retained and deteriorated the properties of the panels [44]. It was determined that when using a traditional hot-pressing cycle with first high and subsequent decreasing pressure, the addition of lignin as a substitute of formaldehyde-based adhesives leads to a deterioration of the panel properties [54]. Good results were achieved by using HL with a very small content of PF resin as an auxiliary binder. PF resin was mainly used to improve the retention of lignin in the pulp, and to enhance the binding reactions of between lignin and wood fibers [55]. The cited study did not fully clarify the effect of the total binder content and the effect of the ratio of lignin to PF resin in the adhesive system.

This work aimed to investigate the effect of the total binder content and the optimal ratio between hydrolysis lignin and PF resin in the production of fiberboards using a modified hot-pressing regime. The main novelty is a derivation of the optimal composition of the adhesive system with maximum HL content when using a modified hot-pressing regime. The modification of the hot-pressing regime aims for maximal utilization of the adhesive abilities of the lignin, in contrast to the classic hot-pressing cycles, which are developed for formaldehyde synthetic resins.

2. Materials and Methods

In this study, PF resin was used as an auxiliary binder to retain the technical HL until the condensation and activation process occurred. The PF resin was chosen because of its better lignin convergence and higher temperature resistance than UF and MUF resins [56–59], Figure 1.

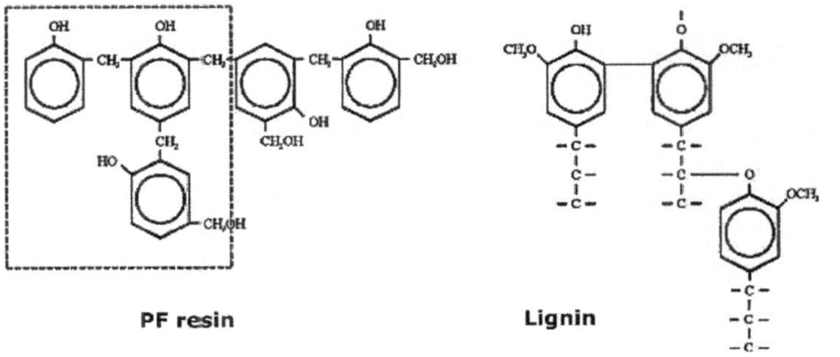

Figure 1. Structure of PF resin and lignin [59].

The present study used a classical approach for mixing PF resin and lignin. Still, a modified hot-pressing cycle was applied to make optimal use of the adhesion capabilities of lignin.

Industrially produced pulp obtained by the thermomechanical refining method in the factory Kronospan-Bulgaria EOOD (Veliko Tarnovo, Bulgaria) was used. The pulp was composed of 40% hardwood (beech and oak) and 60% softwood (pine). The pulp was characterized by a bulk density of 29 kg·m^{-3}, fiber lengths varying from 1120 to 1280 µm, and a moisture content of 11.2% (factory data). The PF resin used was manufactured by Prefere Resins Romania SRL (Rasnov, Romania) and provided by Welde Bulgaria PLC (Troyan, Bulgaria). The PF resin had the following characteristics: dry solids content 46%, viscosity—358 MPa·s; brix 72.7 and acid factor (pH)—6.8.

The technical HL was produced from high temperature diluted sulfuric acid hydrolysis of sawdust and softwood and hardwood chips to sugars. The chemical composition of lignin was determined by the standard TAPPI methods [60,61]. The C, N, S and H analysis was performed by using Elemental Analyzer Euro EA 3000 (EuroVector, Pavia, Italy). After fractionation, only HL from the fraction below 100 µm was used.

In interior design and furniture production, thin and ultra-thin wood-based composites are increasingly used in a variety of end uses [62,63]. Therefore, the target thickness of the laboratory panels was set to 4 mm. Twelve fiberboard panels with dimensions 200 mm × 200 mm × 4 mm were manufactured in laboratory conditions, divided into two series with different adhesive systems. The target density of the laboratory-made fiberboard panels was 850 kg·m^{-3}.

Due to the use of a bio-based binder, i.e., HL, the total binder content was higher compared to the conventional adhesive systems, composed only from thermosetting formaldehyde-based resins [64]. Thus, the total binder content used in this work was 10% and 12%. The higher content of binders when an adhesive system with the participation of lignin is used leads to a significant deterioration in the appearance of the panels, the need to extend the press factor and a very slight improvement in the properties of fiberboards [65,66]. The amount of PF resin in the adhesive system also varied and was 10%, 20% and 30%. The manufacturing parameters of fiberboards bonded with HL and PF resin are given in Table 1.

Table 1. Manufacturing parameters of laboratory-fabricated fiberboard panels bonded with adhesive systems composed of HL and PF resin.

Panel Type	Total Binders Content, %	PF Resin Content in the Adhesive System, %	Technical Hydrolysis Lignin Content in the Adhesive System, %	PF Resin Content Relative to Dry Fibers, %	Technical Hydrolysis Lignin Content Relative to Dry Fibers, %
A	10	10	90	1.0	9.0
B	10	20	80	2.0	8.0
C	10	30	70	3.0	7.0
D	12	10	90	1.2	10.8
E	12	20	80	2.4	9.6
F	12	30	70	3.6	8.4

The HL and PF resin were adjusted to a concentration of 30%. Then they were mixed and almost immediately sprayed into the pulp. A high-speed glue blender at 850 rpm (laboratory prototype, University of Forestry, Sofia, Bulgaria) with needle-shaped blades was used. The adhesive formulation was injected through a nozzle with a diameter of 1.5 mm at a pressure of 0.4 MPa. The whole gluing process had a duration of 60 s.

The hot-pressing was performed on a laboratory press "Servitec-Polystat 200 T" (Servitec Maschinenservice GmbH, Wustermark, Germany). The hot-pressing temperature applied was 200 °C. The pressing was carried out in two stages with subsequent cooling. The first stage was performed at a pressure of 1.2 MPa and lasted 360 s. The second stage was carried out at a pressure of 4.0 MPa for 120 s. Cooling was carried out while maintaining the high pressure (4.0 MPa) up to a temperature below 100 °C. The cooling time was 360 s. That hot-pressing regime was chosen due to a previous study aimed at optimizing the pressing time using HL and PF resin [55]. In that study, an increase in the press factor for the second stage above 30 s·mm^{-1} is unjustified. Preliminary studies indicate that when an adhesive system from HL and PF resin is used, a deterioration in the waterproof properties of the panels is observed at hot-pressing temperatures below 200°. That is confirmed by other similar studies [65,66]. The optimal parameters of the hot-pressing pressure have also been established experimentally. During the first stage, at a pressure above 1.0 MPa, it is difficult to separate the gas mixture from the panels. At a pressure below 4.0 MPa, it is difficult to compress the fibers to the final thickness and density of the panels.

The fiberboard panels were conditioned for 10 days at a room temperature of 20 ± 2 °C and a relative humidity of 65%.

The physical and mechanical properties of the laboratory-fabricated fiberboard panels (Figure 2) were determined according to the EN standards [67–70]. Eight test samples were used for each property. A precision laboratory balance Kern (Kern & Sohn GmbH, Balingen, Germany) with an accuracy of 0.01 g was used to measure the mass of the test specimen. Digital calipers with a 0.01 mm accuracy were used to determine the dimensions of the test samples. Thickness swelling (TS) and water absorption (WA) tests were performed by the weight method after 24 h of immersion in water. A universal testing machine, Zwick/Roell Z010 (ZwickRoell GmbH & Co. KG, Ulm, Germany), was used to determine the mechanical properties of the panels.

Regression analysis was used to analyze the effect of the total binder content and the participation of PF resin in the adhesive system on the properties of the composites, and the following regression model was derived (Equation (1)):

$$\hat{Y} = B_0 + B_1X_1 + B_2X_2 + B_{12}X_1X_2 \quad (1)$$

where $\hat{Y}$ is the predicted value of the given property;

B_0, B_1, B_2, B_{12}—regression coefficients;

X_1, X_2—the studied factors.

Stepwise regression with 1000 iterations was applied to perform optimization. For this purpose, specialized software "QstatLab", version 6.0, was used.

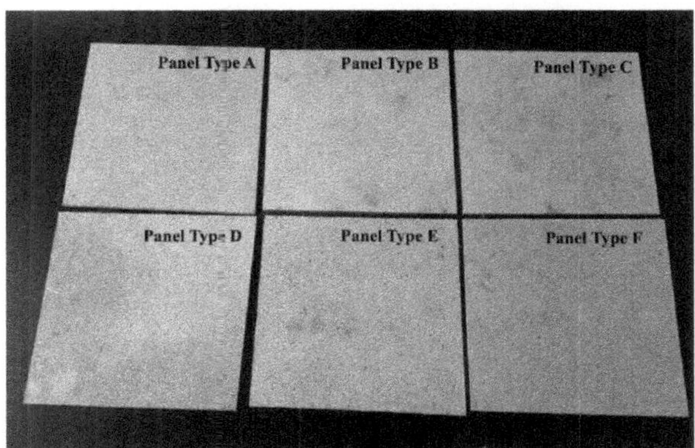

Figure 2. Fiberboard panels bonded with HL and PF resin: Type A—10% of binders, from which 10% was PF resin; Type B—10% of binders, from which 20% was PF resin; Type C—10% of binders, from which 30% was PF resin; Type D—12% of binders, from which 10% was PF resin; Type E—12% of binders, from which 20% was PF resin; Type F—12% of binders, from which 30% was PF resin.

3. Results and Discussion

The technical HL had the following chemical composition: 72.6% lignin, 25.5% cellulose, 2.8% ash content, 55.54 C, 0.74 S, 7.10 H, and 0.26 N. The ash content is relatively low, which leads to improved adhesion properties of technical HL. That is, the ash will not significantly impair the adhesive bonds. The presence of cellulose in the HL contributes to its properties as a binder [71,72].

The results obtained for the density of laboratory-fabricated fiberboard panels are presented in Table 2.

Table 2. Density of fiberboard panels bonded with HL and PF resin.

Panel Type	Average/Mean/ Value, kg·m^{-3}	Standard Deviation, kg·m^{-3}	Standard Error, kg·m^{-3}	Coefficient of Variation, %	Probability, %
A	856	34	4	12	1.42
B	852	43	5	15	1.77
C	862	26	3	9	1.08
D	842	24	3	8	0.99
E	860	24	3	8	0.99
F	850	19	2	7	0.78

The density of the panels varied from 842 to 862 kg·m^{-3}. The difference between the maximum and minimum density values was 2.4%, i.e., significantly below the permissible statistical error of 5%. This was also confirmed by the conducted ANOVA (Table 3). The test performed included the density data obtained for each test sample.

Table 3. ANOVA for the density of fiberboard panels bonded with HL and PF resin.

Source of Variation	SS	df	MS	F	*p*-Value	F_{crit}
Total binder content	247.52	1	247.52	0.356	0.553	4.06
PF resin content in the adhesive system	393.16	2	196.58	0.282	0.755	3.21
Error	30,617.29	44	695.84			
Total	31,257.97	47				

Therefore, it can be concluded that there was no statistically significant difference between the density of the individual fiberboards, thus it will not affect the other physical and mechanical properties.

The derived regression models, in explicit form, are presented in Table 3.

Table 4 shows that the obtained models are statistically significant. In all models, the calculated value of the Fisher's criterion (F_{cal}) is greater than the critical value ($F_{(0.05,3,2)}$).

Table 4. Regression models for the effect of total binder content and PF resin content in the adhesive system on the properties of the fiberboard panels produced in this work.

Regression Coefficient \ Property	Modulus of Elasticity (MOE)	Bending Strength (MOR)	Internal Bond (IB) Strength	Thickness Swelling (24 h)	Water Absorption (24 h)
B_0	−6920.667	−51.867	−2.260	60.538	111.152
B_1	917.333	7.438	0.285	−2.493	−3.613
B_2	230.450	0.395	0.018	−0.410	−0.530
B_{12}	−17.550	–	−0.001	–	–
F_{cal}	147.89	5882.68	974.77	37.892	29.039
$F_{(0.05,3,2)}$	19.164	19.164	19.164	19.164	19.164

A graphical representation of the results obtained for the modulus of elasticity (MOE) of the fiberboard panels fabricated in this work is presented in Figure 3.

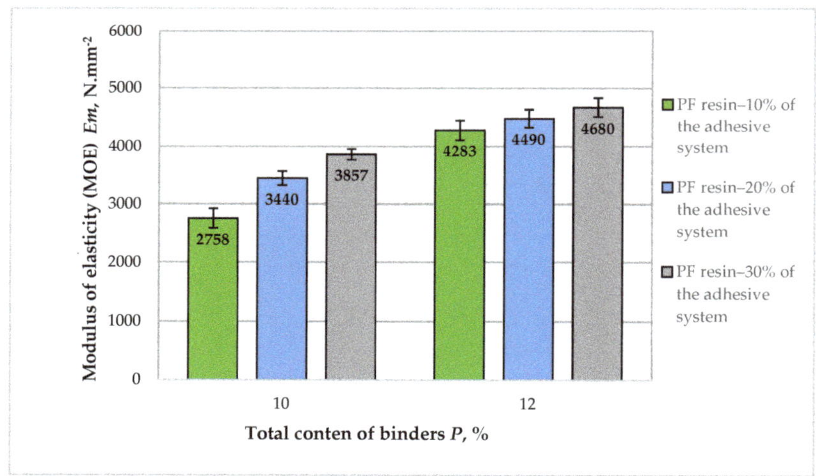

Figure 3. Modulus of elasticity (MOE) of fiberboard panels bonded with HL and PF resin.

The overall improvement of the MOE values with the variation of the adhesive system used was 1.7 times. All manufactured panels bonded with an adhesive composition of HL and PF resin as an auxiliary binder met the standard requirements for dry-process fiberboards for general purpose and use in dry conditions—2700 N·mm^{-2} [70]. All panels except for the fiberboards produced with 10% total binder content and 10% PF resin in the adhesive system, fulfilled the strictest requirements for this property, i.e., for fiberboards for load-bearing applications—3000 N·mm^{-2}.

At a total binder content of 10% with an increase in the PF resin content from 10% to 30% in the adhesive system, an improvement in the MOE values of 1.4 times was observed. That improvement was most significant (by 1.3 times) when the PF resin content in the total adhesive system was increased from 10% to 20%. According to previous studies on the application of lignin as a binder in fiberboards, its incorporation could increase the stiffness of the panels, resulting in higher MOE values [19,71].

The subsequent increase in the content of PF resin from 20% to 30% in the adhesive system had a slight effect, i.e., MOE values improved by 1.2 times. Therefore, it can be concluded that at 10% total binder content, the content of PF resin in the adhesive system should be above 10%. Otherwise, the retention of HL in wood fiber mass is not complete, which results in decreased MOE values of the panels [54,55].

Fiberboard panels fabricated with 12% total binder content had higher MOE values than fiberboards bonded with 10% adhesives. Even at 10% content of PF resin in the adhesive system, a higher MOE was observed compared with the panels manufactured with 10% binder content, of which 30% was PF resin. With the increased PF resin content from 10% to 30% of the total adhesive system used, an improvement of MOE values by 1.09 times was observed. Similarly, more significant increase in MOE values by 1.05 times was determined when the percentage of PF resin content on the adhesive system was increased from 10% to 20%.

The optimal, maximum, MOE value of 4680 N·mm^{-2} was obtained at the upper limit values of the factors—12% total binder content and 30% PF resin content in the adhesive system (Figure 4).

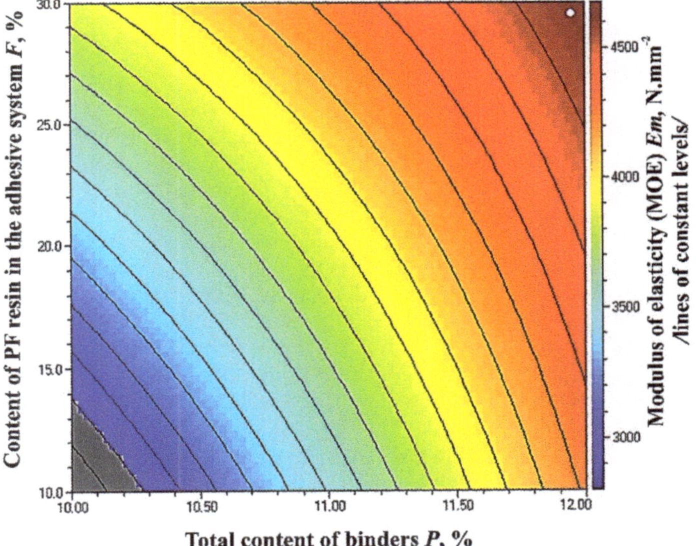

Figure 4. Effect of binder content and PF resin content in the adhesive system on the MOE values of fiberboard panels.

The grey zone in Figure 4 represents the limitation, i.e., the standard requirement for MOE values of fiberboard panels used in load-bearing applications in humid conditions (3000 N·mm^{-2}) [70]. Markedly, fiberboard panels, fulfilling this requirement, can be manufactured with a total binder content of at least 10.3% and a PF resin content in the adhesive system of at least 14%.

The determined MOE values of laboratory-fabricated fiberboard panels with a hydrolysis lignin content of more than 10% in the adhesive system are in accordance with the results obtained by Yotov et al. [44]. Comparable MOE values were also reported by other authors [54,71], who investigated the effect of lignin incorporation in adhesive systems, used for fiberboard manufacturing.

The results obtained for the bending strength (MOR) values of the laboratory-produced fiberboard panels bonded with HL and PF resin are presented in Figure 5.

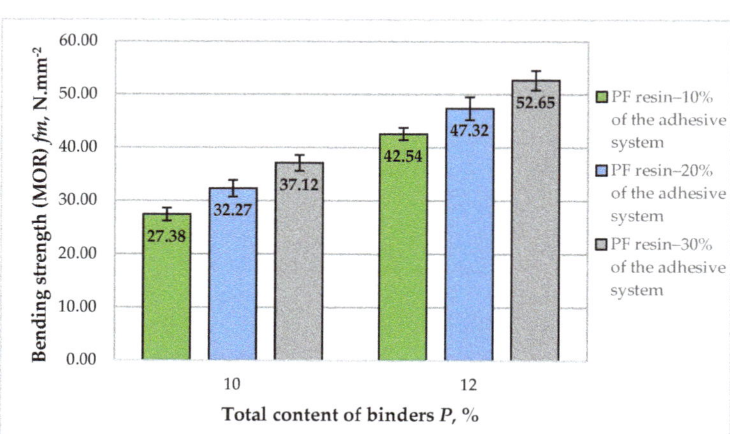

Figure 5. Bending strength (MOR) of fiberboard panels bonded with HL and PF resin.

The fiberboard panels produced in this work exhibited MOR values, varying from 27.38 to 52.65 N·mm^{-2}. The improvement in MOR values of the panels with increasing total binder content was more significant compared to MOE, i.e., an overall improvement of 1.93 times was determined. All laboratory panels met the standard requirements for use in dry conditions—MOR value of at least 27 N·mm^{-2} [70]. All panels, except for the fiberboards manufactured with a total binder content of 10% and 10% PF resin in the adhesive system, also fulfilled the requirements for load-bearing applications and use in humid conditions—30 N·mm^{-2}.

At the total binder content of 10%, the increased PF resin content in the adhesive system from 10% to 30%, resulted in improved MOR values by 1.36 times. The respective increase in MOR was 1.18 times when the PF resin content in the adhesive system was increased from 10% to 20%, and 1.15 times when it was increased to 30%, respectively. MOR values depend on individual fiber strength and bonding strength among wood fibers, i.e., better inter-fiber bonds will result in improved MOR of the panels. The positive effect of HL addition on the fiber surface on MOR values of fiberboards was also confirmed by other authors [19,66,73].

Fiberboard panels fabricated with 12% total binder content exhibited higher MOR values than the panels produced with 10% total binder content. The fiberboards with 12% binder content, of which 10% was PF resin, had 1.15 times higher MOR values than the panels with 10% total binder content, of which 30% was PF resin. Panels bonded with a 12% total binder content showed improved MOR values by 1.42 times when the PF resin content was increased from 10% to 30%.

The optimal, maximum value of the property of 52.65 N·mm^{-2} was obtained at the upper limit values of the factors—total binder content of 12% and 30% PF resin content in the adhesive system (Figure 6).

The grey zone in Figure 6 represents MOR values below 30 N·mm^{-2}. In all other cases, the strictest standard requirement for this property for load-bearing applications and use in humid conditions was fulfilled [70]. To note, this requirement can be achieved with a minimum binder content of 10.22% and PF resin content in the adhesive system of at least 13.5%.

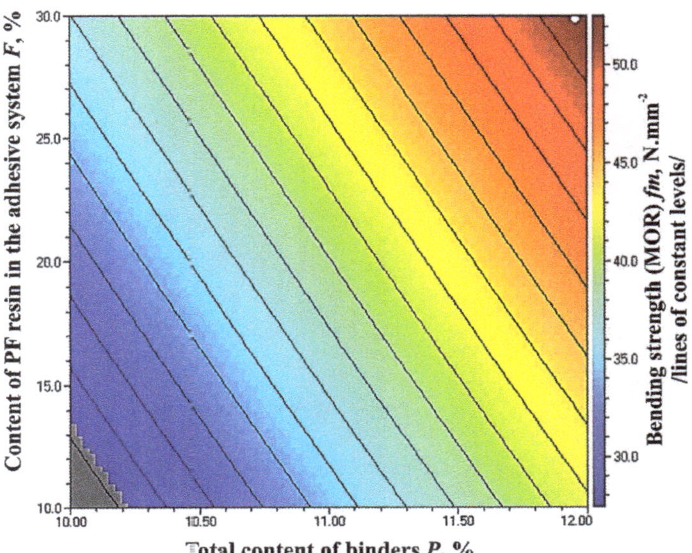

Figure 6. Effect of binder content and PF resin content in the adhesive system on the MOR values of fiberboard panels.

The results show that at the selected levels of variation of the total binder content and the content of PF resin in the adhesive system, the increase in the content of PF resin resulted in improved MOE and MOR values of the panels. However, it should be noted that in all experimental series, the HL is the main binder, and that is, a waste bio-based material is utilized. The improvement in the properties of fiberboards with increasing PF resin content should be due to better retention of lignin in the pulp and the interaction of lignin with PF resin. Which of the two causes dominates should be a subject of subsequent studies.

The results obtained for the MOR values of the panels, fabricated in this work, are in accordance with the findings of other authors [44], where an improvement by 1.7 times in this property was reported with an increase in the binder content from 6% to 10%. In another study [71], the increased lignin content in the adhesive system resulted in a significantly improved (almost twice) MOR values. The beneficial effect of HL addition on MOR values of fiberboards, produced by modifying the hot-pressing regime was also reported by Valchev et al. [55].

The results obtained for the IB strength of the fiberboard panels, bonded with adhesive system composed of HL and PF resin, is presented in Figure 7.

The internal bond (IB) refers to the bonding strength between individual fibers, which is of great importance because it ensures that the panels will not delaminate during post-processing. Internal bonding between wood fibers without synthetic resins is due to the hydrogen binding between fibers, condensation reaction of lignin [74–76], and crosslinking reactions between lignin and polysaccharides [77]. The formed covalent bonds between lignocellulosic polymers contribute to the formation of intermolecular forces which are stronger compared to the ones due to hydrogen bonds [78]. Furthermore, fibers with lignin-rich surfaces have a positive effect on the mechanical properties of the panels due to entanglement of the softened lignin caused by applied pressure and temperature, and supplemented by covalent bond formation [74–76,79].

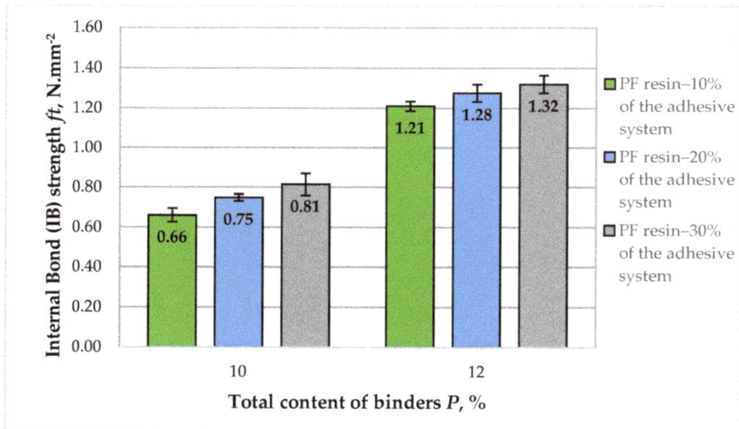

Figure 7. Internal bond (IB) of fiberboard panels bonded with HL and PF resin.

The IB values of the fiberboard panels produced in this work varied from 0.66 to 1.32 N·mm^{-2}. Overall, the IB values of fiberboards were improved with the addition of HL. The increased total binder content and PF resin content in the adhesive system resulted in an almost twice increase in IB strength. For the panels bonded with a 10% total binder content, the IB values increased from 0.66 to 0.81 N·mm^{-2} with the increased PF resin content in the adhesive system from 10% to 30%, i.e., an improvement of 1.23 times was determined. More significant improvement of the IB values was observed with the increased total binder content from 10% to 12%. The tendency of increased IB of the panels with increasing PF resin content in the adhesive system was also confirmed. However, the overall improvement of this property achieved by increasing the PF resin content in the adhesive system from 10% to 30% was only 9%.

All fiberboard panels bonded with HL as the main binder and PF resin as an auxiliary binder fulfilled the European standard requirements for general purpose fiberboards used in dry conditions—0.65 N·mm^{-2} [70]. With the exception of the panels manufactured with 10% total binder content, of which 10% was PF resin, all other panels met the strictest requirements for the property, i.e., for load-bearing applications and use in humid conditions—IB strength of at least 0.70 N·mm^{-2} [70].

The optimal, maximum value of the property of 1.32 N·mm^{-2} was obtained at the upper limit values of the factors—total binder content of 12% and PF resin content of 30% of the adhesive system (Figure 8). In order to meet the most stringent requirement of 0.70 N·mm^{-2}, the total binder content should be at least 10.15%, and the PF resin content in the adhesive system should be at least 15%.

The improvement of the mechanical properties of the panels, MOE, MOR, and IB strength, with the increased total binder content and lignin content, was consistent with previous studies using lignin as a bio-based binder for fiberboard manufacturing [80–83].

However, the predominant effect of improving the properties of fiberboard panels is the increase in total binder content, which increases both the content of HL and PF resin. Although, the results show that the ratio of HL to PF resin is also significant. If the PF resin is the main binder, then in the used hot-pressing regime, the resin polymerization will occur at low pressure, i.e., before the final compression of the material has occurred. That will lead to fabricating panels with deteriorated properties.

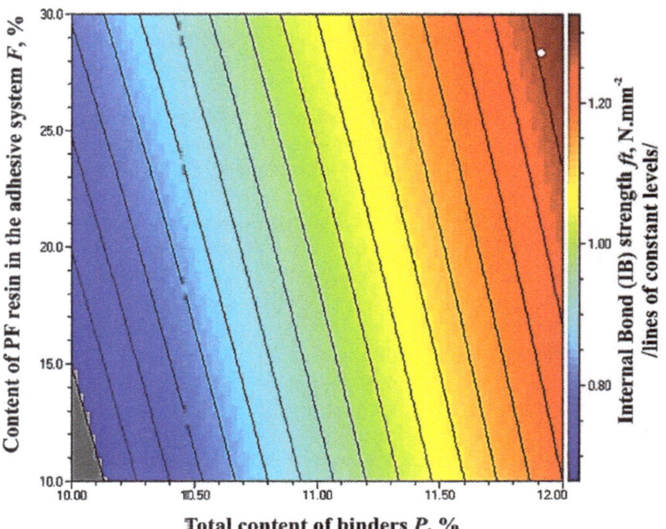

Figure 8. Effect of binder content and PF resin content in the adhesive system on the IB strength of fiberboard panels.

Thickness swelling (TS) and water absorption (WA) are important physical properties of wood composites, strongly related to the dimensional stability of the panels, and providing an insight of panel behavior when used in humid conditions, especially in outdoor applications [84,85].

The results obtained for the TS (24 h) of the fiberboard panels bonded with HL and PF resin are presented in Figure 9.

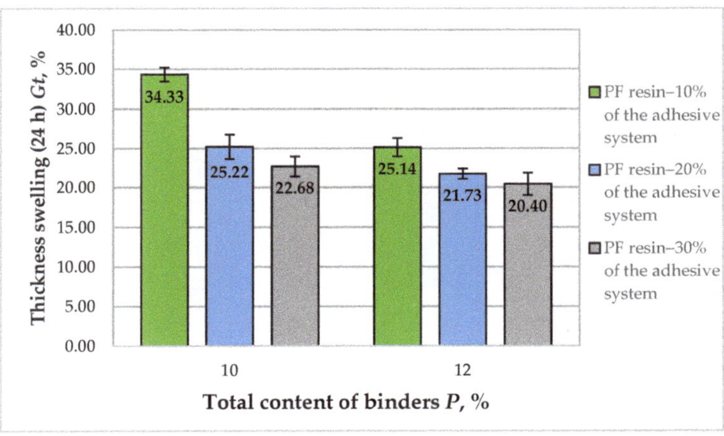

Figure 9. Thickness swelling (24 h) of fiberboard panels bonded with HL and PF resin.

As seen in Figure 8, the TS values of the laboratory-produced fiberboards varied from 34.33% to 20.40%. The increased total binder content resulted in reduced TS values by 1.68 times. All manufactured fiberboards fulfilled the standard requirement for panels used in dry conditions—TS value of 35% [70]. Except for the panels manufactured with a 10% binder content, of which 10% was PF resin, all other fiberboards met the standard requirement for use in humid environments, i.e., 30%. The main reasons for TS in fiberboards are the breakage of bonded areas among wood fibers and the recovery of compressed fibers [74]. Results from previous research works indicated that the addition of HL on the fiber surface resulted in improved dimensional stability of the panels, due to the hydrophobic nature of HL [19,54].

For the panels fabricated with 10% total binder content, the increased PF resin content in the adhesive system, from 10% to 30%, resulted in decreased TS values by 1.51 times. Much more significant improvement of TS was observed when the PF resin content in the adhesive system was increased from 10% to 20%, i.e., 1.36 times. The subsequent increase in the PF resin content resulted in an improvement of 1.11 times.

With regards to the panels, manufactured with 12% total binder content, the increased PF resin content in the adhesive system from 10% to 30% resulted in improved TS values of the panels by 1.23 times. Markedly, the improvement of the studied property was more significant when the PF resin content was increased from 10% to 20% in the adhesive system, i.e., 1.16 times.

It can be concluded that the increased binder content from 10% to 12%, and PF resin content in the adhesive composition from 20% to 30%, respectively, had a limited effect on the TS of the laboratory-made fiberboards bonded with HL and PF resin.

The optimal (minimum) value of the property was obtained again at the upper limit values of the factors (Figure 10). The grey zone in the figure represents TS values higher than 30%, i.e., values that do not meet the standard requirement for use in humid conditions [70]. To fulfill the stringent standard requirements, the panels should be fabricated with a total binder content of at least 10.6%, and the PF resin content in the adhesive system should be at least 14%.

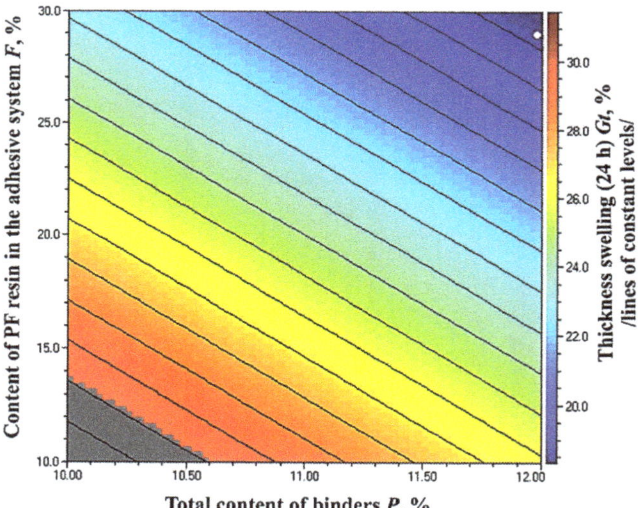

Figure 10. Effect of binder content and PF resin content in the adhesive system on the TS of fiberboard panels.

The improvement of TS by increasing the total binder content is consistent with the results obtained in previous studies [44,54,55,66].

The results of the WA (24 h) of fiberboard panels, bonded with HL and PF resin, are presented in Figure 11.

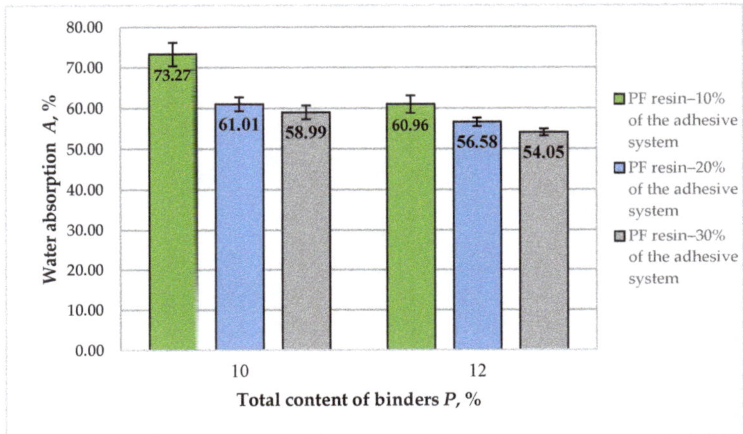

Figure 11. Water absorption (24 h) of fiberboard panels bonded with HL and PF resin.

Lignocellulosic materials absorb water by creating hydrogen bonds between water and hydroxyl groups of lignin, cellulose, and hemicellulose present in the cell wall [86,87]. The addition of lignin in the adhesive system reduces the WA of fiberboards due to the presence of aromatic rings and non-polar hydrocarbon chains in the lignin structure [88].

WA values of the fiberboard panels produced in this work varied from 73.27% to 54.05%. The variation of the total binder content and adhesive system composition resulted in reduced WA of the panels by 1.36 times. The reduced WA values with the incorporation of lignin in the adhesive formulation might be attributed to bulking the cell wall with lignin, which makes it hydrophobic [19,29].

For the panels manufactured with 10% total binder content, the most significant improvement of WA (1.21 times) was determined when the PF resin content in the adhesive system was increased from 10% to 20%. The subsequent increase in the PF resin content in the adhesive system from 20% to 30% had a more negligible effect, and the observed improvement was only 1.04 times.

The increased total binder content from 10% to 12% resulted in improved WA values of the laboratory panels. However, fiberboards fabricated with 10% total binder content, of which 20% was PF resin, exhibited WA values, comparable with the panels bonded with 12% total binder content, of which PF resin was 10%.

The optimal WA value was obtained at the upper limit values of the factors (Figure 12). As WA is not a standardized property of wood-based composites, the strictest restrictions were imposed on the other properties of the panels. Thus, it was found that fiberboards complying with the most stringent requirements, namely for load-bearing applications and use in humid conditions, can be manufactured with a minimum total binder content of 10.6%, where PF resin should be at least 14%. Therefore, the HL addition may constitute 86% of the adhesive system.

The results obtained for improving the waterproof properties of the fiberboard panels by increasing the lignin content are consistent with the findings reported in similar studies [65,80–83]. The results for the physical and mechanical properties of fiberboards manufactured with a modified adhesion system and hot-pressing cycle are comparable or better than the results reported in similar studies on the application of lignin as a bio-based wood adhesive [65,66,80,83]. Compared to the cited studies, an advantage of the technology used in this work is the absence of lignin modification, lower total lignin content, and lower density of the panels.

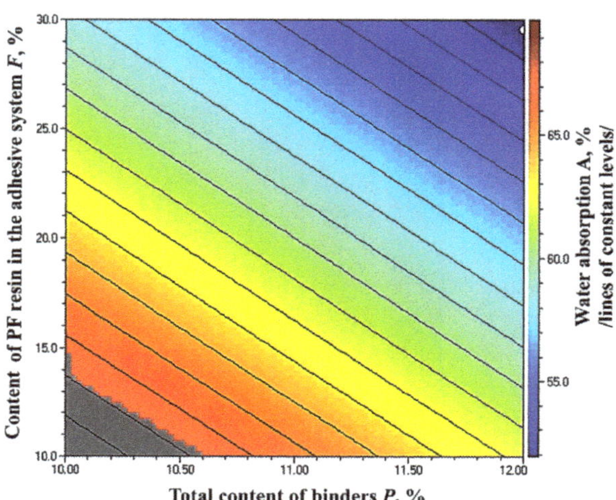

Figure 12. Effect of binder content and PF resin content in the adhesive system on WA (24 h) of fiberboard panels.

4. Conclusions

The HL is activated when modifying the hot-pressing regime, most likely by plasticization and subsequent condensation processes involving the PF resin. That overcomes the disadvantage of low lignin retention in the wood fiber mass. The proposed hot-pressing technology is easily feasible in industrial conditions with continuous presses. These presses have autonomous heating of the individual sections, so cooling will not lead to significant heat losses and additional costs. It should be emphasized that continuous presses with a cooling zone are already in operation. The fiberboard panels produced in this work from industrial wood fiber mass, bonded with an adhesive system composed of HL and PF as an auxiliary binder, exhibited excellent mechanical and hydrophobic properties, fulfilling the requirements of the relevant standard for application in humid conditions. Markedly, fiberboard panels meeting the most stringent standard requirements can be manufactured with a minimum total binder content of 10.6%, composed of at least 14% PF resin content, and at least 86% HL, respectively. It can be concluded that further increase in total binder content and PF resin content is not justified unless for manufacturing special purpose panels. The main disadvantage of the proposed technological solution is the long first stage of hot-pressing, which can be overcome by reducing the moisture content of the fiber mat. Future research should be focused on determining the optimal moisture content of the mat to enable lignin activation at reduced hot-pressing duration.

Author Contributions: Conceptualization, V.S. and I.V.; methodology, V.S. and I.V.; investigation, I.Y. and Z.P.; resources, P.A. and V.S.; writing—original draft preparation, V.S.; writing—review and editing, I.V. and P.A.; visualization, V.S.; supervision, I.V. and P.A.; project administration, I.V. and P.A.; funding acquisition, I.V. All authors have read and agreed to the published version of the manuscript.

Funding: This research was supported by Project No. КП-06-КОСТ/1 at CA 17128 "Study of the potential of lignin raw materials in Bulgaria and development of technologies for their modification and effective application in industry" carried out at the Bulgarian National Science Fund, Sofia, Bulgaria. This research was also supported by project No. НИС-Б-1145/04.2021 "Development, Properties, and Application of Eco-Friendly Wood-Based Composites" carried out at the University of Forestry, Sofia, Bulgaria.

Institutional Review Board Statement: Not applicable.

Informed Consent Statement: Not applicable.

Data Availability Statement: The data presented in this study are available on request from the corresponding authors.

Acknowledgments: The authors would like to thank Kronospan Bulgaria EOOD for supplying the industrial wood fibers; and Welde Bulgaria AD (Troyan, Bulgaria) for providing the phenol-formaldehyde resin.

Conflicts of Interest: The authors declare no conflict of interest.

References

1. Lee, T.C.; Puad, N.A.D.; Selimin, M.A.; Manap, N.; Abdullah, H.Z.; Idris, M.I. An overview on development of environmental friendly medium density fibreboard. *Mater. Today* **2020**, *29*, 52–57. [CrossRef]
2. Mirski, R.; Banaszak, A.; Bekhta, P. Selected Properties of Formaldehyde-Free Polymer-Straw Boards Made from Different Types of Thermoplastics and Different Kinds of Straw. *Materials* **2021**, *14*, 1216. [CrossRef]
3. Popovic, M.; Momčilović, M.D.; Gavrilović-Grmuša, I. New standards and regulations on formaldehyde emission from wood-based composite panels. *Zast. Mater.* **2020**, *61*, 152–160. [CrossRef]
4. Valyova, M.; Ivanova, Y.; Koynov, D. Investigation of free formaldehyde quantity in production of plywood with modified urea-formaldehyde resin. *Int. J. Wood Des. Technol.* **2017**, *6*, 72–76.
5. Kristak, L.; Antov, P.; Bekhta, P.; Lubis, M.A.R.; Iswanto, A.H.; Reh, R.; Sedliacik, J.; Savov, V.; Taghiayri, H.; Papadopoulos, A.N.; et al. Recent Progress in Ultra-Low Formaldehyde Emitting Adhesive Systems and Formaldehyde Scavengers in Wood-Based Panels: A Review. *Wood Mater. Sci. Eng.* **2022**. [CrossRef]
6. FAO. Forest Product Statistics. Available online: http://www.fao.org/forestry/statistics/ (accessed on 24 March 2022).
7. Kibleur, P.; Aelterman, J.; Boone, M.; Bulcke, J. Deep learning segmentation of wood fiber bundles in fiberboards. *Compos. Sci. Technol.* **2022**, *221*, 109287. [CrossRef]
8. Puspaningrum, T.; Haris, Y.H.; Sailah, I.; Yani, M.; Indrasti, N.S. Physical and mechanical properties of binderless medium density fiberboard (MDF) from coconut fiber. *IOP Conf. Ser. Earth Environ. Sci.* **2020**, *472*, 012011. [CrossRef]
9. Akgül, M.; Uner, B.; Çamlibel, O.; Ayata, Ü. Manufacture of Medium Density Fiberboard (MDF) Panels from Agri Based Lignocellulosic Biomass. *Wood Res.* **2017**, *62*, 615–624.
10. Ferdosian, F.; Pan, Z.; Gao, G.; Zhao, B. Bio-Based Adhesives and Evaluation for Wood Composites Application. *Polymers* **2016**, *9*, 70. [CrossRef]
11. Hemmilä, V.; Adamopoulos, S.; Karlsson, O.; Kumar, A. Development of sustainable bio-adhesives for engineered wood panels—A Review. *RSC Adv.* **2017**, *7*, 38604–38630. [CrossRef]
12. Widsten, P.; Hummer, A.; Heathcote, C.; Kandelbauer, A. A preliminary study of green production of fiberboard bonded with tannin and laccase in a wet process. *Holzforschung* **2009**, *63*, 545–550. [CrossRef]
13. Khalaf, Y.; El Hage, P.; Mihajlova, J.D.; Bargeret, A.; Lacroix, P.; El Hage, R. Influence of agricultural fibers size on mechanical and insulating properties of innovative chitosan-based insulators. *Constr. Build. Mater.* **2021**, *278*, 123071. [CrossRef]
14. Wang, Z.; Kang, H.; Liu, H.; Zhang, C.; Wang, Z.; Li, J. Dual-Network Nanocross-linking Strategy to Improve Bulk Mechanical and Water-Resistant Adhesion Properties of Biobased Wood Adhesives. *ACS Sustain. Chem. Eng.* **2020**, *8*, 16430–16440. [CrossRef]
15. Pizzi, A.; Papadopoulus, A.N.; Policardi, F. Wood Composites and Their Polymer Binders. *Polymers* **2020**, *12*, 1115. [CrossRef]
16. Arias, A.; González-Rodríguez, S.; Vetroni Barros, M.; Salvador, R.; de Francisco, A.C.; Piekarski, C.M.; Moreira, M.T. Recent developments in bio-based adhesives from renewable natural resources. *J. Clean. Prod.* **2021**, *314*, 127892. [CrossRef]
17. Tisserat, B.; Eller, F.J.; Mankowski, M.E. Properties of Composite Wood Panels Fabricated from Eastern Redcedar Employing Furious Bio-based Green Adhesives. *BioResources* **2019**, *14*, 6666–6685.
18. Ghahri, S.; Pizzi, A.; Hajihassani, R. A Study of Concept to Prepare Totally Biosourced Wood Adhesives from only Soy Protein and Tannin. *Polymers* **2022**, *14*, 1150. [CrossRef]
19. Mancera, C.; Mansouri, N.E.; Pelach, M.A.; Francesc, F.; Salvadó, J. Feasibility of incorporating treated lignins in fiberboards made from agricultural waste. *Waste Manag.* **2012**, 1962–1967. [CrossRef]
20. Santos, J.; Pereira, J.; Escobar-Avello, D.; Ferreira, I.; Vieira, C.; Magalhães, F.D.; Martins, J.M.; Carvalho, L.H. Grape Canes (*Vitis vinifera* L.) Applications on Packaging and Particleboard Industry: New Bioadhesive Based on Grape Extracts and Citric Acid. *Polymers* **2022**, *14*, 1137. [CrossRef]
21. Kristak, L.; Ruziak, I.; Tudor, E.M.; Barbu, M.C.; Kain, G.; Reh, R. Thermophysical Properties of Larch Bark Composite Panels. *Polymers* **2021**, *13*, 2287. [CrossRef]
22. Pizzi, A. Tannins: Prospectives and Actual Industrial Applications. *Biomolecules* **2019**, *9*, 344. [CrossRef]
23. Dunky, M. Wood Adhesives Based on Natural Resources: A Critical Review Part III. Tannin- and Lignin-Based Adhesives. *Rev. Adhes. Adhes.* **2020**, *8*, 379–525.
24. Anderson, I.; Anderson, B.; Avramidis, G.; Frichart, C.R. *Handbook of Wood Chemistry and Wood Composites*, 2nd ed.; Part III—Wood Composites; Taylor Francis Group: Abingdon, UK, 2013; pp. 255–511.
25. Balakshin, M.; Capanema, E.A.; Zhu, X.; Sulaeva, I.; Potthast, A.; Rosenau, T.; Rojas, O.J. Spruce milled wood lignin: Linear, branched or cross-linked? *Green Chem.* **2020**, *22*, 3985–4001. [CrossRef]
26. Iravani, S.; Varma, R.S. Greener synthesis of lignin nanoparticles and their applications. *Green Chem.* **2020**, *22*, 612. [CrossRef]

27. Westin, M.; Simonson, R.; Östman, B. Kraft lignin wood fiberboards—The effect of kraft lignin addition to wood chips or board pulp prior to fiberboard production. *Holz. Als. Roh.-Und Werkst.* **2001**, *58*, 393–400. [CrossRef]
28. Saffian, H.A.; Yamaguchi, M.; Ariffin, H.; Abdan, K.; Kassim, N.K.; Lee, S.H.; Lee, C.H.; Shafi, A.R.; Humairah Alias, A. Thermal, Physical and Mechanical Properties of Poly(Butylene Succinate)/Kenaf Core Fibers Composites Reinforced with Esterified Lignin. *Polymers* **2021**, *13*, 2359. [CrossRef]
29. Solihat, N.N.; Sari, F.P.; Falah, F.; Ismayati, M.; Lubis, M.A.R.; Fatriasari, W.; Santoso, E.B.; Syafii, W. Lignin as an Active Biomaterial: A Review. *J. Sylva Lestari* **2021**, *9*, 1–22. [CrossRef]
30. Parray, J.; Mir, M.; Shameen, N. Lignin Nanoparticles: Synthesis and Applications. In *Nano-Technology Intervention in Agricultural Productivity*; John Wiley & Sons Ltd.: Hoboken, NJ, USA, 2021; pp. 97–108.
31. Euring, M.; Ostendorf, K.; Rühl, M.; Kües, U. Enzymatic Oxidation of Ca-Lignosulfonate and Kraft Lignin in Different Lignin-Laccase-Mediator-Systems and MDF Production. *Front. Bioeng. Biotechnol.* **2022**, *9*, 788622. [CrossRef]
32. Antov, P.; Savov, V.; Mantanis, G.I.; Neykov, N. Medium-density fibreboards bonded with phenol-formaldehyde resin and calcium lignosulfonate as an eco-friendly additive. *Wood Mater. Sci. Eng.* **2021**, *16*, 42–48. [CrossRef]
33. Antov, P.; Valchev, I.; Savov, V. Experimental and Statistical Modeling of the Exploitation Properties of Eco-Friendly MDF Trough Variation of Lignosulfonate Concentration and Hot-Pressing Temperature. In Proceedings of the 2nd International Congress of Biorefinery of Lignocellulosic Materials (IWBLCM 2019), Córdoba, Spain, 4–7 June 2019; pp. 104–109.
34. Hemmila, V.; Hosseinpourpia, R.; Adamopoulos, S.; Eceiza, A. Characterization of Wood-based Industrial Biorefinery Lignosulfonates and Supercritical Water Hydrolysis Lignin. *Waste Biomass Valor.* **2020**, *11*, 5835–5845. [CrossRef]
35. Savov, V.; Mihajlova, J. Influence of the Content of Lignosulfonate on Physical Properties of Medium Density Fiberboards. *ProLigno* **2017**, *13*, 247–251.
36. Gonçalves, S.; Ferra, J.; Paiva, N.; Martins, J.; Carvalho, L.H.; Magalhães, F.D. Lignosulphonates as an Alternative to Non-Renewable Binders in Wood-Based Materials. *Polymers* **2021**, *13*, 4196. [CrossRef]
37. Savov, V.; Valchev, I.; Antov, P. Processing Factors for Production of Eco-Friendly Medium Density Fibreboards Based on Lignosulfonate Adhesives. In Proceedings of the 2nd International Congress of Biorefinery of Lignocellulosic Materials (IWBLCM 2019), Córdoba, Spain, 4–7 June 2019; pp. 165–169.
38. Savov, V.; Antov, P. Engineering the Properties of Eco-Friendly Medium Density Fibreboards Bonded with Lignosulfonate Adhesive. *Drv. Ind.* **2020**, *71*, 157–162. [CrossRef]
39. Antov, P.; Savov, V.; Krišťák, Ľ.; Réh, R.; Mantanis, G.I. Eco-Friendly, High-Density Fiberboards Bonded with Urea-Formaldehyde and Ammonium Lignosulfonate. *Polymers* **2021**, *13*, 220. [CrossRef]
40. Antov, P.; Krišťák, Ľ.; Réh, R.; Savov, V.; Papadopoulos, A.N. Eco-Friendly Fiberboard Panels from Recycled Fibers Bonded with Calcium Lignosulfonate. *Polymers* **2021**, *13*, 639. [CrossRef]
41. Antov, P.; Savov, V.; Trichkov, N.; Krišťák, Ľ.; Réh, R.; Papadopoulos, A.N.; Taghiyari, H.R.; Pizzi, A.; Kunecová, D.; Pachikova, M. Properties of High-Density Fiberboard Bonded with Urea–Formaldehyde Resin and Ammonium Lignosulfonate as a Bio-Based Additive. *Polymers* **2021**, *13*, 2775. [CrossRef]
42. Bekhta, P.; Noshchenko, G.; Réh, R.; Kristak, L.; Sedliačik, J.; Antov, P.; Mirski, R.; Savov, V. Properties of Eco-Friendly Particleboards Bonded with Lignosulfonate-Urea-Formaldehyde Adhesives and pMDI as a Crosslinker. *Materials* **2021**, *14*, 4875. [CrossRef]
43. Karthäuser, J.; Biziks, V.; Mai, C.; Militz, H. Lignin and Lignin-Derived Compounds for Wood Applications—A Review. *Molecules* **2021**, *26*, 2533. [CrossRef]
44. Yotov, N.; Savov, V.; Valchev, I.; Petrin, S.; Karatotev, V. Study on possibility for utilization of technical hydrolysis lignin in composition of medium density fiberboard. *Innov. Wood. Ind. Eng. Des.* **2017**, *6*, 69–74.
45. Limayem, A.; Ricke, S.C. Lignocellulosic biomass for bioethanol production: Current perspectives, potential issues and future prospects. *Prog. Energy Combust. Sci.* **2012**, *38*, 449–467. [CrossRef]
46. Falah, F.; Lubis, M.A.R.; Fatriasari, W.; Sari, F.P. Utilization of Lignin from the Waste of Bioethanol Production as a Mortar Additive. *J. Sylva Lestari* **2020**, *8*, 326–339. [CrossRef]
47. Gopalakrishnan, K.; Kim, S.; Ceylan, H. Lignin Recovery and Utilization. In *Bioenergy and Biofuel from Biowastes and Biomass*; American Society of Civil Engineers: Reston, VA, USA, 2010; pp. 247–278.
48. Li, H.; Qu, Y.; Xu, J. Microwave-Assisted Conversion of Lignin. *Prod. Biofuels Chem. Microwave. Biofuels Biorefineries* **2015**, *3*, 61–68.
49. Aracri, E.; Blanco, C.D.; Tzanov, T. An enzymatic approach to develop a lignin-based adhesive for wool floor coverings. *Green Chem.* **2014**, *16*, 2597. [CrossRef]
50. Ariton, A.M.; Greanga, S.; Trinca, L.M.; Bors, S.I.; Ungureanu, E.; Malutan, T.; Popa, V. Valorization of lignin modified by hydroxymethylation to ensure birch veneer bioprotection. *Cellul. Chem. Technol.* **2015**, *49*, 765–774.
51. El Mansouri, N.E.; Yuan, O.; Huang, F. Study of chemical modification of alkaline lignin by the glyoxalation reaction. *BioResources* **2011**, *6*, 4523–4536.
52. Ghorbani, M.; Liebner, F.W.G.; van Herwijnen, H.; Pfungen, L.; Krahofer, M.; Budjav, E.; Konnerth, J. Lignin Phenol Formaldehyde Resoles: The Impact of Lignin Type on Adhesive Properties. *BioResources* **2016**, *11*, 6727–6741. [CrossRef]
53. Nasir, M.; Asim, M.; Singh, K. Fiberboard Manufacturing from Laccase Activated Lignin Based Bioadhesive. In *Eco-Friendly Adhesives for Wood and Natural Fiber Composites. Composites Science and Technology*; Springer: Singapore, 2021; pp. 51–83.

54. Valchev, I.; Savov, V.; Yordanov, I. Reduction of Phenol Formaldehyde Resin Content in Dry-Processed Fibreboards by Adding Hydrolysis Lignin. In Proceedings of the 2020 Society of Wood Science and Technology International Convention Renewable Resources for Sustainable and Healthy Future, Portorož, Slovenia, 12–15 July 2020; pp. 592–602.
55. Valchev, I.; Yordanov, Y.; Savov, V.; Antov, P. Optimization of the Hot-Pressing Regime in the Production of Eco-Friendly Fibreboards Bonded with Hydrolysis Lignin. *Period. Polytech. Chem. Eng.* **2022**, *66*, 125–134. [CrossRef]
56. Kumar, R.N.; Pizzi, A. *Adhesives for Wood and Lignocellulosic Materials*; Wiley Publishing House: Hoboken, NJ, USA, 2019; p. 491.
57. Aristri, M.A.; Lubis, M.A.R.; Yadav, S.M.; Antov, P.; Papadopoulos, A.N.; Pizzi, A.; Fatriasari, W.; Ismayati, M.; Iswanto, A.H. Recent Developments in Ligninand Tannin-Based Non-Isocyanate Polyurethane Resins for Wood Adhesives—A Review. *Appl. Sci.* **2021**, *11*, 4242. [CrossRef]
58. Handika, S.O.; Lubis, M.A.R.; Sari, R.K.; Laksana, R.P.B.; Antov, P.; Savov, V.; Gajtanska, M.; Iswanto, A.H. Enhancing Thermal and Mechanical Properties of Ramie Fiber via Impregnation by Lignin-Based Polyurethane Resin. *Materials* **2021**, *14*, 6850. [CrossRef]
59. Malutan, M.; Nicu, R.; Popa, V.I. Contribution to the study of hydroxymetilation reaction of alkali lignin. *Bioresurces* **2008**, *3*, 13–20.
60. Technical Association of the Pulp and Paper Industry. *T 222 om-11 Acid-Insoluble Lignin in Wood and Pulp*; Technical Association of the Pulp and Paper Industry: Peachtree Corners, GA, USA, 2011.
61. Technical Association of the Pulp and Paper Industry. *T 211 om-12 Ash in Wood, Pulp, Paper and Paperboard: Combustion at 525 Degrees C*; Technical Association of the Pulp and Paper Industry: Peachtree Corners, GA, USA, 2012.
62. Jivkov, V.; Elenska-Valchanova, D. Mechanical Properties of Some Thin Furniture Structural Composite Materials. In Proceedings of the 30th International Conference on Wood Science and Technology, Paris, France, 29–30 March 2019; pp. 86–94.
63. Jivkov., V.; Petrova, B. Challenges for furniture design with thin structural materials. In Proceedings of the IFC2020, Trabzon, Turkey, 2–4 November 2020; pp. 113–123.
64. Bertaud, F.; Tapin-Lingua, S.; Pizzi, A.; Navarrete, P.; Petit-Conil, M. Development of green adhesives for fibreboard manufacturing, using tannins and lignin from pulp mill residues. *Cellul. Chem. Technol.* **2012**, *46*, 449–455.
65. Thaeng, D.; Mansouri, N.E.; Arbat, G.; Ngo, B.; Delgado-Aguilar, M.; Pelach, M.A.; Fullana-i-Palmer, P.; Matje, M. Fiberboards made from corn stalks thermomechanical pulp and kraft lignin as green adhesive. *BioResources* **2017**, *12*, 2379–2393. [CrossRef]
66. Zhou, X.; Tang, L.; Zhang, W.; Lv, C.; Zheng, F.; Zhang, R.; Du, G.; Tang, B.; Liu, X. Enzymatic hydrolysis lignin derived from corn stover as an intrinsic binder for bio-composites manufacture: Effect of fiber moisture content and pressing temperature on boards' properties. *BioResources* **2011**, *6*, 253–264. [CrossRef]
67. *EN 310:1999*; Wood-Based Panels—Determination of Modulus of Elasticity in Bending and of Bending Strength. European Committee for Standardization (CEN): Brussels, Belgium, 1999.
68. *EN 317:1998*; Particleboards and Fibreboards—Determination of Swelling in Thickness after Immersion in Water. European Committee for Standardization (CEN): Brussels, Belgium, 1998.
69. *EN 323:2001*; Wood-Based Panels—Determination of Density. European Committee for Standardization (CEN): Brussels, Belgium, 2001.
70. *EN 622-5:2010*; Fibreboards—Specifications—Part 5: Requirements for Dry Process Boards. European Committee for Standardization (CEN): Brussels, Belgium, 2010.
71. Vineeth, S.K.; Gadhave, R.V.; Gadekar, P.T. Chemical Modification of Nanocellulose in Wood Adhesive: Review. *Open J. Polym. Chem.* **2019**, *9*, 86–99. [CrossRef]
72. Karagiannidis, E.; Markessini, C.; Athanassiadou, E. Micro-Fibrillated Cellulose in Adhesive Systems for the Production of Wood-Based Panels. *Molecules* **2020**, *25*, 4846. [CrossRef]
73. Velasquez, J.A.; Ferrando, F.; Salvado, J. Effects of kraft lignin addition in the production of binderless fiberboard from steam exploded *Miscanthus Sinensis*. *Ind. Crops Prod.* **2003**, *18*, 17–23. [CrossRef]
74. Bouajila, A.; Limare, A.; Joly, C.; Dole, P. Lignin plasticization to improve binderless fiberboard mechanical properties. *Polym. Eng. Sci.* **2005**, *45*, 809–816. [CrossRef]
75. Okuda, N.; Hori, K.; Sato, M. Chemical changes of kenaf core binderless boards during hot pressing (I): Influence of the pressing temperature condition. *J. Wood. Sci.* **2006**, *52*, 244–248. [CrossRef]
76. Okuda, N.; Hori, K.; Sato, M. Chemical changes of kenaf core binderless boards during hot pressing (II): Effects on the binderless board properties. *J. Wood. Sci.* **2006**, *52*, 249–254. [CrossRef]
77. Suzuki, S.; Hiroyuki, S.; Park, S.-Y.; Saito, K.; Laemsak, N.; Okuma, M.; Iiyama, K. Preparation of binderless boards from steam exploded pulps of oil palm (*Elaeis guneensis* Jaxq.) fronds and structural characteristics of lignin and wall polysaccharides in steam exploded pulps to be discussed for self-bindings. *Holzforschung* **1998**, *52*, 417–426.
78. Back, E. Oxidative activation of wood surfaces for glue bonding. *For. Prod. J.* **1991**, *41*, 30–36.
79. Quintana, G.; Velasquez, J.; Betancourt, S.; Ganan, P. Binderless fiberboard from steam exploded banana bunch. *Ind. Crops Prod.* **2009**, *29*, 60–66. [CrossRef]
80. Nasir, M.; Gupta, A.; Beg, M.D.H.; Chua, G.K.; Kumar, A. Physical and Mechanical Properties of Medium-Density Fibreboards Using Soy-Lignin Adhesives. *J. Trop. For. Sci.* **2014**, *26*, 41–49.
81. Ammar, M.; Mechi, N.; Hidouri, A.; Elaloui, E. Fiberboards based on filled lignin resin and petiole fibers. *J. Indian Acad. Wood Sci.* **2018**, *15*, 120–125. [CrossRef]

82. Hoareau, W.; Oliveira, F.B.; Grelier, S.; Siegmund, B.; Frollini, E.; Castellan, A. Fiberboards Based on Sugarcane Bagasse Lignin and Fibers. *Macromol. Mater. Eng.* **2006**, *291*, 829–839. [CrossRef]
83. Tupciauskas, R.; Gravitis, J.; Abolins, J.; Veveris, A.; Andzs, M.; Liitia, T.; Tamminen, T. Utilization of lignin powder for manufacturing self-binding HDF. *Holzforschung* **2017**, *71*, 555–561. [CrossRef]
84. Sihag, K.; Yadav, S.M.; Lubis, M.A.R.; Poonia, P.K.; Negi, A.; Khali, D.P. Influence of needle-punching treatment and pressureon selected properties of medium density fiberboard made of bamboo (*Dendrocalamus strictus* Roxb. Nees). In *Wood Material Science & Engineering*; Taylor and Francis Ltd.: London, UK, 2021. [CrossRef]
85. Zakaria, R.; Bawon, P.; Lee, S.H.; Salim, S.; Lum, W.C.; Al-Edrus, S.S.O.; Ibrahim, Z. Properties of Particleboard from Oil Palm Biomasses Bonded with Citric Acid and Tapioca Starch. *Polymers* **2021**, *13*, 3494. [CrossRef]
86. Rowell, R.; Gutzmer, D.; Sachs, I.; Kinney, R. Effects of alkylene oxide treatments on dimensional stability of wood. *Wood Sci.* **1976**, *9*, 51–54.
87. Mantanis, G.I.; Young, R.; Rowell, R. Swelling of Wood. Part 1. Swelling in water. *Wood Sci. Technol.* **1994**, *28*, 119–134.
88. Rozman, H.; Tan, K.; Kumar, R.; Abubakar, A.; Ishak, Z.; Ismail, H. The effect of lignin as a compatibilizer on the physical properties of coconut fiber-polypropylene composites. *Eur. Polym. J.* **2000**, *36*, 1483–1494. [CrossRef]

Article

Novel Biodegradable Poly (Lactic Acid)/Wood Leachate Composites: Investigation of Antibacterial, Mechanical, Morphological, and Thermal Properties

Mohammad Hassan Shahavi [1,*], Peyman Pouresmaeel Selakjani [1], Mohadese Niksefat Abatari [1], Petar Antov [2,*] and Viktor Savov [2]

1 Faculty of Engineering Modern Technologies, Amol University of Special Modern Technologies (AUSMT), Amol 4615664616, Iran; p.pouresmaeel@ut.ac.ir (P.P.S.); m.n.sefat@alumni.ut.ac.ir (M.N.A.)
2 Faculty of Forest Industry, University of Forestry, 1797 Sofia, Bulgaria; victor_savov@ltu.bg
* Correspondence: m.shahavi@ausmt.ac.ir (M.H.S.); p.antov@ltu.bg (P.A.)

Abstract: This research aimed to investigate the effects of using wood leachate (WL) powder as a cost-effective filler added to novel poly (lactic acid) biocomposites and evaluate their mechanical, thermal, morphological, and antibacterial properties. Fourier transform infrared spectroscopy (FTIR), tensile test, Charpy impact test, Shore hardness, scanning electron microscope (SEM), differential scanning calorimetry (DSC), contact angle, and bacterial growth inhibition tests were employed to characterize the developed biocomposites. The SEM results indicated a proper filler dispersion in the polymer matrix. WL powder improved the hydrophobic nature in the adjusted sample's contact angle experiment. Markedly, the results showed that the addition of WL filler improved the mechanical properties of the fabricated biocomposites. The thermal analysis determined the development in crystallization behavior and a decline in glass transition temperature (Tg) from 60.1 to 49.3 °C in 7% PLA-WL biocomposites. The PLA-WL biocomposites exhibited an antibacterial activity according to the inhibition zone for *Escherichia coli* bacteria. The developed novel PLA-WL composites can be effectively utilized in various value-added industrial applications as a sustainable and functional biopolymer material.

Keywords: antibacterial; biocomposite; biopolymers; mechanical properties; poly (lactic acid); wood leachate

Citation: Shahavi, M.H.; Selakjani, P.P.; Abatari, M.N.; Antov, P.; Savov, V. Novel Biodegradable Poly (Lactic Acid)/Wood Leachate Composites: Investigation of Antibacterial, Mechanical, Morphological, and Thermal Properties. *Polymers* **2022**, *14*, 1227. https://doi.org/10.3390/polym14061227

Academic Editor: Gianluca Tondi

Received: 23 February 2022
Accepted: 15 March 2022
Published: 18 March 2022

1. Introduction

In recent years, the enhanced environmental awareness, the increased need for sustainable and "green" materials, as well as the stringent legislative regulations related to waste management and cascading use of natural resources have forced many researchers to develop novel biodegradable polymers as a viable alternative to conventional polymers [1–5]. Despite the numerous advantages of biocomposites such as the potential to create a sustainable industry as well as enhancement in various properties such as durability, flexibility, high gloss, clarity, and tensile strength, there are certain drawbacks, such as deteriorated physical and mechanical properties, poor interface adhesion, brittleness, lower thermal resistance and water absorption, susceptibility to fungi and insect attacks, etc., limiting their wider application as functional materials [6–9]. Poly (lactic acid) (PLA) is one of the most suitable biodegradable polymers widely employed in many applications ranging from the biomedical field, e.g., in engineered drug delivery systems [10,11], tissue engineering (scaffolds) [12,13], and wound dressing [14,15] to food packaging and disposable plastic bags [16–18] due to its versatility, excellent processability, and biocompatibility. It is also widely used as a 3D printing feedstock for desktop fused filament fabrication 3D printers [19]. PLA is a biodegradable aliphatic semicrystalline polyester synthesized from lactic acid monomer (LA; 2-hydroxypropanoic acid), which is soluble in water and occurs in two enantiomeric forms, i.e., l-(+)-LA and d-(−)-LA. PLA represents a sustainable alternative to fossil-derived products, since it is derived from the fermentation of renewable agricultural biomass resources rich in carbohydrates, such as corn starch and sugar beets. Filler materials have been widely used in the polymer industries for economic reasons. Furthermore, it was reported that the fillers could also improve biocomposites' mechanical and thermal properties [20–24]. Lignocellulosic waste and residues are among the most widely used fillers in polymeric composites. Using these natural materials will produce lighter products and lower the environmental footprint due to their biodegradability and lower cost. Another important benefit of using natural waste and by-products is significantly improving polymers' mechanical properties [25–27].

As in any wood-based industry, the medium-density fiberboard (MDF) production factories usually use wood as their feedstock [28]. In the process of MDF production and after wood chipping, the chips are washed with high-pressure steam. The waste produced by this process is a dark brown liquid called leachate. The wood leachate (WL) has a complex composition, but as mentioned in the literature, the leachate compounds are mostly phenol, tannins, lignin, and volatile fatty acids [29–31]. The structure of the main components present in the WL is shown in Figure 1.

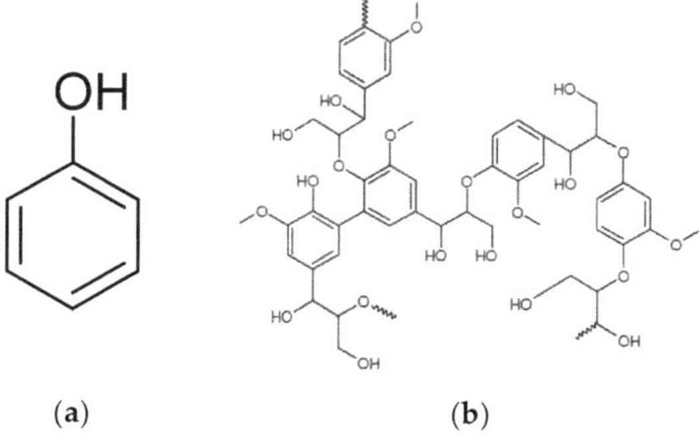

Figure 1. Structure of the main components of the wood leachate (WL) (**a**) phenol, and (**b**) lignin.

The wood-based composite industry has developed rapidly during the last few decades, and its continued development is expected, e.g., the global demand for wood-based materials is projected to increase threefold between 2010 and 2050. In 2019, the annual global production of wood-based composites was estimated to be 357 million m^3 [2]. This generates a massive amount of organic waste, i.e., from acid waste to aromatic leachates. In recent years, due to the increasing global environmental awareness, the need to decrease dependence on petroleum-based products, and new environmental regulations, great attention has been paid to the possibilities of developing biodegradable polymers from renewable, bio-based agricultural waste and residues [32–34]. Considering the great amounts of WL discharged by the wood-based industries, it is feasible to develop new methods for its efficient recycling and reuse in value-added applications. In addition, reusing WL waste solves waste management issues and has remarkable economic and environmental benefits. This acidic waste material could harm the environment and thus needs to be treated before discharge. Leachates may contain large quantities of organic pollutants such as ammonia (NH_3), heavy metals, mineral salts, and phenol derivatives. Moreover, previous work has shown that this leachate can infiltrate and pollute the soil and the surrounding water, creating an environmental problem that needs to be addressed [35–38]. The WL's main components, i.e., lignin, phenol, and tannin, can be used as a natural filler in polymer composites. The lignin in the WL could play a strong bioadhesive role in the polymeric blends, and the phenol and tannin could prevent bacterial growth on the composite surface [39,40].

This research aimed to evaluate the feasibility of using WL powder as an additive in PLA-based biocomposites and evaluate their exploitation properties. The chemical structure of the fabricated composites was characterized using Fourier transform infrared spectroscopy (FTIR). The effects of WL incorporation on the mechanical, thermal, and morphological properties of the PLA matrix were investigated using tensile tests, differential scanning calorimetry (DSC), and field emission-scanning electron microscopy (FE-SEM). Furthermore, the inhibition zone examination was used to detect the antimicrobial behavior of the developed PLA-WL biocomposites using *Escherichia coli (E. coli)* as the model microbe.

2. Materials and Methods

2.1. Materials

The PLA (CAS No: 26100-51-6) with a molecular weight of 80,000 g/mol, a specific gravity of 1.24 g/cm^3, and a melting point of 155–170 °C was purchased from BIOKAS Ö5 (Chemiekas GmbH, Vienna, Austria). The raw WL (with 3% solids content) and fibers used in this work were supplied by the Arian Maryam Incorporation factory (AMI, Rasht, Iran) from the Arian Saeed Industrial Group. The lignin weight percentage of dry WL was determined as the acid-insoluble Klason lignin. The standard approach for determining lignin content is based on Klason's method of hydrolysis of materials using 72% sulfuric acid (Merck 1.00748 98% diluted). Lignin is the substance remaining after hydrolysis in this step, and it is the insoluble residual extracted by filtration and measured by the gravimetric procedure [41,42]. The lignin content of dried WL was determined to be about 22%.

2.2. Biocomposite Preparation

To prepare the novel biocomposites, WL was vacuum dried at 100 °C in a vacuum oven (Memmert® GmbH & Co. KG, Schwabach, Germany) at the first step. Then, the obtained solid material was grounded in powder using a lab-scale mill. The PLA was also vacuum dried at 70 °C to evaporate the moisture content. WL powder was dry blended with PLA and then melted compounded in an internal mixer (Brabender® GmbH & Co. KG, Duisburg, Germany) at 180 °C for 15 min with the screw rotor speed of 50 rpm. The manufacturing parameters of the laboratory-fabricated PLA-WL biocomposites are given in Table 1. The prepared biocomposite blends were placed into a mold of 2 mm height, and the sheets were fabricated by a hot-pressing process in a laboratory press of 5 tons. The press temperature applied was 180 °C. The pressure value used was 1.4 MPa for 5 min [43].

Table 1. Manufacturing parameters of PLA-WL biocomposites produced in this work.

Sample	PLA (%)	WL (%)	Mixing Temp (°C)	Mixing Time (min)	Sheet Thickness (mm)
PLA	100	0	180	15	3.4
PLA-WL-3	97	3	180	15	3.6
PLA-WL-5	95	5	180	15	3.7
PLA-WL-7	93	7	180	15	3.5
PLA-WL-9	91	9	180	15	3.9

2.3. Characterization

2.3.1. Structural and Thermal Analysis

FTIR spectroscopy was employed to identify the composites' functional groups. A Bruker Equinox 55LS 101 series instrument (Bruker, Karlsruhe, Germany) with a resolution of 4 cm^{-1} (50 scans on average) was used for this purpose. The Netzsch DSC Maia 200 F3 facilitated with a low-temperature accessory was employed for performing the DSC analysis. It was performed at a 20 °C/min heating rate at temperatures ranging from −20 to 280 °C through a nitrogen atmosphere. The stepwise specific heat increment midpoint was taken as the glass transition temperature (Tg). All DSC adjustments were made according to ASTM D3418 [44].

2.3.2. Mechanical and Morphological Investigation

The tensile stress–strain experiments were accomplished using a Gotech Universal testing machine AI-7000-LA (Gotech Testing Machines, Inc., Taichung, Taiwan), following the ASTM D638 [45]. The examinations were carried out at room temperature (25 °C). The experiments were carried out at a 5 mm/min cross-head speed. At least five test specimens of any composition were considered for tensile tests. The ASTM D256 [46] standard was followed to perform the Charpy impact strength measurement of the developed composites. The notched test specimens ($80 \times 10 \times 3.8$ mm^3) were tested using FRANK Baldwin-Model-BLI pendulum impact testing equipment (Frank Bacon, Warren, MI, USA). A 45 V-shaped notch was made in the center of the impact sample by razor notching equipment (CEAST). The notch tip radius was 0.25 mm, and the specimen depth remaining under the notch was 10.17 mm. For each composition, at least five samples were examined. Shore A sample hardness was measured using FRANK measuring instruments with ASTM D2240 [47]. Philips (Philips-X130) SEM equipment was employed for identifying the fracture surface of the prepared composite sheets in liquid N_2. The SEM micrographs of the surface were obtained using the cold field emission mode on the surface of samples [48–51].

2.3.3. Surface Hydrophobicity and Antibacterial Activity

The hydrophobicity of the laboratory-fabricated biocomposites was measured using a Kruss G10 contact angle measurement system (KRÜSS GmbH, Hamburg, Germany). The contact angle of the distilled water droplets was calculated on the composite surface [52].

The antimicrobial behavior of the developed composite materials was analyzed using *E. coli* bacteria. The *E. coli* bacteria was inserted into a liquid broth and cultivated under stirring of 120 rpm for 10 h at 37 °C in a shaking incubator. The resulting sample was diluted to around 100 (CFU)/mL using the broth media, and then, the diluted mixture was added to the agar medium. The PLA and PLA-WL samples were sterilized by autoclaving and placed on the plates with the agar's surface in the tangent state and were incubated for 24 h at 37 °C. A comparative analysis of the biocomposites' antimicrobial behavior measured the bacterial growth inhibition region (μm) [53,54].

3. Results and Discussion

3.1. Structural Characterization

The FTIR absorption peaks of the pure PLA and the PLA-WL biocomposites are shown in Figure 2. The C=O stretching absorption peaks related to PLA or ester groups were detected at 1750 cm^{-1}. The C–O bonds in the PLA structure were visible at 1182 cm^{-1}. The peak at 1363 cm^{-1} was related to PLA methyl groups and C–H bonds vibration. The C–C bonds stretching peak of the PLA chemical structure appeared at 836 cm^{-1}.

After incorporating WL to the PLA matrix, the band near 1490 cm^{-1} was attributed to vibration of the phenolic rings, and the peaks at 1450, 1210, 1188 cm^{-1} for phenolic ring or phenolic hydroxyl groups vibrations [55,56] appeared in the spectrums of PLA-WL. That is proved by the WL's phenolic structures (i.e., lignin, tannins, and phenol). The only difference between PLA-WL peaks are in their intensity that is related to the amount of filler and its compounds within the sample.

3.2. Mechanical Properties

The results obtained for the mechanical properties of the laboratory-made PLA-WL biocomposites, i.e., tensile strength, break elongation, and elastic modulus, are presented in Table 2.

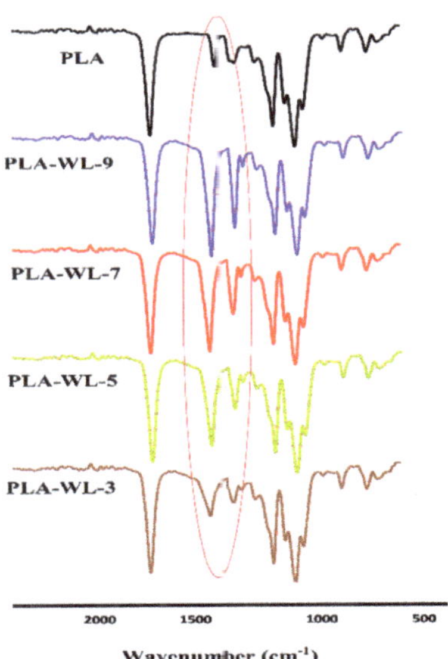

Figure 2. FTIR spectra of PLA and PLA-WL biocomposites.

Table 2. The mechanical characteristics of the fabricated PLA-WL biocomposites.

Sample	Tensile Strength (MPa)	Elongation at Break (%)	Elastic Modulus (MPa)	Charpy Impact Strength (KJ/m^2)	Shore A Hardness
PLA	33.95	4.00	1837	38.62	89.21
PLA-WL-3	58.65	8.80	1553	46.21	91.32
PLA-WL-5	121.75	16.73	1817	62.47	91.85
PLA-WL-7	169.75	18.93	1942	74.56	92.30
PLA-WL-9	66.95	11.71	1427	61.84	86.21

A graphical representation of the stress–strain curves of the developed PLA-WL biocomposites with different addition levels of WL as a filler is shown in Figure 3. Data are in correspondence to previously reported specifications for high molecular weight PLA [57]. The tensile strength of the PLA-WL blends was slightly increased by the filler loading increment (WL powder). The tensile strength was increased by about 400% at PLA-WL-7, which has a 7% concentration of the WL powder in its structure. The higher tensile strength of the filler-induced samples indicated the effect of lignin in the WL structure. The lignin has a paste-like impact within the polymer matrix that causes higher tensile strength values [58]. Lignin penetrates into the gaps of adjacent particles and intramolecular space of the PLA, and it improves the interfacial adhesion between WL filler and PLA matrix (physical adhesion) [59].

Markedly, the elongation at break was also increased in higher WL filler-loaded samples. This might be attributed to the plasticization effect of the lignin. Plasticization is described as the action of plasticizers on the matrix structure of materials, causing it to swell. Plasticizers reduce the T_g and melt viscosity by increasing the free volume by spacing polymer chains and increasing the mobility of chain segments. Lignin's plasticization effect will also help to improve polymer matrix mobility via the effects of OH groups presents on lignin structure, which interacts with the polymer structure and leads to higher mobility as well as the processability and toughness of the resulting composite [60]. It was reported that high filler content increased the agglomeration probability [61], and the regions with stress concentration that requires less energy to crack will exist in this regard. The findings showed a significant decrease in tensile strength at 9% WL powder content. The localization of WL agglomerates between the PLA polymer chains in PLA-WL-9 resulted in decreased biocomposite break elongation. The steric hindrance of lignin, phenol, and tannin bonds in high WL powder concentrations leads to lower lock energy between the PLA chains.

The elastic modulus of the PLA-WL-3 specimen indicated a drop following WL's addition due to lignin's plasticization effect, referring to Table 2. The lignin mechanically bound the PLA chains in higher WL volumes, and the lignin's paste activity exceeded its plasticization impact. Lignin could fit the polymer matrix, causing less mobility than the samples without or with a fewer percentage of WL powder filler [62].

During force loading in the testing machine, partial spaces were made, which barricaded the stress dissemination between the matrix and the filler. The degree of locking increased as the filler loading increased, raising the elastic modulus and stiffness. At the PLA-WL-9 sample, a sudden drop was observed related to the agglomeration of filler materials. The data relating to the Charpy impact strength of the fabricated films are presented in Table 2. The Charpy impact strength of the biocomposites was enhanced by the increment of the filler content from the results. Regular energy transfer in the matrix is substantial for impact persistence [63]. The highest impact resistant composition in this study was the PLA-WL-7 sample with an impact strength of 74.56 KJ/m^2, and the factor has a growing trend by the addition of filler up to this composition. In dispersed phases addition into a polymer matrix, the dispersed phase's good dispersion, filler particle sizes, and proper interfacial incorporation are the most critical factors determining the materials' optimum performance. The blend's impact strength drop with 9% WL powder was due to some unfavorable filler dispersion in the polymer matrix.

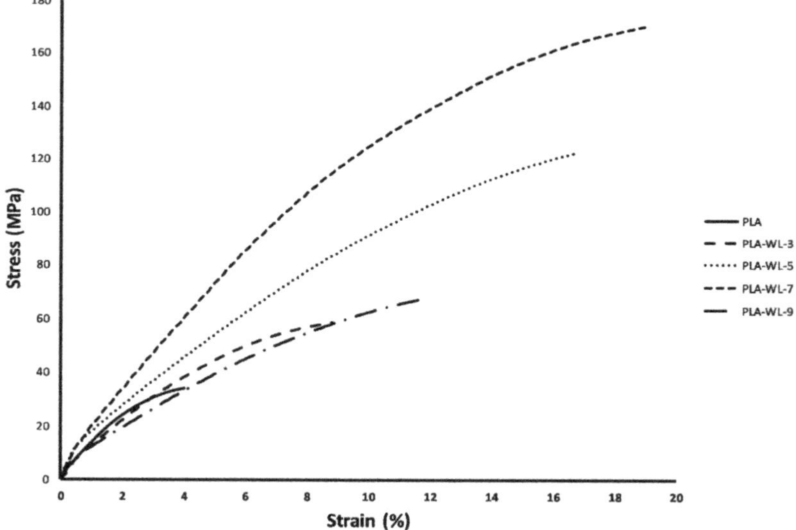

Figure 3. Stress–strain curves of tensile tests for neat PLA and PLA-WL biocomposites.

The composite's hardness results are shown in Table 2. The hardness of the blended sheets increased with the higher filler content. It may be due to the WL powder dispersion and void minimization into the PLA matrix and intense physical engagement between the WL and PLA chains. However, the hardness increment value by filler addition has a slight trend, which is caused by dispersion of the WL within the structure and not just on the surface of the developed biocomposites. The highest hardness value of 92.3 was determined for the PLA-WL-7 sample. It can also be deduced that the filler increased the hardness of polymeric materials by filling the empty micro gaps between the polymer frameworks. Overall, the mechanical properties showed that 3–7% content of WL powder could be applied as filling material, leading to amelioration in the PLA composite's mechanical properties, such as tensile strength, impact strength, and hardness. In addition, the induction of lignin, tannin, and phenol in low concentrations to the polymer blend can make it widely applicable in different industries according to the antibacterial effects of these materials, which are explained in the following section.

3.3. Thermal Behavior

The DSC analysis was employed to investigate the effect of various filler (WL powder) loadings on the thermal characteristics of the developed PLA-WL biocomposites. Understanding the polymers' crystallization behavior under dynamic conditions is considerable because most processing procedures perform in non-isothermal conditions [64,65]. Figure 4 illustrates the crystallization exotherms of the PLA-WL biocomposites. The parameters of the DSC curves for crystallization exotherms, such as the crystallization peak (Tc), glass transition temperature (Tg), and the crystallization enthalpy (dHc), were calculated from the under peak surface area.

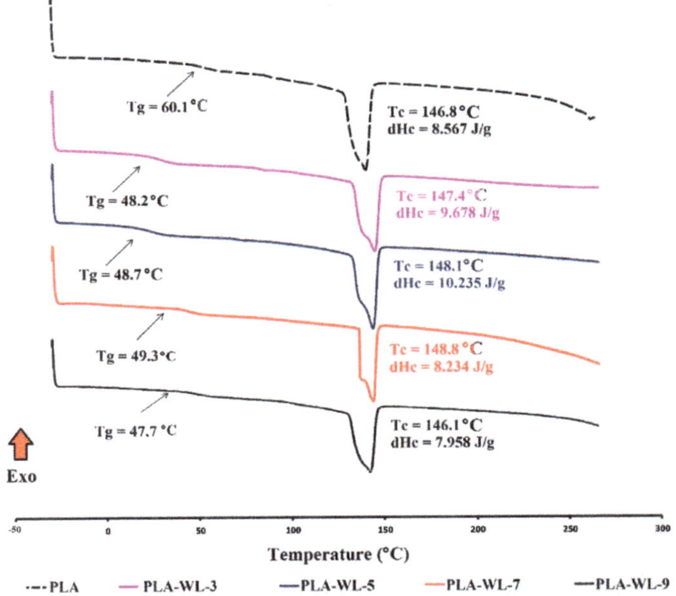

Figure 4. The DSC curves for PLA-WL biocomposites.

The WL had a positive effect on the crystallization behavior of the PLA as the crystallization peak temperature was shifted to higher values after incorporating the WL filler. As expected, it also induced the crystallization enthalpy. The highest crystallization degree was determined for the PLA-WL sample. The higher crystalline structure of the PLA-WL-7 led to a high mechanically strengthened material [66,67], which complies with the results from mechanical testing. The plasticization effect of lignin leads to the free movement of the PLA chains, and as a result, the crystallinity of the modified composites became higher than that of neat PLA. On the other hand, high amounts of lignin and other phenolic compounds (in PLA-WL-9) present in the WL composition significantly reduced the polymer chains mobility caused by the steric hindrance impact associated with their aromatic composition.

Subsequently, the glass transition temperature was lowered by incorporating WL powder into the composite structure. The PLA had a Tg drop at 60.1 °C, and it is evident that WL lowers Tg as it makes the chains more mobile during the matrix due to its plasticization effect. According to the DSC results, the Tg has a positive trend by increasing the WL content in the laboratory-made biocomposites. This might be attributed to the chain locks occurring by increasing the filler amount. Lignin could hinder the polymer chains movement, resulting in greater Tg in higher WL filler ratios. Conversely, lignin could enhance the PLA chains mobility due to its plasticization effect, facilitating crystallization growth and improving crystallinity.

3.4. Morphology

The SEM technique was used to identify the biocomposites' morphological modifications on the surface and cross-sections split in liquid N2. A graphical representation of the samples' SEM images (PLA, PLA-WL-7, and PLA-WL-9 biocomposites) is presented in Figure 5.

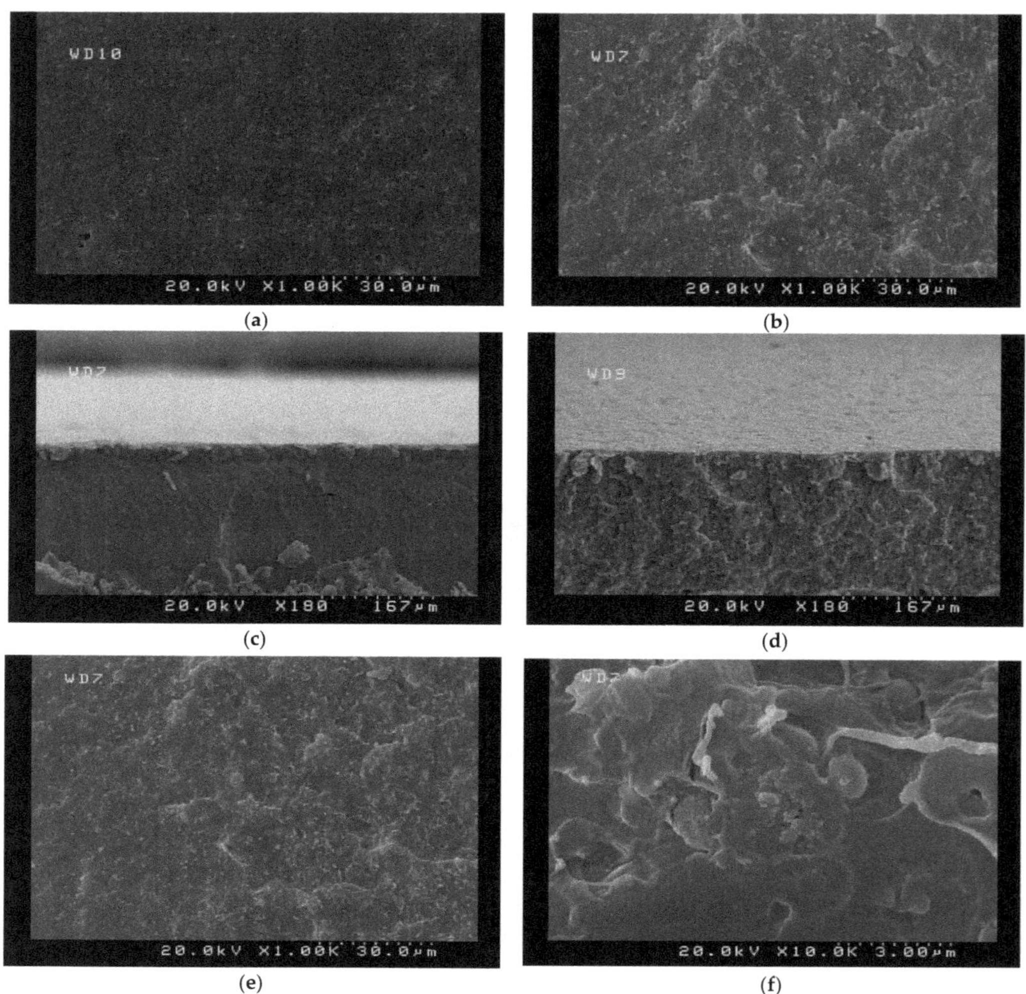

Figure 5. The SEM micrographs of the surface ((**a**) PLA and (**b**) PLA-WL-7 biocomposites), fracture surface ((**c**) PLA and (**d**) PLA-WL-7 biocomposites), and fracture surface of PLA-WL-9 in two different magnifications ((**e**) 1KX and (**f**) 10 KX).

The PLA had a smooth surface with some voids related to the humidity vaporization from the surface during the hot pressing process. The uniform structure was detected for PLA in both surface and cross-section images. The WL powder's presence in the structure of the PLA-WL-7 biocomposite was completely observable from changes in the morphology in the SEM micrographs of this sample. Different phases were detectable, and WL powder was well dispersed between the PLA matrixes. No micro-domains were observed in the developed PLA-WL composites. The homogeneous surface resulted from the obtained good dispersion of WL in the PLA matrix during the process, confirming the composites' enhanced mechanical performance containing 7% WL. The optimal mechanical and thermal characteristics were determined for the PLA-WL-7 sample. These improvements were proved by illustrating WL proper dispersion in the polymer matrix. A powerful filler and matrix engagement due to the lignin's paste impact was evident in the images. The SEM micrographs of the fracture surface for the PLA-WL-9 sample are illustrated in Figure 5e,f. Some agglomerations of WL are detectable within the image. The results correspond with the mechanical testing results due to the reduced mechanical strength of this sample compared with the PLA-WL-7 composite.

3.5. Contact Angle

The developed biocomposites' contact angles determined the surface wettability (Figure 6). The contact angles increased from about 46° to 66° when the WL was incorporated into the PLA matrix at 7 wt % concentration. It was observed that the addition of phenolic compounds which are presented in the WL enhanced the hydrophobicity of the composite surface. The reason for this hydrophobicity is the emplacement of the WL with hydrophobic phenolic functional groups in the micro spaces of the PLA matrix. The data achieved from the contact angle tests agree with the antibacterial activity test, which is reported in the next section. This study may serve as a stepping stone for future investigations to solve some of the critical problems of bioplastic industries by preventing moisture diffusion and inducing the antibacterial activity into a biodegradable packaging material, i.e., PLA.

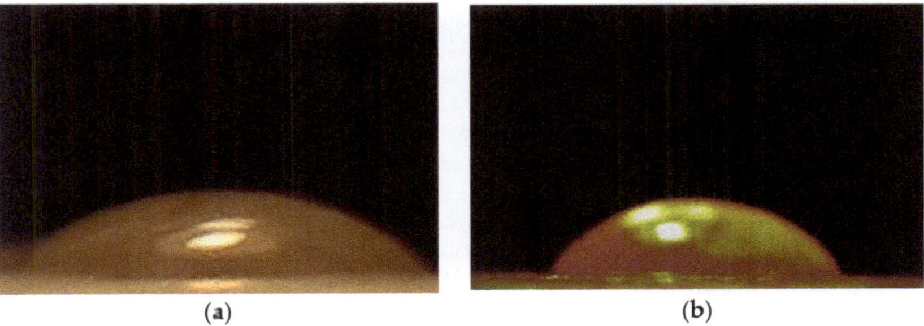

Figure 6. The contact angle images for fabricated biocomposites ((**a**) PLA and (**b**) PLA-WL-7).

3.6. Antibacterial Behavior

The materials' antibacterial activity is associated with their physical and chemical properties such as the functional groups, dispersion of antibacterial agents, and surface roughness. It is well known that bacteria tend to be cultured on hydrophilic surfaces, and thus, hydrophobic surfaces may inhibit bacterial growth [68–70]. For examining the antibacterial characteristics of PLA-WL biocomposites, the association of the *E. coli* bacteria with neat PLA and PLA-WL was mentioned. Figure 7 indicates composites' behavior in the forms of cytotoxic effects. The density of the bacterial growth around the neat PLA indicates that there is no antimicrobial activity in the neat PLA structure. WL powder insertion into the PLA matrix structure led to a lower microbe density across the test sheet, suggesting an inhibitory effect. The results illustrated that the PLA-WL-3 has no bacterial growth inhibition, unlike the neat PLA sample. On the other hand, PLA-WL-5, PLA-WL-7 and PLA-WL-9 have shown growth inhibition zones, and this activity is more indicatable in PLA-WL-9. This result proved the claim of the antibacterial effect of WL on *E. coli* growth. The expressed antimicrobial behavior of the PLA-WL composites proved their potential use in various packaging applications to inhibit bacterial growth. The authors postulate that phenolic compounds in the WL structure caused the PLA-WL composite's antibacterial properties.

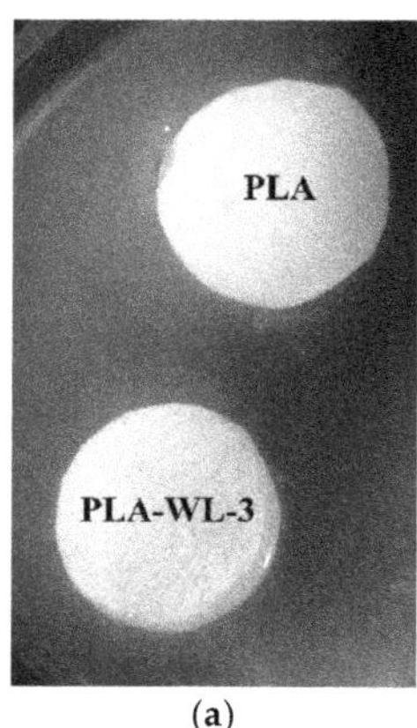

(**a**)

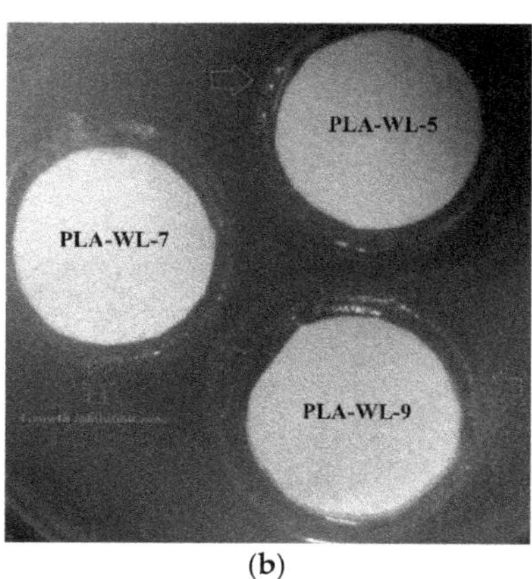

(**b**)

Figure 7. The antibacterial activity of the prepared biocomposites ((**a**) PLA and PLA-WL-3 and (**b**) PLA-WL-5, PLA-WL-7, and PLA-WL-9) against *E. coli*.

4. Conclusions

The results indicated that the WL powder filler induction in the PLA matrix could enhance the polymer's mechanical characteristics. WL significantly improved all of the PLA's mechanical properties, such as tensile strength, elongation at break, and hardness. The addition of WL resulted in an enhanced Charpy impact strength of the biocomposites. The PLA's thermal stability and crystallization behavior were also improved with the addition of WL as a filler. The surface hydrophobicity of the PLA-WL biocomposites was improved compared to the neat PLA sample. The antibacterial activity is another benefit that the WL added to the PLA. In bio-industries, preventing moisture diffusion and bacterial growth are very significant challenges. The induction of these properties to the PLA as a biopolymer could make the WL suitable as a filler for various industrial applications.

Author Contributions: Conceptualization, M.H.S.; methodology, M.H.S.; investigation, P.P.S. and M.N.A.; resources, M.H.S.; writing—original draft preparation, P.P.S.; writing—review and editing, M.H.S., P.A. and V.S.; visualization, P.P.S.; supervision, M.H.S.; project administration, M.H.S.; funding acquisition, M.H.S. All authors have read and agreed to the published version of the manuscript.

Funding: This research work was supported by project No. P/8/995 carried out at the Amol University of Special Modern Technologies, Amol, Iran. This research was also supported by the project No. НИС-Б-1145/04.2021 "Development, Properties, and Application of Eco-Friendly Wood-Based Composites" carried out at the University of Forestry, Sofia, Bulgaria.

Institutional Review Board Statement: Not applicable.

Informed Consent Statement: Not applicable.

Data Availability Statement: Not applicable.

Conflicts of Interest: The authors declare no conflict of interest.

References

1. Vitolina, S.; Shulga, G.; Neiberte, B.; Jaunslavietis, J.; Verovkins, A.; Betkers, T. Characteristics of the Waste Wood Biomass and Its Effect on the Properties of Wood Sanding Dust/Recycled PP Composite. *Polymers* **2022**, *14*, 468. [CrossRef] [PubMed]
2. Antov, P.; Krišťák, L.; Réh, R.; Savov, V.; Papadopoulos, A.N. Eco-friendly fiberboard panels from recycled fibers bonded with calcium lignosulfonate. *Polymers* **2021**, *13*, 639. [CrossRef] [PubMed]
3. Scaffaro, R.; Maio, A.; Sutera, F.; Gulino, E.F.; Morreale, M. Degradation and recycling of films based on biodegradable polymers: A short review. *Polymers* **2019**, *11*, 651. [CrossRef]
4. Silva, A.L.P. Future-proofing plastic waste management for a circular bioeconomy. *Curr. Opin. Environ. Sci. Health* **2021**, *22*, 100263. [CrossRef]
5. Dorieh, A.; Selakjani, P.P.; Shahavi, M.H.; Pizzi, A.; Movahed, S.G.; Pour, M.F.; Aghaei, R. Recent developments in the performance of micro/nanoparticle-modified urea-formaldehyde resins used as wood-based composite binders: A review. *Int. J. Adhes. Adhes.* **2022**, *114*, 103106. [CrossRef]
6. Fonseca-García, A.; Jiménez-Regalado, E.J.; Aguirre-Loredo, R.Y. Preparation of a novel biodegradable packaging film based on corn starch-chitosan and poloxamers. *Carbohydr. Polym.* **2021**, *251*, 117009. [CrossRef]
7. Ramesh, M.; Rajeshkumar, L.; Sasikala, G.; Balaji, D.; Saravanakumar, A.; Bhuvaneswari, V.; Bhoopathi, R. A Critical Review on Wood-Based Polymer Composites: Processing, Properties, and Prospects. *Polymers* **2022**, *14*, 589. [CrossRef]
8. Dorieh, A.; Farajollah Pour, M.; Ghafari Movahed, S.; Pizzi, A.; Pouresmaeel Selakjani, P.; Valizadeh Kiamahalleh, M.; Hatefnia, H.; Shahavi, M.H.; Aghaei, R. A review of recent progress in melamine-formaldehyde resin based nanocomposites as coating materials. *Prog. Org. Coat.* **2022**, *165*, 106768. [CrossRef]
9. Selakjani, P.P.; Dorieh, A.; Pizzi, A.; Shahavi, M.H.; Hasankhah, A.; Shekarsaraee, S.; Ashouri, M.; Movahed, S.G.; Abatari, M.N. Reducing free formaldehyde emission, improvement of thickness swelling and increasing storage stability of novel medium density fiberboard by urea-formaldehyde adhesive modified by phenol derivatives. *Int. J. Adhes. Adhes.* **2021**, *111*, 102962. [CrossRef]
10. Mundel, R.; Thakur, T.; Chatterjee, M. Emerging uses of PLA–PEG copolymer in cancer drug delivery. *3 Biotech* **2022**, *12*, 41. [CrossRef]
11. Vlachopoulos, A.; Karlioti, G.; Balla, E.; Daniilidis, V.; Kalamas, T.; Stefanidou, M.; Bikiaris, N.D.; Christodoulou, E.; Koumentakou, I.; Karavas, E. Poly (Lactic Acid)-Based Microparticles for Drug Delivery Applications: An Overview of Recent Advances. *Pharmaceutics* **2022**, *14*, 359. [CrossRef]
12. Abifarin, J.K.; Prakash, C.; Singh, S. Optimization and significance of fabrication parameters on the mechanical properties of 3D printed chitosan/PLA scaffold. *Mater. Today Proc.* **2021**, *50*, 2018–2025. [CrossRef]
13. Młotek, M.; Gadomska-Gajadhur, A.; Sobczak, A.; Kruk, A.; Perron, M.; Krawczyk, K. Modification of PLA scaffold surface for medical applications. *Appl. Sci.* **2021**, *11*, 1815. [CrossRef]
14. Fan, T.; Daniels, R. Preparation and characterization of electrospun polylactic acid (PLA) fiber loaded with birch bark triterpene extract for wound dressing. *Aaps Pharmscitech* **2021**, *22*, 1–9. [CrossRef]
15. Liu, J.; Shi, R.; Hua, Y.; Gao, J.; Chen, Q.; Xu, L. A new cyanoacrylate-poly (lactic acid)-based system for a wound dressing with on-demand removal. *Mater. Lett.* **2021**, *293*, 129666. [CrossRef]
16. Khosravi, A.; Fereidoon, A.; Khorasani, M.M.; Naderi, G.; Ganjali, M.R.; Zarrintaj, P.; Saeb, M.R.; Gutiérrez, T.J. Soft and hard sections from cellulose-reinforced poly (lactic acid)-based food packaging films: A critical review. *Food Packag. Shelf Life* **2020**, *23*, 100429. [CrossRef]
17. Backes, E.H.; Pires, L.D.N.; Beatrice, C.A.G.; Costa, L.C.; Passador, F.R.; Pessan, L.A. Fabrication of biocompatible composites of poly (lactic acid)/hydroxyapatite envisioning medical applications. *Polym. Eng. Sci.* **2020**, *60*, 636–644. [CrossRef]
18. Sin, L.T.; Rahmat, A.R.; Rahman, W.A.W.A. 3—Applications of Poly(lactic Acid). In *Handbook of Biopolymers and Biodegradable Plastics, Ebnesajjad, S., Ed.*; William Andrew Publishing: Boston, MA, USA, 2013; pp. 55–69.
19. Alexandre, A.; Cruz Sanchez, F.A.; Boudaoud, H.; Camargo, M.; Pearce, J.M. Mechanical properties of direct waste printing of polylactic acid with universal pellets extruder: Comparison to fused filament fabrication on open-source desktop three-dimensional printers. *3D Print. Addit. Manuf.* **2020**, *7*, 237–247. [CrossRef]
20. Giorcelli, M.; Khan, A.; Pugno, N.M.; Rosso, C.; Tagliaferro, A. Biochar as a cheap and environmental friendly filler able to improve polymer mechanical properties. *Biomass Bioenergy* **2019**, *120*, 219–223. [CrossRef]
21. Arrigo, R.; Bartoli, M.; Malucelli, G. Poly (lactic acid)–biochar biocomposites: Effect of processing and filler content on rheological, thermal, and mechanical properties. *Polymers* **2020**, *12*, 892. [CrossRef]
22. Ebrahimpour, M.; Shahavi, M.H.; Jahanshahi, M.; Najafpour, G. Nanotechnology in Process Biotechnology: Recovery and Purification of Nanoparticulate Bioproducts Using Expanded Bed Adsorption. *Dyn. Biochem. Process Biotechnol. Mol. Biol.* **2009**, *3*, 57–60.
23. Kuang, T.; Ju, J.; Liu, T.; Hejna, A.; Saeb, M.R.; Zhang, S.; Peng, X. A facile structural manipulation strategy to prepare ultra-strong, super-tough, and thermally stable polylactide/nucleating agent composites. *Adv. Compos. Hybrid Mater.* **2022**. [CrossRef]
24. Barczewski, M.; Hejna, A.; Aniśko, J.; Andrzejewski, J.; Piasecki, A.; Mysiukiewicz, O.; Bąk, M.; Gapiński, B.; Ortega, Z. Rotational molding of polylactide (PLA) composites filled with copper slag as a waste filler from metallurgical industry. *Polym. Test.* **2022**, *106*, 107449. [CrossRef]

25. Scaffaro, R.; Maio, A.; Gulino, E.F.; Pitarresi, G. Lignocellulosic fillers and graphene nanoplatelets as hybrid reinforcement fo polylactic acid: Effect on mechanical properties and degradability. *Compos. Sci. Technol.* **2020**, *190*, 108008. [CrossRef]
26. Wolski, K.; Cichosz, S.; Masek, A. Surface hydrophobisation of lignocellulosic waste for the preparation of biothermoelastoplasti composites. *Eur. Polym. J.* **2019**, *118*, 481–491. [CrossRef]
27. Singh, T.; Lendvai, L.; Dogossy, G.; Fekete, G. Physical, mechanical, and thermal properties of Dalbergia sissoo wood waste-fille poly (lactic acid) composites. *Polym. Compos.* **2021**, *42*, 4380–4389. [CrossRef]
28. Moreno-Anguiano, O.; Cloutier, A.; Rutiaga-Quiñones, J.G.; Wehenkel, C.; Rosales-Serna, R.; Rebolledo, P.; Hernández-Pacheco C.E.; Carrillo-Parra, A. Use of Agave durangensis Bagasse Fibers in the Production of Wood-Based Medium Density Fiberboar (MDF). *Forests* **2022**, *13*, 271. [CrossRef]
29. Scussel, R.; Feltrin, A.C.; Angioletto, E.; Galvani, N.C.; Fagundes, M.Í.; Bernardin, A.M.; Feuser, P.E.; de Ávila, R.A.M.; Pich, C.T Ecotoxic, genotoxic, and cytotoxic potential of leachate obtained from chromated copper arsenate-treated wood ashes. *Environ Sci. Pollut. Res.* **2022**, 1–14. [CrossRef]
30. Raclavská, H.; Růžičková, J.; Juchelková, D.; Šafář, M.; Brťková, H.; Slamová, K. The quality of composts prepared in automati composters from fruit waste generated by the production of beverages. *Bioresour. Technol.* **2021**, *341*, 125878. [CrossRef]
31. Kannepalli, S.; Strom, P.F.; Krogmann, U.; Subroy, V.; Giménez, D.; Miskewitz, R. Characterization of wood mulch and leachate/runoff from three wood recycling facilities. *J. Environ. Manag.* **2016**, *182*, 421–428. [CrossRef]
32. Zhu, Y.; Romain, C.; Williams, C.K. Sustainable polymers from renewable resources. *Nature* **2016**, *540*, 354–362. [CrossRef] [PubMed]
33. Shahavi, M.H.; Esfilar, R.; Golestani, B.; Sadeghabad, M.S.; Biglaryan, M. Comparative study of seven agricultural wastes fo renewable heat and power generation using integrated gasification combined cycle based on energy and exergy analyses. *Fue* **2022**, *317*, 123430. [CrossRef]
34. Kazemeini, H.; Azizian, A.; Shahavi, M.H. Effect of Chitosan Nano-Gel/Emulsion Containing Bunium Persicum Essential Oil and Nisin as an Edible Biodegradable Coating on Escherichia Coli O157:H7 in Rainbow Trout Fillet. *J. Water Environ. Nanotechnol* **2019**, *4*, 343–349. [CrossRef]
35. Maresca, A.; Krüger, O.; Herzel, H.; Adam, C.; Kalbe, U.; Astrup, T.F. Influence of wood ash pre-treatment on leaching behaviou liming and fertilising potential. *Waste Manag.* **2019**, *83*, 113–122. [CrossRef]
36. Berger, F.; Gauvin, F.; Brouwers, H. The recycling potential of wood waste into wood-wool/cement composite. *Construct Build Mater.* **2020**, *260*, 119786. [CrossRef]
37. Hosseini, M.; Shahavi, M.H.; Yakhkeshi, A. AC & DC-currents for separation of nano-particles by external electric field. *Asian J Chem.* **2012**, *24*, 181–184.
38. de Klerk, S.; Ghaffariyan, M.R.; Miles, M. Leveraging the Entrepreneurial Method as a Tool for the Circular Economy: The Case of Wood Waste. *Sustainability* **2022**, *14*, 1559. [CrossRef]
39. Che, X.; Wu, M.; Yu, G.; Liu, C.; Xu, H.; Li, B.; Li, C. Bio-inspired water resistant and fast multi-responsive Janus actuator assembled by cellulose nanopaper and graphene with lignin adhesion. *Chem. Eng. J.* **2021**, *433*, 133672. [CrossRef]
40. Pizzi, A. Bioadhesives for wood and fibres. *Rev. Adhes. Adhes.* **2013**, *1*, 88–113. [CrossRef]
41. Gaff, M.; Kačík, F.; Gašparík, M. Impact of thermal modification on the chemical changes and impact bending strength of European oak and Norway spruce wood. *Compos. Struct.* **2019**, *216*, 80–88. [CrossRef]
42. Mohammadi-Rovshandeh, J.; Pouresmaeel-Selakjani, P.; Davachi, S.M.; Kaffashi, B.; Hassani, A.; Bahmeyi, A. Effect of lignin removal on mechanical, thermal, and morphological properties of polylactide/starch/rice husk blend used in food packaging. *J Appl. Polym. Sci.* **2014**, *131*, 41095. [CrossRef]
43. Hwang, S.W.; Shim, J.K.; Selke, S.; Soto-Valdez, H.; Rubino, M.; Auras, R. Effect of Maleic-Anhydride Grafting on the Physical and Mechanical Properties of Poly (L-lactic acid)/Starch Blends. *Macromol. Mater. Eng.* **2013**, *298*, 624–633. [CrossRef]
44. *ASTM D3418-21*; Standard Test Method for Transition Temperatures and Enthalpies of Fusion and Crystallization of Polymers by Differential Scanning Calorimetry. ASTM International: West Conshohocken, PA, USA, 2021. [CrossRef]
45. *ASTM D638-14*; Standard Test Method for Tensile Properties Of Plastics. ASTM International: West Conshohocken, PA, USA, 2014; p. 18. [CrossRef]
46. *ASTM D256-10*; Standard Test Methods for Determining the Izod Pendulum Impact Resistance of Plastics. ASTM International: West Conshohocken, PA, USA, 2018. [CrossRef]
47. *ASTM D2240-15*; Standard Test Method for Rubber Property—Durometer Hardness. ASTM International: West Conshohocken, PA, USA, 2021. [CrossRef]
48. Abatari, M.N.; Emami, M.R.S.; Jahanshahi, M.; Shahavi, M.H. Superporous pellicular κ-Carrageenan–Nickel composite beads, morphological, physical and hydrodynamics evaluation for expanded bed adsorption application. *Chem. Eng. Res. Des.* **2017**, *125*, 291–305. [CrossRef]
49. Rad, A.S.; Shahavi, M.H.; Esfahani, M.R.; Darvishinia, N.; Ahmadizadeh, S. Are nickel- and titanium- doped fullerenes suitable adsorbents for dopamine in an aqueous solution? Detailed DFT and AIM studies. *J. Mol. Liq.* **2021**, *322*, 114942. [CrossRef]
50. Pérez Quiñones, J.; Brüggemann, O.; Kjems, J.; Shahavi, M.H.; Peniche Covas, C. Novel brassinosteroid-modified polyethylene glycol micelles for controlled release of agrochemicals. *J. Agric. Food Chem.* **2018**, *66*, 1612–1619. [CrossRef] [PubMed]
51. Jahanshahi, M.; Shahavi, M.H. Chapter 17—Advanced Downstream Processing in Biotechnology. In *Biochemical Engineering and Biotechnology*, 2nd ed.; Najafpour, G.D., Ed.; Elsevier: Amsterdam, The Netherlands, 2015; pp. 495–526.

52. Sauerbier, P.; Köhler, R.; Renner, G.; Militz, H. Surface activation of polylactic acid-based wood-plastic composite by atmospheric pressure plasma treatment. *Materials* **2020**, *13*, 4673. [CrossRef]
53. Lashkenari, A.S.; Najafi, M.; Peyravi, M.; Jahanshahi, M.; Mosavian, M.T.H.; Amiri, A.; Shahavi, M.H. Direct filtration procedure to attain antibacterial TFC membrane: A facile developing route of membrane surface properties and fouling resistance. *Chem. Eng. Res. Des.* **2019**, *149*, 158–168. [CrossRef]
54. Shahavi, M.H.; Hosseini, M.; Jahanshahi, M.; Meyer, R.L.; Darzi, G.N. Clove oil nanoemulsion as an effective antibacterial agent: Taguchi optimization method. *Desalination Water. Treat.* **2016**, *57*, 18379–18390. [CrossRef]
55. Halder, P.; Kundu, S.; Patel, S.; Parthasarathy, R.; Pramanik, B.; Paz-Ferreiro, J.; Shah, K. TGA-FTIR study on the slow pyrolysis of lignin and cellulose-rich fractions derived from imidazolium-based ionic liquid pre-treatment of sugarcane straw. *Energy Convers. Manag.* **2019**, *200*, 112067. [CrossRef]
56. Hejna, A.; Barczewski, M.; Andrzejewski, J.; Kosmela, P.; Piasecki, A.; Szostak, M.; Kuang, T. Rotational molding of linear low-density polyethylene composites filled with wheat bran. *Polymers* **2020**, *12*, 1004. [CrossRef]
57. Rahaman, M.N.; Brown, R.F. *Materials for Biomedical Engineering: Fundamentals and Applications*; John Wiley & Sons: Hoboken, NJ, USA, 2021.
58. Davachi, S.M.; Bakhtiari, S.; Pouresmaeel-Selakjani, P.; Mohammadi-Rovshandeh, J.; Kaffashi, B.; Davoodi, S.; Yousefi, A. Investigating the effect of treated rice straw in PLLA/starch composite: Mechanical, thermal, rheological, and morphological study. *Adv. Polym. Technol.* **2018**, *37*, 5–16. [CrossRef]
59. Börcsök, Z.; Pásztory, Z. The role of lignin in wood working processes using elevated temperatures: An abbreviated literature survey. *Eur. J. Wood Wood Prod.* **2021**, *79*, 511–526. [CrossRef]
60. Park, C.-W.; Youe, W.-J.; Kim, S.-J.; Han, S.-Y.; Park, J.-S.; Lee, E.-A.; Kwon, G.-J.; Kim, Y.-S.; Kim, N.-H.; Lee, S.-H. Effect of lignin plasticization on physico-mechanical properties of lignin/poly (lactic acid) composites. *Polymers* **2019**, *11*, 2089. [CrossRef] [PubMed]
61. Ferrández-Montero, A.; Lieblich, M.; Benavente, R.; González-Carrasco, J.L.; Ferrari, B. Study of the matrix-filler interface in PLA/Mg composites manufactured by Material Extrusion using a colloidal feedstock. *Addit. Manuf.* **2020**, *33*, 101142. [CrossRef]
62. Khan, T.A.; Lee, J.-H.; Kim, H.-J. Chapter 9—Lignin-Based Adhesives and Coatings. In *Lignocellulose for Future Bioeconomy, Ariffin, H., Sapuan, S.M., Hassan, M.A., Eds.*; Elsevier: Amsterdam, The Netherlands, 2019; pp. 153–206.
63. Nofar, M.; Sacligil, D.; Carreau, P.J.; Kamal, M.R.; Heuzey, M.-C. Poly (lactic acid) blends: Processing, properties and applications. *Int. J. Biol. Macromol.* **2019**, *125*, 307–360. [CrossRef]
64. Tang, J.; Li, L.; Wang, X.; Yang, J.; Liang, X.; Li, Y.; Ma, H.; Zhou, S.; Wang, J. Tailored crystallization behavior, thermal stability, and biodegradability of poly (ethylene adipate): Effects of a biocompatible diamide nucleating agent. *Polym. Test.* **2020**, *81*, 106116. [CrossRef]
65. Spinelli, G.; Guarini, R.; Kotsilkova, R.; Ivanov, E.; Romano, V. Experimental, Theoretical and Simulation Studies on the Thermal Behavior of PLA-Based Nanocomposites Reinforced with Different Carbonaceous Fillers. *Nanomaterials* **2021**, *11*, 1511. [CrossRef]
66. Tábi, T. The application of the synergistic effect between the crystal structure of poly (lactic acid)(PLA) and the presence of ethylene vinyl acetate copolymer (EVA) to produce highly ductile PLA/EVA blends. *J. Therm. Anal. Calorim.* **2019**, *138*, 1287–1297. [CrossRef]
67. Vanaei, H.R.; Shirinbayan, M.; Costa, S.F.; Duarte, F.M.; Covas, J.A.; Deligant, M.; Khelladi, S.; Tcharkhtchi, A. Experimental study of PLA thermal behavior during fused filament fabrication. *J. Appl. Polym. Sci.* **2021**, *138*, 49747. [CrossRef]
68. Kachur, K.; Suntres, Z. The antibacterial properties of phenolic isomers, carvacrol and thymol. *Crit. Rev. Food Sci. Nutr.* **2020**, *60*, 3042–3053. [CrossRef]
69. Sonseca, A.; Madani, S.; Rodríguez, G.; Hevilla, V.; Echeverría, C.; Fernández-García, M.; Muñoz-Bonilla, A.; Charef, N.; López, D. Multifunctional PLA blends containing chitosan mediated silver nanoparticles: Thermal, mechanical, antibacterial, and degradation properties. *Nanomaterials* **2020**, *10*, 22. [CrossRef]
70. Chong, W.J.; Shen, S.; Li, Y.; Trinchi, A.; Pejak, D.; Kyratzis, I.L.; Sola, A.; Wen, C. Additive manufacturing of antibacterial PLA-ZnO nanocomposites: Benefits, limitations and open challenges. *J. Mater. Sci. Technol.* **2022**, *111*, 120–151. [CrossRef]

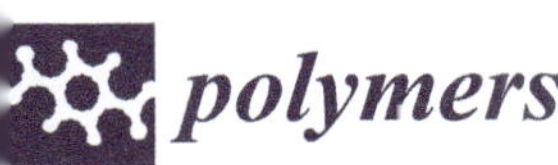

Review

Recent Advances in the Development of Fire-Resistant Biocomposites—A Review

Elvara Windra Madyaratri [1], Muhammad Rasyidur Ridho [1,2], Manggar Arum Aristri [1,2], Muhammad Adly Rahandi Lubis [2], Apri Heri Iswanto [3,4,*], Deded Sarip Nawawi [1,*], Petar Antov [5], Lubos Kristak [6], Andrea Majlingová [6] and Widya Fatriasari [2,*]

1 Department of Forest Products, Faculty of Forestry and Environment, IPB University, Bogor 16680, Indonesia; elvarawindra@yahoo.com (E.W.M.); rasyidmuhammad0505@gmail.com (M.R.R.); arumaristri@gmail.com (M.A.A.)
2 Research Center for Biomaterials BRIN, Jl Raya Bogor KM 46, Cibinong 16911, Indonesia; marl@biomaterial.lipi.go.id
3 Department of Forest Product, Faculty of Forestry, Universitas Sumatera Utara, Medan 20155, Indonesia
4 JATI-Sumatran Forestry Analysis Study Center, Jl. Tridharma Ujung No. 1, Kampus USU, Medan 20155, Indonesia
5 Faculty of Forest Industry, University of Forestry, 1797 Sofia, Bulgaria; p.antov@ltu.bg
6 Faculty of Wood Sciences and Technology, Technical University in Zvolen, 96001 Zvolen, Slovakia; kristak@tuzvo.sk (L.K.); majlingova@tuzvo.sk (A.M.)
* Correspondence: apri@usu.ac.id (A.H.I.); dsnawawi@apps.ipb.ac.id (D.S.N.); widya.fatriasari@biomaterial.lipi.go.id or widy003@brin.go.id (W.F.)

Abstract: Biocomposites reinforced with natural fibers represent an eco-friendly and inexpensive alternative to conventional petroleum-based materials and have been increasingly utilized in a wide variety of industrial applications due to their numerous advantages, such as their good mechanical properties, low production costs, renewability, and biodegradability. However, these engineered composite materials have inherent downsides, such as their increased flammability when subjected to heat flux or flame initiators, which can limit their range of applications. As a result, certain attempts are still being made to reduce the flammability of biocomposites. The combustion of biobased composites can potentially create life-threatening conditions in buildings, resulting in substantial human and material losses. Additives known as flame-retardants (FRs) have been commonly used to improve the fire protection of wood and biocomposite materials, textiles, and other fields for the purpose of widening their application areas. At present, this practice is very common in the construction sector due to stringent fire safety regulations on residential and public buildings. The aim of this study was to present and discuss recent advances in the development of fire-resistant biocomposites. The flammability of wood and natural fibers as material resources to produce biocomposites was researched to build a holistic picture. Furthermore, the potential of lignin as an eco-friendly and low-cost FR additive to produce high-performance biocomposites with improved technological and fire properties was also discussed in detail. The development of sustainable FR systems, based on renewable raw materials, represents a viable and promising approach to manufacturing biocomposites with improved fire resistance, lower environmental footprint, and enhanced health and safety performance.

Keywords: advanced biocomposites; biopolymers; natural fibers; green flame retardants; fire retardancy; product safety

Citation: Madyaratri, E.W.; Ridho, M.R.; Aristri, M.A.; Lubis, M.A.R.; Iswanto, A.H.; Nawawi, D.S.; Antov, P.; Kristak, L.; Majlingová, A.; Fatriasari, W. Recent Advances in the Development of Fire-Resistant Biocomposites—A Review. *Polymers* **2022**, *14*, 362. https://doi.org/10.3390/polym14030362

Academic Editor: Bob Howell

Received: 18 December 2021
Accepted: 15 January 2022
Published: 18 January 2022

1. Introduction

Increased environmental awareness and the scarcity of natural resources, as well as recent stringent environmental regulations and the unsustainable consumption of fossil-derived resources, have forced many manufacturing industries to search for new eco-friendly materials from renewable feedstocks to substitute conventional materials in several end uses. This growing need for "green" materials has led to the increased utilization of natural fibers in the production of biobased composite materials. The development of high-performance biocomposites fabricated from natural resources is increasing worldwide, and the greatest challenge in working with natural biobased composites is the large variations in their properties and characteristics [1] and the resulting variations in final composites. Natural fiber-reinforced composites are innovative composite materials consisting of a polymer matrix reinforced with high-strength natural fibers. The matrices currently used in the development of biobased composites are petroleum-based (thermoplastics or thermosets) or biobased (polylactide acid, polyhydroxybutyrate, starch, etc.) [2]. Thermoplastics include polymers such as polypropylene, polyethylene, polystyrene, and polyvinyl chloride, while polyesters, epoxy and polyurethane represent examples of thermosets used for manufacturing natural fiber reinforced composites. Wood, woody, and non-woody lignocellulosic biomass, all of which are renewable and sustainable materials, can be used as natural resources in biocomposites and have gained great attention in many value-added applications due to their excellent properties, such as their low cost, flexibility during processing, and highly specific stiffness, etc. [3–5]. Green biocomposites can be viable alternatives to the conventional synthetic fiber-reinforced composites as structural or semi-structural components, especially in lightweight applications [6–8]. Some of the most common applications of biobased composites include automotive panels and upholstery, window and door frames, furniture, railroad sleepers, packaging, and other applications that do not require very high mechanical properties but significantly reduce production and maintenance costs [9,10]. Biobased composite materials in the form of panels and sandwich structures have been used to replace wooden furniture, fittings, and noise-insulating panels [11]. When woody or non-woody fibers are combined with thermoplastic matrices, such as polyethylene, polypropylene, or polyvinyl chloride, wood plastic composites (WPC) are produced. Due to their excellent properties, such as high strength, durability, stiffness, and resistance to wear, these engineered materials have found a wide range of applications [12]. Despite the numerous advantages of natural fibers, there are also some drawbacks limiting their potential as a natural feedstock for the development of biobased composites, such as their insufficient adhesion and incompatibility with the matrices, lower water and thermal resistance, and their susceptibility to insect and fungi attacks, etc. In addition, they belong to the group of highly-combustible polymer materials [13]. To inhibit or suppress the burning process, refractory additives function by chemically or physically inhibiting particular stages of the burning process and lowering the amount of heat emitted during the early stages of fire by slowing its spread [14].

Although complete fire protection of biocomposites for indefinite periods is unachievable, appropriate flame retardants (FRs) can turn these materials into hard-to-ignite materials, thus extending the range of their applications [15–17]. FRs are used to reduce the risk of fire in items by preventing ignition and delaying the spread of the fire and the flashover time, while also protecting the lifetime of the item and offering environmental protection by preventing local pollution and long-term environmental effects [18]. The most important parameters are a time to ignite, spontaneous ignition and flash point temperature, rate of heat release, thermal stability index, smoke toxicity, extinction flammability index, mass loss, limiting oxygen index (LOI), flame propagation on the surface, and fire resistance. Because of their high calorific capacity, polymers burn quickly. However, by adding FRs, it is possible to improve their fire behavior (e.g. by neutralizing or decreasing heat and smoke) [19]. Regarding these purposes, FRs have been introduced in the manufacturing of many goods to meet fire safety requirements.

Since the 1980s, there has been a growth in polymeric material utilization, which has enhanced the risk of fires caused by the flammability of polymeric materials [20]. To address this weakness, several FR treatments and techniques have been introduced, such as halogenated and non-halogenated FRs, layered silicates, nano fillers, copolymerization, grafting, and the synergistic use of natural fiber and FRs [21]. The two main categories of additive FRs are halogenated and non-halogenated refractory materials. Because they are inexpensive and effective, halogen-based compounds are the most used FR additions on the market. Several halogen compounds, however, have been banned due to their toxicity and environmental issues related to halogen-based refractory additives. The use of halogen-based compounds in the industrial sector of wood products in Europe has been prohibited since 2006 [22].

As a result, non-halogen refractory materials are becoming more widely used [23]. Environmental issues, mechanical/physical attributes, and processing constraints all necessitate a narrow range of options in the development of FR biocomposite materials. Nowadays, the need for unique FR solutions has been increased, with companies realizing the need for a product that is not only environmentally friendly but also long-lasting and cost-effective [18]. Polymer-based FRs are undergoing research and development. Because of their high availability and annual renewability, biobased FRs from animal origins, including chitin, DNA, and biomass sources (e.g., those that are cellulose based, such as lyocell fibre, saccharide based, and those based on polyphenolic compounds, etc.), hold promise in terms of their potential as "green" FRs in the development of biobased composites [24–29]. Aromatic compounds such as lignin and tannin are well known for their capability for producing char in combination with phosphorous [30]. Furthermore, the FRs employed must be safe for humans and animals, i.e., they must not emit hazardous compounds during normal material use. Using non-toxic nanofillers in polymers to achieve flame retardancy is a viable option [31]. Markedly, the addition of FRs in the matrix can result in the compromised physical and mechanical properties of the fabricated composites [18,19]. The aim of this research work was to present and discuss the recent advances in the development of fire-resistant biocomposites. The flammability of wood and natural fibers as material resources to produce biocomposites was evaluated to build a holistic picture. Furthermore, the potential of lignin as an eco-friendly and low-cost FR additive in the matrix of biocomposites with improved technological and fire properties was investigated. The limitations and perspectives of the economic and environmental elements of FRs were also highlighted for future implementation.

2. Flammability of Biocomposites

2.1. Woody Biomass

Biomass is the richest natural resource on the planet. Lignocellulosic biomass has gained increasing research interest because of its renewable nature [32]. Lignocellulosic biomass refers to both non-woody and wood biomass, which differ in their chemical and physical composition [33]. Holocellulose (a mixture of hemicellulose and cellulose) and lignin make up the category of lignocellulosic biomass. The composition of lignocellulose highly depends on its source, i.e., whether it is derived from woody or non-woody biomass [34,35]. Woody biomass is denser, stronger, and physically larger than non-woody. Furthermore, wood fibers can be collected throughout the year, minimizing the need for long-term storage [36].

One of the main disadvantages of wood as a structural material is its dimensional instability in conditions of environmental change. Wood is also susceptible to wood-destroying organisms such as insects and fungi, not to mention the fact that wood fibers are flammable. However, because wood has many advantages, such as excellent mechanical strength and insulation properties, and a pleasing appearance, it is commonly employed as a building material [37]. Furthermore, wood biomass is renewable and can be used as an organic matter-based sustainable energy source. A range of energy sources, including renewable energy sources, fossil fuels, solar energy, and nuclear power, can be utilized to generate electricity or other forms of power [38].

Several factors can influence the moisture content of wood biomass used as a source of energy [39]. In woody biomass, there are two forms of water, i.e., free water and bound water [40]. The cell cavity contains free water, whereas the cell walls of wood (cellulose and hemicellulose) contain bound water. Bound water, on the other hand, is repressed in wood's chemical constituents, which contain hydroxyl groups that generate strong intermolecular hydrogen bonds. As a result, drying is necessary to lower the moisture content of wood [41]. The moisture content of wood biomass lowers its overall calorific value.

Meanwhile, the anatomy of wood has an important impact on the pace of combustion [42]. Wood is mostly treated with FRs [43], which are usually inorganic salts, e.g., mono-diammonium phosphate, zinc chloride, ammonium sulfate, boric acid, sodium tetraborate, and other compounds. The use of refractory salt is applied to materials intended for interior applications only because it is not stable to washing with water [37]. Furthermore, some lignocellulosic plants have developed refractory behavior [30]. Due to their intrinsic capacity to form a thermally stable charred residue when engaging with fire, cellulose and lignin, which are the major constituents of lignocellulosic plants, have certain potential when it comes to their use as FRs additives [44].

Wood, due to its organic nature, is a combustible material. The burning rate of wood is determined by its density, air oxygen concentration, wood moisture content, and heat flux, and it is one of the most vital aspects of fire behavior [45]. The combustion rate refers to the rate at which a specific material is reacted by fire. It can be expressed in terms of mass loss, heat release, or char generation [46]. The process of heat transport in a charred wood sample is presented in Figure 1. When wood is subjected to heat, the surface temperature rises to the point where moisture content is removed, and the constituents of wood (lignin, cellulose, and hemicellulose) begin to decompose at a temperature of 160–180 °C. The pyrolysis and flame combustion of wood occur at temperatures greater than 225–275 °C. If given a spark, wood can burn at 350–360 °C, and the deterioration process begins with the development of a charred layer [30], while carbonization occurs within the range of 500–800 °C.

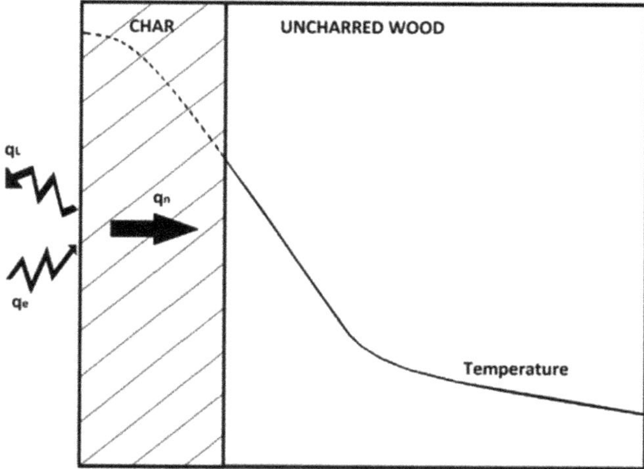

Figure 1. Heat transport in a charred wood sample [30]. Copyright @ 2017 Elsevier, License Number: 5157971091655.

2.2. Non-Woody Biomass

Non-wood fibers and wood fibers are the two types of natural fibers (Figure 2). Material obtained from agricultural waste or non-wood plant fibers is known as lignocellulosic biomass. The worldwide availability and biodegradability of lignocellulosic fibers, their low cost compared to synthetic fibers, and good mechanical properties, have resulted in increased industrial and scientific interest in the context of their wider utilization in the production of biocomposite materials. In addition, the use of lignocellulosic biomass has created new business development opportunities in countries with deficient fossil fuel stocks, which has provided conditions for sustainable development. Biomass obtained from crop residues on farmland or material leftovers after crops have been processed into usable goods is referred to as "agricultural residues". Most agricultural waste is used as fertilizer or animal feed. Meanwhile, to save time and effort, some may be disposed of by burning or landfilling [47].

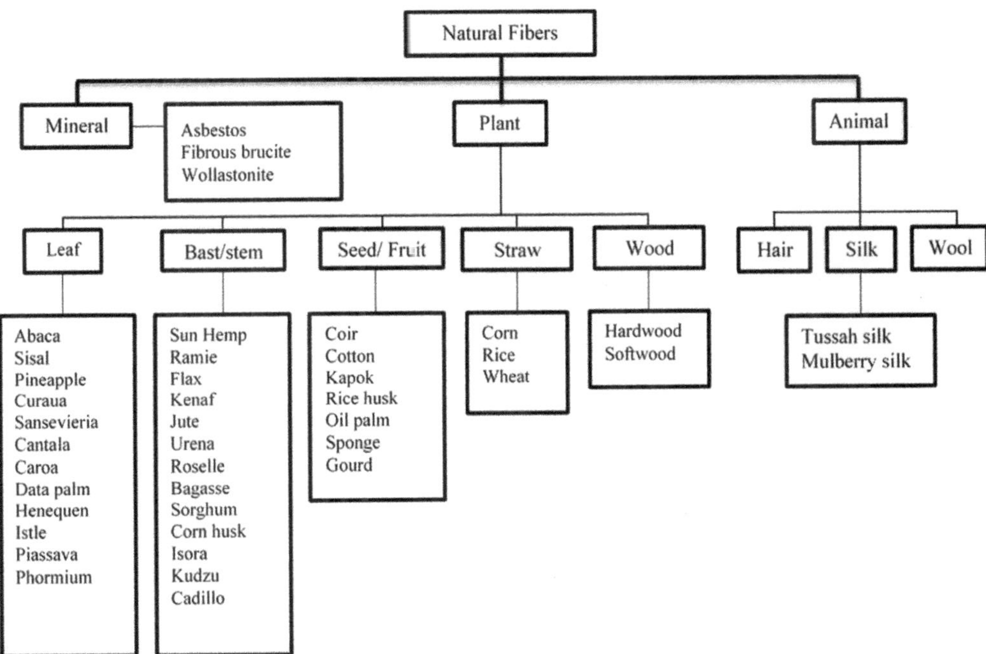

Figure 2. Classification of natural fibers modified from [48]. Copyright @ 2014 Elsevier, License Number: 5212860548284.

Non-wood biomass is a great potential raw material because of its better characteristics and endurance, as well as its ease of modification [49]. Textiles, paper, fabrics, biofuels, and composite reinforcing materials can all be made from natural fibers or non-wood plant species. In the automotive sector, composite reinforcement can be used for packaging, construction, and use [26,50]. Because non-wood fibers are more readily available than wood fibers, they are gaining increased attention as biomass feedstock for bioproducts. Non-wood fibers also have a more open structure, making them easier to process, which results in less processing energy. Furthermore, non-wood fibers are less expensive than wood fibers due to the fact that the majority of non-wood fibers are derived from perennial plants with a predictable supply [51].

Hemicellulose, cellulose, lignin, and pectin are all components of lignocellulosic biomass, which includes both non-wood and wood [52], with the proportion amount varying depending on plant species, tissue, growth stage [53], growth location [54], and axial position [55], as illustrated in Figure 3. Other constituents include extractives, ash, pectin, and waxes [56,57]. Plants are made up of several types of cells with varying physical properties, which are represented in proteins, structural components (polyphenolic compounds, and polysaccharides), and lipids. The presence of stiff cell walls with thicknesses varying from 0 to 10 μm in all plant cells determines their mechanical strength, their resistance to disease, while also influencing cell adhesion properties and the crucial interactions that allow plants to adapt to a variety of environments [58,59]. Natural fibers have a fiber diameter of 10–30 μm and are separated into three main layers: the outside primary cell wall, the inside secondary cell wall, and the outside secondary cell wall [60]. Plant cell walls can govern organ growth as well as the ability to withstand tensile or compressive stresses [59,61].

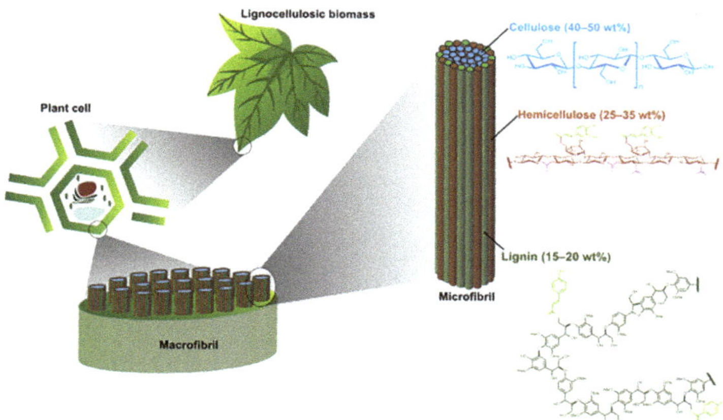

Figure 3. The cell wall structure of lignocellulosic biomass [62]. Copyright @ 2020 Elsevier, License number: 5157980402686.

Cellulose fibers are hydrophilic, which means they absorb water. The moisture level of the fiber can range from 5% to 10%. This can result in dimensional variances in the composite, as well as a change in its mechanical characteristics. Hemicelluloses are responsible for fiber biodegradation, water absorption, and thermal deterioration, while lignin, which is thermally stable, is responsible for UV degradation. Lignin works as a natural adhesive, providing a protective barrier that prevents water and enzymes from accessing cellulose, increasing a plant's resilience to pathogens and biomass breakdown. Some studies have summarized the variation of chemical components including lignin, hemicellulose, and cellulose in natural fiber [1]. Generally, fibers are made up of 40–60% cellulose, 10–25% lignin, and 20–40% hemicellulose [25]. Even though natural fibers have many advantages when it comes to the reinforcement of biocomposites, including annual renewability, lower production costs, good specific mechanical properties, reduced energy consumption during manufacturing, biodegradability, etc., their hydrophilic nature and poor fire resistance has become a limitation when it comes to expanding their range of uses [63]. Due to its low molecular weight, hemicellulose degrades quickly in the presence of heat. Lignin, meanwhile, has a unique highly aromatic structure and a high charring capacity upon heating at elevated temperatures, which decreases the heat release rate and combustion heat of polymeric materials, making it a feasible FR additive option [64]. Together with hemicellulose, lignin contributes to flame degradation properties [65].

The flame retardancy of natural fibers is primarily affected by their chemical composition, as well as their crystallinity and orientation. The characteristics of the resulting natural fiber reinforced composites (NFRCs) are affected by the fiber content, matrix types, filler concentration, compatibilizer, and fiber surface treatment [65].

At temperatures of 200–260 °C and 260–350 °C, respectively, hemicellulose and cellulose begin to degrade. During thermal decomposition, char, volatiles, and gases such as CO, ethylene, and methane are generated. Levoglucosan is generated at temperatures ranging from 280 to 350 °C. As the temperature rises, decomposition produces combustible volatiles, fumes, and carbonaceous char. Lignin is thermally degraded at temperatures of 160 to 400 °C. Bond cleavage takes place at lower temperatures, whereas aromatic ring bond cleavage takes place at higher temperatures [66]. Plant biomaterials have a high degree of biochemical and physical complexity due to the variety in the composition and varying numbers of structural constituents in plant cell walls of diverse species and tissues, which makes the physicochemical characterization of plant biomass difficult [53].

Due to their abundant availability, biomass chemicals hold promise in term of their potential as FRs in polymers. The chemical reaction of cellulose during heat degradation that results in char formation is exceedingly complex and perplexing, and is therefore disputed [36]. Natural fibers can be used as a fuel source, are susceptible to ignition and combustion, and are strongly consumed during combustion [63,67]. Natural fibers have a significant amounts of carbon, hydrogen, and oxygen, making them highly combustible [68]. They are an insulator with high mechanical qualities and a low thermal conductivity of 0.29–032 W/mK. Bark fibers are much less flammable than leaf fibers [68]. Increased thermal stability can be achieved by coating or adding chemicals [69]. The flammability of fibers is affected by their intermediate surroundings, which include the composition of the polymer matrix and other FRs present, the existence of a coupling agent, and the method used to produce the NFRCs. Horizontal and vertical burning tests, cone calorimeter testing, the LOI test, thermogravimetric analysis (TGA), differential scanning calorimetry (DSC), and dynamic mechanical analysis (DMA) have all been used to examine the flammability and thermal behavior of NFRCs [24,70–72].

Carbonization, followed by enhanced char production, is the mechanism of FR treatment of natural fibers [63]. Non-wood fibers are projected to play a larger part in the energy portfolio in the future, despite accounting for the bulk of biomass utilized in fuel generation [47]. Due to their thermoplastic and thermosetting properties, jute, sisal, coir, hemp, banana, bamboo, kenaf, sugarcane, flax, and a range of other natural fibers are used as a reinforcement alternative in polymer composites [64]. Due to their low lignin content, flax fibers have the highest thermal resistance among natural fibers, as measured by a long period before flashover and the duration to ignition. Meanwhile, jute fiber composites have the shortest duration but the fastest spreading fire with the least amount of smoke emission. The reduced smoke is a significant advantage because it diminishes the principal hazards of fire [73].

Vahabi et al. [24,28] have described a general mechanism for FR polymers in which they decompose with some activities in the condensed and/or phase phases, depending on the chemical composition of the polymer matrix and its chemical interaction with it. Modifying the decomposition pathway of polymers to create fewer combustible volatiles and more char, resulting in the production of a barrier or protective layer on the polymer's surface, the cooling effect, and melt dripping are all achievable in the condensed phase. Some FRs aid in the production of polyaromatic structures and intermolecular processes during burning, resulting in carbonaceous char. The barrier effect is a well-known property of condensed phase solutions. In another piece of research, Nah et al. [74] have stated that FRs can act in both chemical and physical ways, e.g., by reducing flame spread, by raising the ignition temperature, by reducing the rate of burning, by cooling, and by forming a protective layer.

Some techniques, such as the chemical alteration of polymer matrices, have been used to provide flame retardancy. A phosphorus-containing reagent was used to chemically modify poly (vinyl alcohol) [75]. Aside from that, the FR coating of composites can be done in a variety of ways, such as using UV-curable boron in hybrid coatings or by using plasma coating techniques. Micro or nano FR incorporation in materials has also been reported to improve the flame retardancy and thermal properties of polymers [65]; however, the mechanical properties of the composites decreased [76]. To manage these qualities, suitable FR filler distribution, surface treatment, and compatibilizer addition are used [77,78].

2.3. Development of Lignin-Based FRs

Due to sustainability and environmental issues, the use of bio-based and renewable polymers and additives to improve fire retardancy has significantly evolved in recent years. There are two types of bio-based FRs: those that arise from biomass, such as lignin, starch, phytic acid, cellulose, tannins, proteins, and oils, and those obtained from animal

DNA and chitosan [24,25], as presented in Figure 4. In recent years, lignin has attracted considerable attention in the context of promoting the FRs of polymers [79]. Lignin has a high thermal resistance, so it has great potential as an FR additive. It can also be effectively used as a carbon source for the design of intumescent systems in combination with other FR additives [80]. Numerous studies have demonstrated that using lignin or lignin derivatives can enhance the mechanical and thermal properties of polymeric materials [81]. The capacity of lignin to act as a flame retardant additive for polymers is highly dependent on its heat stability and ability to generate char [80]. The types and sources of lignin have a direct effect on their thermal decomposition behavior, which is usually characterized by a primary decomposition temperature range between 160–500 °C [82] and the fact that it produces a thermally stable product (char) at 700 °C [83]. The combination of starch or lignin with ammonium polyphosphate (APP) decreases LOI to an acceptable value (above 32%) [84]. Due to the fact that aromatic functional groups have varied thermal characteristics, observations regarding the thermal degradation of lignin cover a varied temperature range [85]. Table 1 summarizes a review of the literature on the development of lignin-based additives.

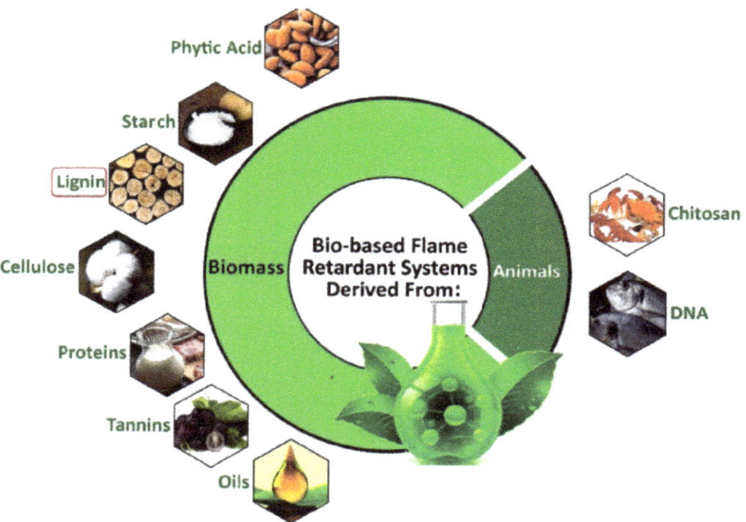

Figure 4. Schematic of various bio-based flame retardants from two main sources; biomass and animals.

Table 1. Development of lignin-based refractory additives from various biomass sources.

No	Method	Findings	Citation
1	Lignin in the form of LL and DL was combined with FR phosphinate (zinc phosphinates (ZnP) and aluminum phosphinate (AIP)) in Polyamide 11 (PA 11).	The mixtures PA80–LL7–ZnP13 and PA80–LL10–ZnP10 showed that these mixtures had the best fire resistance. Sulfonates were identified to play a role in the synthesis of thermally stable molecules in LL.	[85]
2	KL was soaked in a solution of ammonium dihydrogen phosphate (ADP) and urea in a ratio of 1:10 at 70 °C for 1 h.	Phosphorylated KL (PKL) can increase the value of Ti (ignition time), which is 399 °C, which means that PKL has a superior thermal stability and FR qualities to kraft lignin	[86]

Table 1. *Cont.*

No	Method	Findings	Citation
3	A mixture of LS and ZnP in a polyamide 11 (PA11) matrix was melted under pressure to form pellets and plates with a thickness of 3 mm.	A mixture of alkaline lignin with low sulfonate groups and ZnP can effectively improve the refractory properties of PA 11. Mixtures containing 10% alkaline lignin with low sulfonate groups and 10% ZnP showed the best performance.	[87]
4	Lignin was modified for phosphorus and nitrogen grafting. Melting procedure: a mixture of lignin and molten polylactic acid (PLA) was mixed at 160 °C for 7 min at 700 rpm per minute. Previously, before usage, the lignin and PLA were dried overnight in a vacuum oven at 60 °C (control PLA and composite PLA with 20% lignin type, i.e., KL and OL).	The thermal stability of OL is relatively small compared to KL in PLA-lignin composites because OL contains a lot of carboxylic acids and phenolic acid groups, which cause the high degradation of PLA during the melting process. The use of lignin as a refractory ingredient in PLA was proved to reduce the heat generated during combustion by forming a charcoal layer on the burned sample's surface. The flammability of the PLA composite was lowered by lignin that had been changed by grafting phosphorus and nitrogen.	[88]
5	Alkaline lignin, ammonium polyphosphate (APP) and zein powder were tested. Lignin was dissolved in polyethylene glycol (PEG), then zein powder was added and mixed. Plastic was thermally molded (70 °C, 500 rpm for 10 min).	Lignin and APP additives can improve mechanical properties and refractory properties. By performing a vertical burning test and simultaneously degrading lignin, lignin can increase the FR characteristics of thermo-plasticized zein (TPZ). The best biocomposite is 10% APP and 3% Alkali lignin (AL), which has good tensile strength and stiffness	[89]
6	Functionalized lignin (F-Lignin) was created by dissolving 5.0 g lignin, 0.081 mL formaldehyde and 1.38 g DEP in 50 mL DMF following the addition of HCl, then put in 5 g of Cu $(Ac)_2$ and washed and dried at 80 °C. Composite wood-polypropylene (Wood/PP) and wood-PP-lignin were made by melt compounding at 180 °C for 10 min at 60 rpm.	In the presence of FRs components (P and N), F-lignin is considerably better in improving the thermal stability and fire resistance of plastic composites, as well as the catalytic activity of Cu^{2+} on char production. F-lignin lowers the rate of smoke during combustion. The char residue increases and compact char is formed continuously during firing which is responsible for the increased fire resistance.	[90]
	Phenolization AL was made by mixing lignin alkali with H_2SO_4. Phenolated lignin (Ph-lignin) and melamine ammonium polyphosphate (MELAPP) were mixed to produce F-lignin-APP. FRs of EP composites were made by mixing F-lignin with an EP homogeneous, then adding diaminodiphenylmethan (DMM). A vacuum was applied until the resin is mixed.	F-lignin@APP applied to epoxy resins (EP) has strong fire resistance, while AL added to EP has less good fire resistance. F-lignin@APP can increase LOI values on EP, improve smoke suppression (9.9 m^2) and lower the HRR (53.1 MJ/m^2).	[91]
	Enzymatic hydrolysis lignin-based FR (Lig) synthesis was modified with nitrogen and phosphorus. Lig was added to EP to make Lig/EP composites	Composite Lig-F/EP with high phosphorus content has the best fire resistance, and the ul-94 test results achieved an excellent V-0 smoke suppression rating with a significant reduction of THR, 46.6%, and smoke production, 53%. The increase in fire retardant quality was caused by the synergistic action of nitrogen and phosphorus.	[92]

In the forced combustion test, the interaction between lignin and zinc phosphinates dramatically reduced the peak of heat release rate (PHRR), by 74%, and the total heat release (THR) by 22% in the mixture of lignosulfonate (LS) and kraft lignin (KL) [85]. Based on DSC and TGA test findings, KL impregnated with $NH_4H_2PO_4$ and urea solutions raised the main degradation temperature (T_{max}) from 541 °C to 620 °C and glass transition temperature (Tg) from 176 °C to 265 °C. The ignition time (T_i) values increased by 339 °C, suggesting weaker thermal stability and fire resistance than KL itself [86]. After cone calorimetry testing, the combination of alkaline lignin with low sulfonic groups (LS) and ZnP in the polyamide matrix 11 (PA11) may reduce the PHRR value by approximately 50%, the THR value by about 13%, and the maximum value (MARHE) by 35% [87].

The chemical modifications of two lignins, kraft lignin (KL) and organosolv lignin (OL), by grafting phosphorus and nitrogen reduce the PHRR and THR by around 21% and 23% for KL with poly lactic acid (PLA) of 20%, respectively [88]. Functionalized lignin (F-lignin) with phosphorus-nitrogen grafting and a metal element (Cu^{2+}) was used in the study of Liu et al. [90] to increase fire resistance and thermal stability. This reveals that PHRR values declined by 9%, THR values decreased by 25%, and average mass loss rate (AMLR) values reduced by 19%. Lignin can be added to the poly propylene (PP) matrix to act as an FR and toughening agent [90]. The use of alkali lignin (AL) in epoxy resins did not demonstrate a good FR in a study [91] since the carbon supply is solely in AL. The application of lignin as an FR in polymer composites is still falling short of industrial standards, such as its high LOI value > 28%. Traditional FRs such as APP, boric acid, and ammonium dihydrogen phosphate (ADP) can be employed as additives in the polymer matrix with lignin to achieve a high LOI while lowering the PHRR [93].

3. Challenges

3.1. Fire Retardant Agents

Compounds containing halogens (i.e., bromine or chlorine), nitrogen, phosphorus, borax, boric acid, or inorganic metal compounds are often employed in refractory materials in the wood [94]. Ammonium chloride, ammonium sulfate, mono- and di-ammonium phosphate, boric acid, borax, calcium, zinc, aluminum chloride, and magnesium are among the key chemical constituents found in most FRs chemicals now on the market [94]. Coating, thermoplastics, thermosets, rubbers, and fabrics all have FR additive qualities that provide fire resistance. FR additives can prevent, minimize, and stop combustion, with the additive breaks the combustion cycle thereby reducing the burning rate of the fiber and, in cases, extinguishing the flame [95]. FRs are chemicals introduced to materials to decrease fires, increase thermal stability, control the spread of fires, or put a stop to combustion [96]. Table 2 shows the mechanism of action of FRs [97] when subjected to heat or when a fire occurs.

Due to its chemical makeup, polypropylene is readily flammable. Incorporating FRs into the polymer is one way to minimize its flammability. A neutralizing intumescent flame-retardant agent (NIFR) was produced and tested in polypropylene (PP) and found to be extremely effective as a fire retardant [98]. In situ liquid ring-opening polymers of both the ε caprolactam with a combination of ED (6,6′-(Ethane-1,2-diylbis(azanediyl)) bis(dibenzo[c,e][1,2]oxaphosphinine-6-oxide), NED (6,6′-(1-(2-Naphthyl)ethane-1,2-diyl)bis (dibenzo[c,e][1,2]-oxaphosphinine-6-oxide), and PHED (6,6′-(1-Phenylethane-1,2-diyl)bis (dibenzo[c,e][1,2]-oxaphosphinine-6-oxide) materials yielded the polyamide 6 (PA6) system, and the schematic for the preparation of PA6/FR nano-dispersed system can be seen in Figure 5. The advantage of using PA6 which is synthesized in situ from the three materials is that it has a uniform distribution, and the UL 94 test results for PHED material in the PA6 matrix show that it has an excellent fire-resistant effect that can decrease and prevent ignition [99].

Table 2. Important mechanisms of action on FRs.

Flame Retardants	Mechanism
Halogen-based flame retardant	The flame retardant mechanism is based on a radical reaction that acts on the vapor phase
Decabromodiphenly oxide, hexabromocyclododecane, and other bromine-based compounds	Bromine gases, FRs that frequently synergizes with antimony trioxide, can protect materials from oxygen and heat exposure while also limiting the chemical processes that occur during the condensed phase of combustion
Non-halogenated FRs	The char formation mechanism, acting on the flame inhibition of the condensed phase
Ammonium polyphosphate, sodium phosphate, and other phosphorus compounds	The material forms charcoal mainly as an automatic extinguisher by inhibiting oxygen contact and can protect against flammable gases gas
Compounds containing antimony, such as antimony trioxide	Capable of increasing charcoal production and scavenging free radicals through synergism with halogens
Metal hydroxide-based compounds, such as magnesium and aluminum	It works better as an insulator, slows the flame at high temperatures, and can dissolve smoke, making it safer for humans and the environment
Zinc borate, for example, is a boron-based chemical	Smoke output is reduced, and chars are produced
Melamine and its salts are nitrogen-based chemicals	This shows that phosphorus and nitrogen work together to generate char
Silicon-based compounds such as silica etc.	It provides a synergistic effect with APP by forming a layer of charcoal and silicon, also known as inert diluents

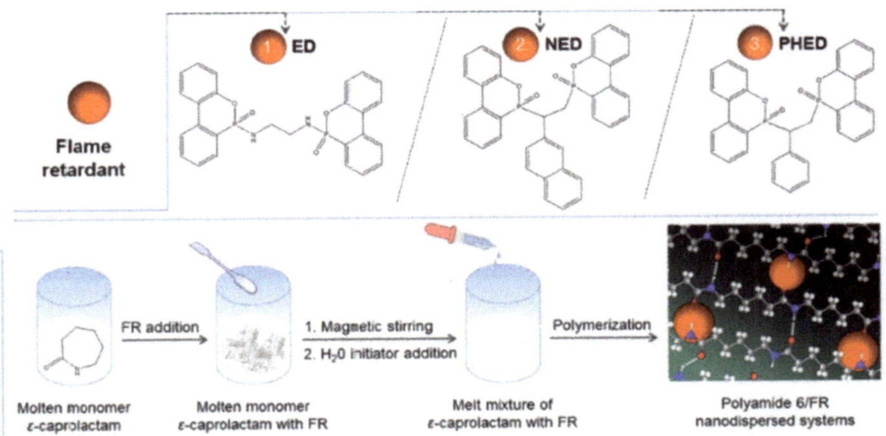

Figure 5. The fabrication method for the FR polyamide 6 (PA6)/FR nano-dispersed systems is based on the molecular structures of the flame retardants (FRs), i.e., ED, NED, and PHED [99] Copyright @ 2020 MDPI under CC by 4.0.

Wood flour in wood plastic composite (WPC) with 30% ethanolamine (ETA-APP) (modified ammonium polyphosphate) can provide fire-resistant characteristics while also increasing flexural properties. Figure 6 presents the probable creation method of WPC/30 wt.% ETA-APP. The ETA-APP produces phosphoric acid, which catalyzes cellulose chain scissions into many tiny molecules that may be destroyed quickly by heat. The release of NH_3 and H_2O results in the creation of stable molecules such as P-O-C, P-N-C, and CN, which react with hydroxyl cellulose to create a stable carbon layer and oxygen barrier. Polyoses, hemicelluloses, and lignin are all cross-linked [100].

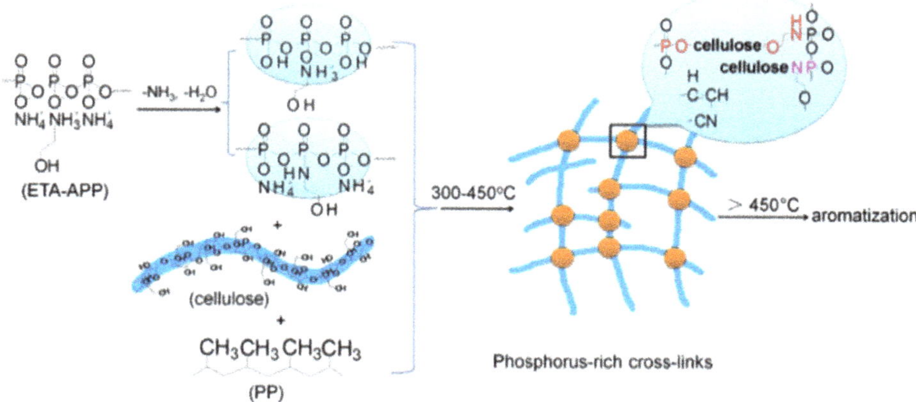

Figure 6. Pyrolysis process of WPC/ETA-APP biocomposites (30 wt.%) [100]. Copyright @ 2015 American Chemical Society under CC.

In some reports, Cu_2O acts as a synergist EDA-APP for epoxy/ethanediamine-modified ammonium polyphosphate (EP/EDA-APP) systems. Cu_2O also acts as an adjuvant with EDA-APP in terms of increasing the quantity of char produced, intumescent degree, and compactness [101]. The EDA-APP may be utilized to make refractory neat EP composites. Cu_2O can help to decrease smoke and carbon monoxide emissions. Figure 7 shows a schematic of the FR and suppression of smoke mechanism of EP/EDA-APP/Cu_2O. Up to a temperature of 450 °C, phosphoric acids are formed and Cu_2O hastens the breakdown of EP and EDA-APP, which can aid in the production of charcoal as a flammable barrier and reduce smoke and hazardous gas emissions [101].

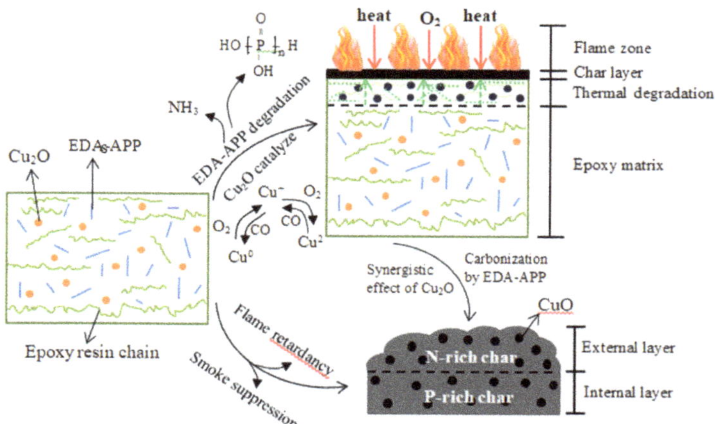

Figure 7. EP/EDA-APP/Cu_2O composite's potential flame-retardant and smoke-suppressant mechanisms. Modified from Chen et al. [102]. Copyright @ 2017 CC BY-NC 3.0.

The use of APP can be used as an FR; the material reacts with carbon compounds that can form charcoal as a protective layer which can prevent the further spread of fire [103].

Magnesium hydroxide ($Mg(OH)_2$) is a chemical compound. It is also an FR substance which is capable of slowing the flame by releasing water at a temperature of 360 °C [104]. The addition of APP or APP combinated with zinc borate applied to sisal/polypropylene composites was able to increase fire resistance, thermal stability, and did not decrease their mechanical properties [105]. This coating technique can assist in increasing the composite's fire resistance. It is applied during the finishing step or by impregnation [106]. Several chemicals, such as boron phosphate and silicon, have been found to improve the fire resistance of epoxy resin systems [107–109].

A method that is widely used in the context of adding active compounds to polymers is the addition of thermal (hydrated oxides) or inert fillers (silica, talc) to make less-flammable composite reinforced natural fiber [97]. The creation of several types of FRs from natural fiber reinforced polymer composites is depicted in Table 3.

Table 3. Examples of fire-retardant natural fiber-reinforced polymer composites.

Polymers or Reinforcement Materials	Flame Retardants	Property Improvement	References
Wood fibers/PP composite	Silica and APP	APP and silica are excellent fire retardants for wood fiber/PP composite. Apart from tensile strength, the mechanical characteristics of the composites degraded after flame retardants were introduced.	[110]
Sisal/PP composites	Zinc borate and $Mg(OH)_2$	The addition of FRs to sisal/PP can slow down the process while raising the temperature. The addition pf $Mg(OH)_2$ and zinc borate to the sisal/PP composite can improve its fire resistance while not affecting its mechanical characteristics.	[77]
Cotton fabric/epoxy	Montmorillonite (MMT)	The thermal properties and flammability of the cotton fabric composite improved after treatment based on TGA study, vertical flame, and oxygen index analysis. There was no residue from the combustion on the control cloth, but on the MMT-treated cloth there was still some residue left.	[86]
Binder (flax short fibers/pea protein)	Some of the materials utilized in the manufacturing of aluminum tri-hydroxide include melamine phosphate (MMP), zinc borate (ZB), and melamine borate (MMB)	Using a protein binder, fire-resistant chemicals were integrated into insulating materials made from flax short fibers. MMB with 20 wt.% shows an increase in flame retardancy behavior.	[111]

3.2. Manufacturing FRs

Methods for preparing FR-treated natural fibers include the insertion of FR into adhesive and the mixing of fibers with FR before the addition of an adhesive [63] during the preparation process for biocomposites. Xiong et al. studied the development and application of refractory adhesives. In this research, a new design was studied to optimize the quality of fire-resistant adhesives as decorative panels for household needs. The orthogonal test determines the proportion of the adhesive with fire-resistant additives and the number of layers so that the fire-resistant adhesive can improve the performance of fire-resistant film-coated wood boards in the production process. The techniques of dyeing, peeling, strengthening the surface bond, and releasing formaldehyde follow national and industry standards. These techniques are also superior to the previous conventional technique, which used a panel wood-based veneer soaking technique by first using fire-resistant additives and then pasting them. The advanced technique can be used simultaneously with the production and treatment of refractory materials and their adhesives. Other advantages of this method are the emission of fewer materials used so that environmental pollution can be minimized, also the reduced use of FR additives and adhesives. Hence, the costs used are cheaper, and the production process is simple, easy, and eco-friendly [112].

Polybrominated diphenyl ethers (PBDEs) are materials originally utilized in consumer products such as foam furniture, padding, and electronics [113]. PBDEs are FRs that have been utilized in everyday products to prevent the spread of fire. They are added to numerous consumer products, including electrical circuits, building materials, thermoplastics, polyurethane foams, and other products as one of the most often used brominated flame retardant (BFR) classes [114].

As a result of the introduction of the flammability requirement, and the rising usage of synthetic materials, halogenated FRs have been in use since the 1940s, with a fast increase in demand and manufacturing since then. The growing demand has been satisfied by the development of new compounds with improved fire-resistant qualities [115]. PBDEs are one of the most frequently employed organic FRs, and they are found in a wide range of polymers used in construction products, consumer goods, and automobiles [116]. Furthermore, with the discontinuance of penta-, octa-, and decabrominated diphenyl ethers (BDEs), other "new" FRs (NFRs) were utilized in greater quantities to satisfy flammability regulations [117]. Table 4 lists a variety of commercial polymer refractory materials used in the fabrication of additives that meet the majority of fire safety criteria [23].

Table 4. Flame-retardant polymers for additive manufacturing.

Name	Material	FR Additives	Machine
PA 2210 FR	PA 12	Phosphorus	P 385, P380i, P 380, P 360, P 350/2, P 700
PA 2241 FR	PA 12	Halogen	EOSINT P395/760/390/730
PA 606-FR	PA 12	Not identified	Unspecified
FR-106	PA 11	Not identified	Unspecified
DuraForm ProX FR 1200	PA 12	Not identified	ProX SLS 500
DuraForm FR 1200	PA 12	Unknown	sPro 60 HD-HS
DuraForm FR 100	Unspecified	Halogen free	3D systems
PEEK HP3	PEEK	-	EOSINT P800
ULTEM 9085	PEI	-	Fortus 400 mc/450 mc/900 mc
ULTEM 1010	PEI	-	Fortus 450 mc/900 mc
PPSF	PPSF/PPSU	-	Fortus 400 mc/900 mc

Note: Poly ether ether ketone (PEEK), polyphenylsulfone (PPSF), polyethylenimine (PEI): selective laser sintering (SLS), polyamide (nylon) (PA): polyetherimide (ULTEM): polyphenylsulfone (PPSU).

Other thermally stable polymers can be employed in fire-prone situations, in addition to refractory materials. According to the flow chart in Figure 8, refractory additives (chemical FR) and materials with good heat stability are identified. There are three groups of chemical FRs, including halogen-based, phosphorous-based, and nitrogen-based FRs. Because flammability is not dependent on thermal stability, thermally stable polymers are not necessarily refractory [23].

3.3. *Safety of FRs*

Non-halogenated and halogenated FRs are the two types of FRs. Due to their durability, bioaccumulation, and potential human health impacts, halogenated flame retardants (HFRs) containing bromine or chlorine linked to carbon,] have gotten a lot of attention [96]. Halogen-based compounds are the most prevalent refractory additives on the market because they are cheap and effective. However, due to the environmental and toxicity problems connected with halogen-based refractory additives, several halogen compounds have been banned. The chemical interference with a radical chain mechanism in the gas phase during burning was used to prevent fires [118]. HFRs are used to lower the flammability of produced materials such as textiles, plastics, furniture, and polyurethane foam to inhibit the spread of fire [5].

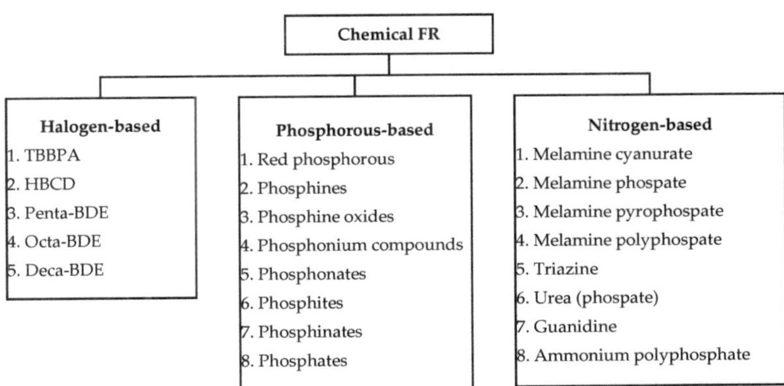

Figure 8. Flame-retardant additives. Modified from [23]. Copyright @ 2020 Elsevier, License number: 5157990641808.

Many home and commercial products contain FRs to minimize the fire consequences and losses, but the commonly used FRs, polybrominated diphenyl ethers (PBDEs), have negative consequences for human health and the environment [79]. To replace PBDEs, alternative flame retardants (AFRs) have been developed. The two primary components of AFRs, for example, are di-(2-Ethylhexyl)-tetrabromophthalate (TBPH) and 2-Ethylhexyl-2,3,4,5-tetrabromobenzoate (TBB), which is currently one of the most extensively used commercial fire-retardant mixes. A 1,2-bis (2,4,6-tribromophenoxy) ethane (TBE), pentabromobenzene (PBBZ), pentabromoethyl benzene (PBEB), and hexabromobenzene (HBB) are some of the other AFRs [117]. According to recent studies, the alternative level of FRs in the air has risen to levels comparable to PBDE [119]. Furthermore, due to their low cost, refractory additions to polymers containing halogen elements are frequently employed. Halogens, on the other hand, are reactive, as they can produce harmful and corrosive fumes. Due to their availability and low cost, phosphate-containing compounds such as phosphate are also utilized as an alternative to halogenated FRs. However, following further investigation, it was discovered that this refractory addition also possesses hazardous qualities [120].

At this time, a substitute for dangerous compounds in FR additives is required that performs similarly and may be employed in a safe and environmentally responsible manner [96]. The usage of polymeric composite materials that have not been treated with FR is hazardous to human safety [121]. Natural fiber/polymer composites are becoming more popular, and FR development must be evaluated in terms of human safety, human health, and environmental friendliness. Natural fiber has been more widely employed in the automobile industries and packaging [122], where fire safety regulations are less severe than in the aerospace industry. The flammability qualities and fire retardance function of biocomposites must be examined to widen their range of applicability for additional advanced manufacturing applications such as aerospace, electronics equipment, and building [63]. Several types of FRs are utilized in refractory goods, including aluminum, antimony, phosphorus, chlorides, bromides, and boron-containing compounds [123]. Metallic hydroxides, including magnesium hydroxide ($Mg(OH)_2$) and aluminum hydroxide ($Al(OH)_3$), and are commonly employed materials that are both safe for humans and environmentally benign [124]. Magnesium hydroxide (MgH), and aluminum hydroxide (ATH) were the refractory fillers utilized by Mohapatra et al. [120]. Non-halogen and non-phosphorous refractory additives were employed in this study. Inorganic fillers are becoming more important in industry due to their favorable mixture of low smoke, low cost, and reasonably good refractory efficiency.

4. Potential

4.1. Economic

Fire is one of humanity's greatest creations, but every year, approximately 4000 people in the United States and 5000 people in Europe pass away in fires and approximately 0.3% of the gross domestic product is lost as a result of fires [125]. In other countries, such as South Korea, the National Emergency Management Agency has reported a six-fold increase in large-scale fires (10 casualties, 5 lives lost, and $4 million in destruction of property) and in related damage (varying from 45 deaths to 232 deaths in 2019) and property costs (ranging from $5 million and $330 million) from 2010 to 2020. Mechanical and electrical problems are the primary causes of fires, contributing to large-scale fires [126,127], in addition to human error when ensuring the safety of electrical equipment in domestic buildings. Poor quality inexpensive electrical plugs and outlets are widely available in electrical stores and are primarily purchased by householders and small home industries due to their low cost [127]. During a fire, those who were near charred furniture may suffer burns, smoke, and poisonous gas inhalation, such as CO and other gasses (found in the soot in their airways), which may result in them becoming serious victims [128]. Aside from that, in several major cities in Indonesia, the construction of houses in densely populated areas is still largely based on wood [129] that vulnerable to fire. Residential settlements with very tight building distances make it easier for fire to spread and the type of flammable building (Regulation No. 14, 2012) affects fire vulnerability [127]. Throughout 2018, there were 698 cases of fires in Jakarta (DKI Jakarta Provincial Fire and Rescue Service, 2017) which counts as evidence for the high number of fires in Indonesia. Based on some data, the fire protection on the materials used in building construction, including furniture, becomes crucial even though it is very challenging to manage. The adverse effects of fire hazards can be mitigated by providing fire safety in buildings. The aim is to enhance fire safety through the establishment of rational fire design approaches, cost-effective fire removal technologies identifying innovative materials, defining performance codes, and evaluating wildfire fire hazard [130].

FRs, which are widely used additives in the plastics industry, are in high demand in the current global market. The market share of these agents is expected to be $2.3 billion [131]. Since the 1930s, halogenated FRs technology has been in use, demonstrating that FRs technology prices are quite low. Furthermore, boosting the fire resistance of composite materials could minimize the severity of incidents like aviation accidents [121], because the usage of composites without FR treatment is harmful to human safety. Typical biocomposites must be adjusted in several ways, such as chemical modification, to meet the high requirements and environmental demands of the circular economy. However, the higher flammability of biocomposites reinforced with lignocellulosic biomass when attacked by a heat flux or flame source can limit the wide uses of biocomposites. At 300–500 °C, cellulose-based polymers decompose into gas and condensed phases, resulting in incombustible liquids, char, gases, and smoke with potentially hazardous dripping. Therefore, some FRs have been introduced to decrease this negative effect associated with biocomposites' easily flammable properties.

Nano-sized FR agents with a low or non-toxic environmental impact are increasingly used to improve the fire performance of combustible biocomposites [66]. They greatly limit the heat release rate, delay ignition, and slow flame propagation by adding 2–10% to the total material weight [121]. The challenge in developing FR biocomposites is how to maintain mechanical strength performance [132]. Furthermore, the usage of FRs raises the cost of the finished product [133]. Halogen-based FRs and nano-sized FRs have been known to advance the flame retardancy of composite even at lower concentrations compared to metal hydroxide compounds. However, because of their environmentally hazardous effect, their utilization is being banned. Intumescent FRs are new kinds of FR materials that considerably enhance flame retardancy, although further study is needed to improve this type of material [121].

4.2. Environmental

The flammability and post-ignition fire behavior of material and goods are increasingly regulated. The toxicity of FRs and FR-treated materials' thermal composite products was not paid much attention in the context of them being heated, burned, or combusted for waste disposal, and there is a huge and growing tonnage of FR compounds in all stages of their life cycle. In recent years, toxicity issues concerning FRs have been a growing concern. One problem is that the small amounts of very dangerous combustibles that are emitted during unintentional fires and garbage incineration may cause environmental contamination. Further concerns relate to the occupational and environmental dangers in processing and recycling them, and technological problems in the recycling of particular materials [134].

Developing FR polymers with the potential to generate novel fire-safe materials is crucial. Some current FRs are subject to some regulations due to multiple environmental harmful effects. Despite the challenges, however, FR technology is progressing, and future fire-resistant polymer materials can reduce fire risks [126]. Globally, living standards and fire safety, including the prevention of ignition and flame propagation and extended escape times, are increasing because of the increased consumption of FRs [135]. Recently, the tendency has been towards them being more environmentally compatible. Concerning FR utilization, persistence, bioaccumulation, and toxicity (PBT), chemicals are a source of concern because they are degradation-resistant and can remain this way in the environment [136] and have been proven to be hazardous to humans and wildlife in recent years [137]. They have been known to be very effective and inexpensive for FR additives into the polymer. However, they pose toxicity issues during fires due to the formation of smoke, asphyxiants, irritants, and the direct discharge of irritating acid gases. Some environmental contaminations may have happened as a result of the emission of halogenated dioxins and dibenzofurans [134]. Halogenated FRs have greater levels of emission of corrosive fumes and gases during fires, possible environmental leaching, and are difficult to recycle [138]. Polymer composites and/or mixes have been used in some applications where fire danger and harm to humans and structures are key issues [63]. As a result, some efforts have been made to substitute them with more environmentally friendly FRs, such as non-halogenated FRs.

Melamine polyphosphate (MPP), a non-halogenated FR, has been demonstrated to be effective as an FR for bio-based thermoset polymer systems while having a minimal environmental impact in terms of toxicity [139]. Compounds that contain phosphorus are quite effective at preventing fire, are eco-friendly, and not very hazardous, especially when it comes to high-oxygen polymers [140]. Organophosphorus fire retardants are non-toxic and ecologically benign [141,142]. They have significantly lower toxicity than their organohalogen competitors, and may be made more effective by adding other components like sulfur, boron, nitrogen, or silicon [143]. In matrices which have an oxygen or nitrogen atom backbone, phosphorus and nitrogen-based chemicals are potential solutions when it comes to FR additions [65]. FRs containing nitrogen are one of the most environmentally friendly in terms of chemicals since they produce less smoke and produce no dioxins or halogen by-products during burning [137]. Nitrogen compounds are less effective than organohalogen (since phased out) or organophosphorus compounds at diluting the fuel load in the combustion zone by releasing inert fragments into the gas phase. Because nitrogen-based FRs decompose at a high temperature, they may be successfully integrated into thermoplastic polymers. When mixed with melamine phosphate and ammonium phosphate, they create a synergistic system [144]. Organophosphorus FRs derived from plants represent a promising solution for the development of biobased FRs due to their abundance, renewability, and non-toxicity. Isosorbide, a dihydroxy ether, is included in the category of organophosphorus FRs that can be produced from starch materials via the hydrolysis of glucose following the reduction of glucose followed by double dehydration [145]. It can also be extracted from seed grains [146]. By employing 5-hydoxymethyl-2-furfural (HMF) obtained from renewable resources, furan-based FRs constitute a harmless alternative to organophosphorus. Moreover, furan-based flame retardant (FBF) has a significant char yield and a high LOI [147].

When FR additives can be substituted by chemicals or systems which do not pollute the environment, while the fiber is bio accumulative inactive or toxicologically active, and when prices are comparable to the efficacy of FRs, FR additives can be advantageous [148]. The characteristic features and properties of several standard FR systems and products are summarized in Table 5 [148].

Table 5. Characteristic features and properties of common FR systems.

System	Possibility of Reducing Fire	Reduces Fire Size	FR System or Substance Bioavailability/Environmental Toxic Hazard	Toxic Combustion Product Yields	Toxic Product Yields in theEnvironment
Mineral wool, glass fiber, ceramic fiber, and aramid fiber are examples of inert insulation panels, layers, fillers, and interlinears.	Yes	Yes	Glass fiber and mineral wool are biodegradable if breathed in and have minimal absorption and toxicity. During installation or removal, there are certain minor health risks	There are not or not very harmful or toxic.	There are no or very minor environmental hazards
Coatings that are inflammable	Yes	Yes	There are no detected problems	No	None
Magnesium dihydroxide and aluminum trihydroxide	Yes	Yes, by releasing water.	There are no detected problems	No	None
Boric acid	Yes	Yes, due to the creation of glass.	REACH classification H360FD (may harm fertility): easily released from the substrate. Although there is a chance that the unborn child will be harmed, the risk of exposure is typically low. FR is regarded as a "green" product.	Reduced	Reduced
FRs of phosphorus, and nitrogen combined phosphorus	Yes	Yes, through char formation and in the gas phase.	It is contingent on the compound and the application's durability	Some reduced	Reduced
Phosphorus halogen	Yes	Yes, primarily in the gas phase.	It is contingent on the compound, and the application's durability	Enhanced	Enhanced
Ammonium polyphosphate	Yes	Yes, when it comes to char creation.	No detected issues	Reduced	Reduced
Organic halogens antimony salts	Yes	Yes, helps char in gaseous stages.	Some compounds have been phased out, depending on the chemical.	Enhanced	Enhanced
Inorganic halogens: PVC	Yes	To some degree	There is not any evidence that this is a problem (except VCM during manufacturing).	Enhanced	Enhanced
Fluoropolymers	Yes	Yes	It is unknown whether or not there is a problem.	Under some circumstances, such as HF and PFIB, there are increased severely toxic nanoparticles	Enhanced
Nanoparticle clays and fibers	Possibly	Possibly	Is it possible for nanoparticles to escape during use? What kind of toxicity happens?	There is a reduction in existing toxic compounds, but there is a risk of toxic exposure from aerosolized nanoparticles	Potential issue

Note: Vinyl chloride monomer (VMC), Registration, Evaluation, and Authorization of Chemicals (REACH), Perfluoroisobutylene (PFIB), Hydrogen Fluoride (HF).

5. Conclusions

Increased environmental awareness, the scarcity of non-renewable resources, and recent technological advances have enhanced the industrial development of biobased composites with engineered properties for a wide range of value-added end uses. However, their flammability limits their broader use in more advanced applications. To improve fire protection in biocomposites, a wide variety of FR additives are commonly incorporated in their composition, enabling biocomposites with poor fire characteristics to fulfill regulatory fire performance criteria, widening their range of applications. This paper outlined the recent advancements in the development of fire-resistant biocomposites, providing an analysis of the flammability of woody and non-woody biocomposites. Due to its abundant availability and good thermal properties, the potential of lignin-based FRs in biocomposites as a 'green' alternative to the traditional FR compounds was also highlighted. Manufacturing biocomposites with FR properties, as well as their production properties and safety considerations, were described. Furthermore, the effects of incorporating FRs into biocomposites on the economy as well as their environmental impact were also presented and evaluated. Although it might be difficult to develop effective alternatives to the existing FR additives used in biocomposites for some applications, in most cases an improved fire performance can be obtained using biobased FRs with less environmental impact.

Author Contributions: Conceptualization, W.F., A.H.I. and M.A.R.L.; methodology, W.F., A.H.I. and M.A.R.L.; validation, W.F., A.H.I., D.S.N. and M.A.R.L.; formal analysis, M.R.R., E.W.M. and W.F.; investigation, W.F., M.R.R. and E.W.M.; resources, W.F., A.H.I. and M.A.R.L.; data curation, E.W.M., M.R.R. and W.F.; writing—original draft preparation, review, and editing, W.F., M.A.R.L., A.H.I., P.A., L.K., A.M., M.A.A., M.R.R. and E.W.M.; supervision, W.F. and D.S.N.; project administration, A.H.I. and W.F. All authors have read and agreed to the published version of the manuscript.

Funding: The authors are grateful for a research grant from the Deputy for Strengthening Research and Development, Ministry of Research and Technology in the National Competitive Research grant with the title "The Characteristics of Fire-Resistant Wood Panel Composite Based Home Components" from the Deputy of Strengthening Research and Development, Ministry of Research and Technology/National Research and Innovation Agency 2021 Fiscal Year (95/UN5.2.3.1/PPM/KP-DRPM/2021). Thanks are due to the facilities and the scientific and tech-nical support provided by the Integrated Laboratory of Bioproducts (iLaB), Research Center for Biomaterials, National Research and Innovation Agency, through E-Layanan Sains Lembaga Ilmu Pengetahuan, Indonesia. This study is part of Elvara Windra Madyaratri's master's thesis at IPB University. This research was also supported by the Slovak Research and Development Agency under contracts No. APVV-18-0378, APVV-19-0269, and APVV-20-0159, also by the Scientific Grant Agency of Ministry of Education, Science, Research, and Sport of the Slovak Republic (grant number VEGA 1/0714/21) and by Project No.НИС-Б-1145/04.2021, "Development, Properties and Application of Eco-Friendly Wood-Based Composites", carried out at the University of Forestry, Sofia, Bulgaria.

Institutional Review Board Statement: Not applicable.

Informed Consent Statement: Not applicable.

Data Availability Statement: The data presented in this study are available on request from the corresponding author.

Conflicts of Interest: The authors declare no conflict of interest.

References

1. Karimah, A.; Ridho, M.R.; Munawar, S.S.; Adi, D.S.; Ismadi; Damayanti, R.; Subiyanto, B.; Fatriasari, W.; Fudholi, A. A review on natural fibers for development of eco-friendly bio-composite: Characteristics, and utilizations. *J. Mater. Res. Technol.* **2021**, *13*, 2442–2458. [CrossRef]
2. Faruk, O.; Bledzki, A.K.; Fink, H.-P.; Sain, M. Biocomposites reinforced with natural fibers: 2000–2010. *Prog. Polym. Sci.* **2012**, *37*, 1552–1596. [CrossRef]
3. Wackernagel, M.; Yount, J.D. The Ecological Footprint: An Indicator of Progress Toward Regional Sustainability. *Environ. Monit. Assess.* **1998**, *51*, 511–529. [CrossRef]

4. Mohammed, A.S.; Meincken, M. Properties of Low-Cost WPCs Made from Alien Invasive Trees and rLDPE for Interior Use in Social Housing. *Polymers* **2021**, *13*, 2436. [CrossRef] [PubMed]
5. Stevens, R.S.A.; Van Es, D.; Bezemer, R.C.; Kranenbarg, A. The structure-activity relationship of fire retardant phosphorus compounds in wood. *Polym. Degrad. Stab.* **2006**, *91*, 832–841. [CrossRef]
6. Sathishkumar, T.P.; Navaneethakrishnan, P.; Shankar, S. Tensile and flexural properties of snake grass natural fiber reinforced isophthallic polyester composites. *Compos. Sci. Technol.* **2012**, *72*, 1183–1190. [CrossRef]
7. Yusriah, L.; Sapuan, S.; Zainudin, E.S.; Jaafar, M. Characterization of physical, mechanical, thermal and morphological properties of agro-waste betel nut (Areca catechu.) husk fibre. *J. Clean. Prod.* **2014**, *72*, 174–180. [CrossRef]
8. Sanjay, M.R.; Yogesha, B. Studies on Natural/Glass Fiber Reinforced Polymer Hybrid Composites: An Evolution. *Mater. Today Proc.* **2017**, *4*, 2739–2747. [CrossRef]
9. Al-Oqla, F.M.; Salit, M.S. 2—Natural fiber composites. In *Materials Selection for Natural Fiber Composites*; Al-Oqla, F.M., Salit, M.S., Eds.; Woodhead Publishing: Sawston, UK, 2017; pp. 23–48.
10. La Mantia, F.P.; Morreale, M. Green composites: A brief review. *Compos. Part A Appl. Sci. Manuf.* **2011**, *42*, 579–588. [CrossRef]
11. Ho, M.-P.; Wang, H.; Lee, J.-H.; Ho, C.-K.; Lau, K.-T.; Leng, J.; Hui, D. Critical factors on manufacturing processes of natural fibre composites. *Compos. Part B Eng.* **2012**, *43*, 3549–3562. [CrossRef]
12. Zhang, L.; Chen, Z.; Dong, H.; Fu, S.; Ma, L.; Yang, X. Wood plastic composites based wood wall's structure and thermal insulation performance. *J. Bioresour. Bioprod.* **2021**, *6*, 65–74. [CrossRef]
13. Peng, L.; Hadjisophocleous, G.; Mehaffey, J.; Mohammad, M. Fire Resistance Performance of Unprotected Wood–Wood–Wood and Wood–Steel–Wood Connections: A Literature Review and New Data Correlations. *Fire Saf. J.* **2010**, *45*, 392–399. [CrossRef]
14. Östman, B.; Voss, A.; Hughes, A.; Jostein Hovde, P.; Grexa, O. Durability of fire retardant treated wood products at humid and exterior conditions review of literature. *Fire Mater.* **2001**, *25*, 95–104. [CrossRef]
15. Dasari, A.; Cai, G.P.; Mai, Y.W.; Yu, Z.Z. 10—Flame retardancy of polymer–clay nanocomposites. In *Physical Properties and Applications of Polymer Nanocomposites*; Tjong, S.C., Mai, Y.W., Eds.; Woodhead Publishing: Sawston, UK, 2010; pp. 347–403.
16. Kozłowski, R.; Władyka-Przybylak, M. Flammability and fire resistance of composites reinforced by natural fibers. *Polym. Adv. Technol.* **2008**, *19*, 446–453. [CrossRef]
17. Mahmud, S.; Hasan, K.M.F.; Jahid, M.A.; Mohiuddin, K.; Zhang, R.; Zhu, J. Comprehensive review on plant fiber-reinforced polymeric biocomposites. *J. Mater. Sci.* **2021**, *56*, 7231–7264. [CrossRef]
18. Grover, T.; Khandual, A.; Chatterjee, K.; Jamdagni, R. Flame retardants: An overview. *Colourage* **2014**, *61*, 28–36.
19. Seidi, F.; Movahedifar, E.; Naderi, G.; Akbari, V.; Ducos, F.; Shamsi, R.; Vahabi, H.; Saeb, M.R. Flame Retardant Polypropylenes: A Review. *Polymers* **2020**, *12*, 1701. [CrossRef]
20. Zhuang, J.; Payyappalli, V.M.; Behrendt, A.; Lukasiewicz, K. *Total Cost of Fire in the United States*; Fire Protection Research Foundation: Buffalo, NY, USA, 2017.
21. Rashid, M.; Chetehouna, K.; Cablé, A.; Gascoin, N. Analysing Flammability Characteristics of Green Biocomposites: An Overview. *Fire Technol.* **2021**, *57*, 31–67. [CrossRef]
22. REACH. Regulation on Registration, Evaluation, Authorisation and Restriction of Chemicals. No 1907/2006 2006. Available online: https://osha.europa.eu/en/legislation/directives/regulation-ec-no-1907-2006-of-the-european-parliament-and-of-the-council (accessed on 17 December 2021).
23. Lv, Y.F.; Thomas, W.; Chalk, R.; Singamneni, S. Flame retardant polymeric materials for additive manufacturing. *Mater. Today Proc.* **2020**, *33*, 5720–5724. [CrossRef]
24. Vahabi, H.; Laoutid, F.; Mehrpouya, M.; Saeb, M.R.; Dubois, P. Flame retardant polymer materials: An update and the future for 3D printing developments. *Mater. Sci. Eng. R Rep.* **2021**, *144*, 100604. [CrossRef]
25. Vahabi, H.; Brosse, N.; Latif, N.A.; Fatriasari, W.; Solihat, N.; Hashimd, R.; Hussin, M.; Laoutid, F.; Saeb, M. Chapter 24—Nanolignin in materials science and technology-does flame retardancy matter? In *Biopolymeric Nanomaterials: Fundamental and Applications*; Elsevier: Amsterdam, The Netherlands, 2021; Volume 1, pp. 515–550.
26. Gao, R.; Jing, Y.; Ni, Y.; Jiang, Q. Effects of Chitin Nanocrystals on Coverage of Coating Layers and Water Retention of Coating Color. *J. Bioresour. Bioprod.* **2021**, *in press*. [CrossRef]
27. Jiang, X.; Bai, Y.; Chen, X.; Liu, W. A review on raw materials, commercial production and properties of lyocell fiber. *J. Bioresour. Bioprod.* **2020**, *5*, 16–25. [CrossRef]
28. Vahabi, H.; Jouyandeh, M.; Parpaite, T.; Saeb, M.R.; Ramakrishna, S. Coffee Wastes as Sustainable Flame Retardants for Polymer Materials. *Coatings* **2021**, *11*, 1021. [CrossRef]
29. Hussin, M.H.; Appaturi, J.N.; Poh, N.E.; Abd Latif, N.H.; Brosse, N.; Ziegler-Devin, I.; Vahabi, H.; Syamani, F.A.; Fatriasari, W.; Solihat, N.N.; et al. A recent advancement on preparation, characterization and application of nanolignin. *Int. J. Biol. Macromol.* **2022**, *200*, 303–326. [CrossRef]
30. Costes, L.; Laoutid, F.; Brohez, S.; Dubois, P. Bio-based flame retardants: When nature meets fire protection. *Mater. Sci. Eng. R Rep.* **2017**, *117*, 1–25. [CrossRef]
31. Dasari, A.; Yu, Z.-Z.; Cai, G.-P.; Mai, Y.-W. Recent developments in the fire retardancy of polymeric materials. *Prog. Polym. Sci.* **2013**, *38*, 1357–1387. [CrossRef]
32. Ofori-Boateng, C.; Lee, K.T. Sustainable utilization of oil palm wastes for bioactive phytochemicals for the benefit of the oil palm and nutraceutical industries. *Phytochem. Rev.* **2013**, *12*, 173–190. [CrossRef]

33. Cseke, L.J.; Podila, G.K.; Kirakosyan, A.; Kaufman, P.B. *Plants as Sources of Energy*; Springer: Dordrecht, The Netherlands; Heidelberg, Germany; London, UK; New York, NY, USA, 2009.
34. Anwar, Z.; Gulfraz, M.; Irshad, M. Agro-industrial lignocellulosic biomass a key to unlock the future bio-energy: A brief review. *J. Radiat. Res. Appl. Sci.* **2014**, *7*, 163–173. [CrossRef]
35. Bertero, M.P. Fuels from bio-oils: Bio-oil production from different residual sources, characterization and thermal conditioning. *Fuel* **2012**, *95*, 263–271. [CrossRef]
36. Alvira, P.; Tomás-Pejó, E.; Ballesteros, M.; Negro, M.J. Pretreatment technologies for an efficient bioethanol production process based on enzymatic hydrolysis: A review. *Bioresour. Technol.* **2010**, *101*, 4851–4861. [CrossRef]
37. Pries, M.; Mai, C. Fire resistance of wood treated with a cationic silica sol. *Eur. J. Wood Wood Prod.* **2013**, *71*, 237–244. [CrossRef]
38. Proskurina, S.; Junginger, M.; Heinimö, J.; Tekinel, B.; Vakkilainen, E. Global biomass trade for energy—Part 2: Production and trade streams of wood pellets, liquid biofuels, charcoal, industrial roundwood and emerging energy biomass. *Biofuels Bioprod. Biorefining* **2019**, *13*, 371–387. [CrossRef]
39. Lachowicz, H.; Sajdak, M.; Paschalis-Jakubowicz, P.; Cichy, W.; Wojtan, R.; Witczak, M. The Influence of Location, Tree Age and Forest Habitat Type on Basic Fuel Properties of the Wood of the Silver Birch (Betula pendula Roth.) in Poland. *BioEnergy Res.* **2018**, *11*, 638–651. [CrossRef]
40. Rowell, R.M. *Handbook of Wood Chemistry and Wood Composites*; CRC Press: Boca Raton, FL, USA, 2005.
41. Stenström, S. Drying of paper: A review 2000–2018. *Dry. Technol.* **2020**, *38*, 825–845. [CrossRef]
42. Schaffer, E.L. *Charring Rate of Selected Woods—Transverse to Grain*; Forest Products Lab Madison Wis: Madison, WI, USA, 1967.
43. White, R.H.; Dietenberger, M. Fire safety of wood construction. In *Wood Handbook: Wood as an Engineering Material: Chapter 18*; U.S. Dept. of Agriculture, Forest Service, Forest Products Laboratory: Madison, WI, USA, 2010; pp. 18.1–18.22.
44. Mikkola, E. Charring of wood based materials. *Fire Saf. Sci.* **1991**, *3*, 547–556. [CrossRef]
45. Hugi, E.; Weber, R. Fire Behaviour of Tropical and European Wood and Fire Resistance of Fire Doors Made of this Wood. *Fire Technol.* **2012**, *48*, 679–698. [CrossRef]
46. Tran, H.C.; White, R.H. Burning rate of solid wood measured in a heat release rate calorimeter. *Fire Mater.* **1992**, *16*, 197–206. [CrossRef]
47. Tye, Y.; Lee, K.T.; Wan Nadiah, W.; Peng, L.C. The world availability of non-wood lignocellulosic biomass for the production of cellulosic ethanol and potential pretreatments for the enhancement of enzymatic saccharification. *Renew. Sustain. Energy Rev.* **2016**, *60*, 155–172. [CrossRef]
48. Rowell, R.M. 25—The use of biomass to produce bio-based composites and building materials. In *Advances in Biorefineries*; Waldron, K., Ed.; Woodhead Publishing: Sawston, UK, 2014; pp. 803–818.
49. Dungani, R.; Khalil, H.P.S.A.; Sumardi, I.; Suhaya, Y.; Sulistyawati, E.; Islam, M.N.; Suraya, N.L.M.; Aprilia, N.A.S. Non-wood renewable materials: Properties Improvement and its application. In *Biomass and Bioenergy: Applications*; Hakeem, K., Jawaid, M., Rashid, U., Eds.; Springer International Publishing: Cham, Switzerland, 2014; pp. 1–29.
50. Campilho, R. Chapter 1. Introduction. In *Natural Fiber Composites*; Campilho, R., Ed.; CRC Press; Taylor & Francis Group, LLC: Boca Raton, FL, USA, 2016; pp. 1–34.
51. Chandra, M. Use of nonwood plant fibers for pulp and paper industri in Asia: Potential in China. Master's Thesis, Virginia Polytechnic Institute and State University, Blacksburg, Virginia, 1998.
52. Saxena, M.; Pappu, A.; Sharma, A.; Haque, R.; Wankhede, S. Composite Materials from Natural Resources:Recent Trends and Future Potentials. In *Advances in Composite Materials—Analysis of Natural and Man-Made Materials*; Tesinova, P., Ed.; InTech: London, UK, 2011; pp. 121–162.
53. Moran-Mirabal, J. Advanced-Microscopy Techniques for the Characterization of Cellulose Structure and Cellulose-Cellulase Interactions. In *Cellulose-Fundamental Aspects*; InTechOpen: London, UK, 2013; p. 44.
54. Iswanto, A.H.; Tarigan, F.O.; Susilowati, A.; Darwis, A.; Fatriasari, W. Wood chemical compositions of raru species originating from Central Tapanuli, North Sumatra, Indonesia: Effect of Differences in Wood Species and Log Positions. *J. Korean Wood Sci. Technol.* **2019**, *49*, 416–429.
55. Iswanto, A.H.; Siregar, Y.S.; Susilowati, A.; Darwis, A.; Hartono, R.; Wirjosentono, B.; Rachmat, H.H.; Hidayat, A.; Fatriasari, W. Variation in chemical constituent of Styrax sumatrana wood growing at different cultivation site in North Sumatra, Indonesia. *Biodiversitas* **2019**, *20*, 448–452. [CrossRef]
56. Rowell, R.M.; Han, J.S.; Rowell, J.S. Characterization and factors effecting fiber properties. In *Natural Polymers and Agrofibers Composites*; Frollini, E., Leão, A.L., Mattoso, L.H.C., Eds.; Embrapa Instrumentação Agropecuária: Sãn Carlos, Brazil, 2000; pp. 115–134.
57. Liu, D.; Song, J.; Anderson, D.P.; Chang, P.R.; Hua, Y. Bamboo fiber and its reinforced composites: Structure and properties. *Cellulose* **2012**, *19*, 1449–1480. [CrossRef]
58. Chundawat, S.P.S.; Beckham, G.T.; Himmel, M.E.; Dale, B.E. Deconstruction of Lignocellulosic Biomass to Fuels and Chemicals. *Annu. Rev. Chem. Biomol. Eng.* **2011**, *2*, 121–145. [CrossRef]
59. Lee, K.J.D.; Marcus, S.E.; Knox, J.P. Cell Wall Biology: Perspectives from Cell Wall Imaging. *Mol. Plant* **2011**, *4*, 212–219. [CrossRef]
60. Al-Oqla, F.M.; Salit, M.S. 4—Material selection for composites. In *Materials Selection for Natural Fiber Composites*; Al-Oqla, F.M., Salit, M.S., Eds.; Woodhead Publishing: Sawston, UK, 2017; pp. 73–105.

61. Pauly, M.; Keegstra, K. Cell-wall carbohydrates and their modification as a resource for biofuels. *Plant J. Cell Mol. Biol.* **2008**, *54*, 559–568. [CrossRef]
62. Bertella, S.; Luterbacher, J.S. Lignin Functionalization for the Production of Novel Materials. *Trends Chem.* **2020**, *2*, 440–453. [CrossRef]
63. Mngomezulu, M.E.; John, M.J.; Jacobs, V.; Luyt, A.S. Review on flammability of biofibres and biocomposites. *Carbohydr. Polym.* **2014**, *111*, 149–182. [CrossRef]
64. Wu, Q.; Ran, F.; Dai, L.; Li, C.; Li, R.; Si, C. A functional lignin-based nanofiller for flame-retardant blend. *Int. J. Biol. Macromol.* **2021**, *190*, 390–395. [CrossRef]
65. Prabhakar, M.N.; Shah, A.U.R.; Song, J.-I. A Review on the flammability and flame retardant properties of natural fibers and polymer matrix based composites. *Compos. Res.* **2015**, *28*, 29–39. [CrossRef]
66. Kovačević, Z.; Flinčec Grgac, S.; Bischof, S. Progress in Biodegradable Flame Retardant Nano-Biocomposites. *Polymers* **2021**, *13*, 741. [CrossRef] [PubMed]
67. Bourbigot, S. 2—Flame retardancy of textiles: New approaches. In *Advances in Fire Retardant Materials*; Horrocks, A.R., Price, D., Eds.; Woodhead Publishing: Sawston, UK, 2008; pp. 9–40.
68. Kozlowski, R.M.; Muzyczek, M.; Walentowska, J. Flame retardancy and protection against biodeterioration of natural Fibers: State-of-Art and future prospect. In *Polymer Green Flame Retardants*; Elsevier: Amsterdam, The Netherlands, 2014; pp. 801–836.
69. Saheb, D.N.; Jog, J.P. Natural fiber polymer composites: A review. *Adv. Polym. Technol.* **1999**, *18*, 351–363. [CrossRef]
70. Podkościelna, B.; Wnuczek, K.; Goliszek, M.; Klepka, T.; Dziuba, K. Flammability Tests and Investigations of Properties of Lignin-Containing Polymer Composites Based on Acrylates. *Molecules* **2020**, *25*, 5947. [CrossRef]
71. Stark, N.M.; White, R.H.; Mueller, S.A.; Osswald, T.A. Evaluation of various fire retardants for use in wood flour–polyethylene composites. *Polym. Degrad. Stab.* **2010**, *95*, 1903–1910. [CrossRef]
72. Shumao, L.; Jie, R.; Hua, Y.; Tao, Y.; Weizhong, Y. Influence of ammonium polyphosphate on the flame retardancy and mechanical properties of ramie fiber-reinforced poly(lactic acid) biocomposites. *Polym. Int.* **2010**, *59*, 242–248. [CrossRef]
73. Manfredi, L.B.; Rodríguez, E.S.; Wladyka-Przybylak, M.; Vázquez, A. Thermal degradation and fire resistance of unsaturated polyester, modified acrylic resins and their composites with natural fibres. *Polym. Degrad. Stab.* **2006**, *91*, 255–261. [CrossRef]
74. Nah, C.; Oh, J.; Mensah, B.; Jeong, K.U.; Ahn, D.U.; Kim, S.J.; Lee, Y.; Nam, S.H. Effects of thermal aging on degradation mechanism of flame retardant-filled ethylene–propylene–diene termonomer compounds. *J. Appl. Polym. Sci.* **2016**, *132*, 41324. [CrossRef]
75. Sauca, S.; Giamberini, M.; Reina, J.A. Flame retardant phosphorous-containing polymers obtained by chemically modifying poly(vinyl alcohol). *Polym. Degrad. Stab.* **2013**, *98*, 453–463. [CrossRef]
76. Yotkuna, K.; Chollakup, R.; Imboon, T.; Kannan, V.; Thongmee, S. Effect of flame retardant on the physical and mechanical properties of natural rubber and sugarcane bagasse composites. *J. Polym. Res.* **2021**, *28*, 455. [CrossRef]
77. Suppakarn, N.; Jarukumjorn, K. Mechanical properties and flammability of sisal/PP composites: Effect of flame retardant type and content. *Compos. Part B Eng.* **2009**, *40*, 613–618. [CrossRef]
78. Cárdenas, M.A.; García-López, D.; Gobernado-Mitre, I.; Merino, J.C.; Pastor, J.M.; Martínez, J.d.D.; Barbeta, J.; Calveras, D. Mechanical and fire retardant properties of EVA/clay/ATH nanocomposites@ Effect of particle size and surface treatment of ATH filler. *Polym. Degrad. Stab.* **2008**, *93*, 2032–2037. [CrossRef]
79. Liu, S.-M.; Huang, J.-Y.; Jiang, Z.-J.; Zhang, C.; Zhao, J.-Q.; Chen, J. Flame retardance and mechanical properties of a polyamide 6/polyethylene/surface-modified metal hydroxide ternary composite via a master-batch method. *J. Appl. Polym. Sci.* **2010**, *117*, 3370–3378. [CrossRef]
80. Mandlekar, N.; Cayla, A.; Rault, F.; Giraud, S.; Salaün, F.; Malucelli, G.; Guan, J.-P. *An Overview on the Use of Lignin and Its Derivatives in Fire Retardant Polymer Systems*; IntechOpen: London, UK, 2017.
81. Kubo, S.; Kadla, J.F. The Formation of Strong Intermolecular Interactions in Immiscible Blends of Poly(vinyl alcohol) (PVA) and Lignin. *Biomacromolecules* **2003**, *4*, 561–567. [CrossRef]
82. Brebu, M.; Tamminen, T.; Spiridon, I. Thermal degradation of various lignins by TG-MS/FTIR and Py-GC-MS. *J. Anal. Appl. Pyrolysis* **2013**, *104*, 531–539. [CrossRef]
83. Brodin, I.; Sjöholm, E.; Gellerstedt, G. The behavior of kraft lignin during thermal treatment. *J. Anal. Appl. Pyrolysis* **2010**, *87*, 70–77. [CrossRef]
84. Réti, C.; Casetta, M.; Duquesne, S.; Bourbigot, S.; Delobel, R. Flammability properties of intumescent PLA including starch and lignin. *Polym. Adv. Technol.* **2008**, *19*, 628–635. [CrossRef]
85. Mandlekar, N.; Cayla, A.; Rault, F.; Giraud, S.; Salaün, F.; Guan, J. Valorization of Industrial Lignin as Biobased Carbon Source in Fire Retardant System for Polyamide 11 Blends. *Polymers* **2019**, *11*, 180. [CrossRef] [PubMed]
86. Gao, C.; Zhou, L.; Yao, S.; Qin, C.; Fatehi, P. Phosphorylated kraft lignin with improved thermal stability. *Int. J. Biol. Macromol.* **2020**, *162*, 1642–1652. [CrossRef]
87. Mandlekar, N.; Malucelli, G.; Cayla, A.; Rault, F.; Giraud, S.; Salaün, F.; Guan, J. Fire retardant action of zinc phosphinate and polyamide 11 blend containing lignin as a carbon source. *Polym. Degrad. Stab.* **2018**, *153*, 63–74. [CrossRef]
88. Costes, L.; Laoutid, F.; Aguedo, M.; Richel, A.; Brohez, S.; Delvosalle, C.; Dubois, P. Phosphorus and nitrogen derivatization as efficient route for improvement of lignin flame retardant action in PLA. *Eur. Polym. J.* **2016**, *84*, 652–667. [CrossRef]

89. Verdolotti, L.; Oliviero, M.; Lavorgna, M.; Iannace, S.; Camino, G.; Vollaro, P.; Frache, A. On revealing the effect of alkaline lignin and ammonium polyphosphate additives on fire retardant properties of sustainable zein-based composites. *Polym. Degrad. Stab.* **2016**, *134*, 115–125. [CrossRef]
90. Liu, L.; Qian, M.; Songand, P.a.; Huang, G.; Yu, Y.; Fu, S. Fabrication of green lignin-based flame retardants for enhancing the thermal and fire retardancy properties of polypropylene/wood composites. *ACS Sustain. Chem. Eng.* **2016**, *4*, 2422–2431. [CrossRef]
91. Liang, D.; Zhu, X.; Dai, P.; Lu, X.; Guo, H.; Que, H.; Wang, D.; He, T.; Xu, C.; Robin, H.M.; et al. Preparation of a novel lignin-based flame retardant for epoxy resin. *Mater. Chem. Phys.* **2021**, *259*, 124101. [CrossRef]
92. Dai, P.; Liang, M.; Ma, X.; Luo, Y.; He, M.; Gu, X.; Gu, Q.; Hussain, I.; Luo, Z. Highly Efficient, Environmentally Friendly Lignin-Based Flame Retardant Used in Epoxy Resin. *ACS Omega* **2020**, *5*, 32084–32093. [CrossRef] [PubMed]
93. Yang, H.; Yu, B.; Xu, X.; Bourbigot, S.; Wang, H.; Song, P. Lignin-derived bio-based flame retardants toward high-performance sustainable polymeric materials. *Green Chem.* **2020**, *22*, 2129–2161. [CrossRef]
94. Popescu, C.-M.; Pfriem, A. Treatments and modification to improve the reaction to fire of wood and wood based products—An overview. *Fire Mater.* **2020**, *44*, 100–111. [CrossRef]
95. Shah, A.U.R.; Prabhakar, M.N.; Song, J.-I. Current Advances in the Fire Retardancy of Natural Fiber and Bio-Based Composites—A Review. *Int. J. Precis. Eng. Manufactiring-Green Technol.* **2017**, *4*, 247–262. [CrossRef]
96. Zhang, M.; Buekens, A.; Li, X. Brominated flame retardants and the formation of dioxins and furans in fires and combustion. *J. Hazard. Mater.* **2016**, *304*, 26–39. [CrossRef]
97. Saba, N.; Jawaid, M.; Alothman, O.Y.; Inuwa, I.M.; Hassan, A. A review on potential development of flame retardant kenaf fibers reinforced polymer composites. *Polym. Adv. Technol.* **2017**, *28*, 424–434. [CrossRef]
98. Fontaine, G.; Bourbigot, S.; Duquesne, S. Neutralized flame retardant phosphorus agent: Facile synthesis, reaction to fire in PP and synergy with zinc borate. *Polym. Degrad. Stab.* **2008**, *93*, 68–76. [CrossRef]
99. Vasiljević, J.; Čolović, M.; Čelan Korošin, N.; Šobak, M.; Štirn, Ž.; Jerman, I. Effect of Different Flame-Retardant Bridged DOPO Derivatives on Properties of in Situ Produced Fiber-Forming Polyamide 6. *Polymers* **2020**, *12*, 657. [CrossRef]
100. Guan, Y.-H.; Huang, J.-Q.; Yang, J.-C.; Shao, Z.-B.; Wang, Y.-Z. An Effective Way To Flame-Retard Biocomposite with Ethanolamine Modified Ammonium Polyphosphate and Its Flame Retardant Mechanisms. *Ind. Eng. Chem. Res.* **2015**, *54*, 3524–3531. [CrossRef]
101. Chen, M.J.; Wang, X.; Li, X.L.; Liu, X.Y.; Zhong, L.; Wang, H.Z.; Liu, Z.G. The synergistic effect of cuprous oxide on an intumescent flame-retardant epoxy resin system. *RSC Adv.* **2017**, *7*, 35619–35628. [CrossRef]
102. Zhang, Y.; Chen, X.L.; Fang, Z.P. Synergistic effects of expandable graphite and ammonium polyphosphate with a new carbon source derived from biomass in flame retardant ABS. *J. Appl. Polym. Sci.* **2013**, *128*, 2424. [CrossRef]
103. Samyn, F.; Bourbigot, S.; Duquesne, S.; Delobel, R. Effect of zinc borate on the thermal degradation of ammonium polyphosphate. *Thermochim. Acta* **2007**, *456*, 134–144. [CrossRef]
104. Sain, M.; Park, S.H.; Suhara, F.; Law, S. Flame retardant and mechanical properties of natural fibre–PP composites containing magnesium hydroxide. *Polym. Degrad. Stab.* **2004**, *83*, 363–367. [CrossRef]
105. Jeencham, R.; Suppakarn, N.; Jarukumjorn, K. Effect of flame retardants on flame retardant, mechanical, and thermal properties of sisal fiber/polypropylene composites. *Compos. Part B Eng.* **2014**, *56*, 249–253. [CrossRef]
106. Suardana, N.P.G.; Ku, M.S.; Lim, J.K. Effects of diammonium phosphate on the flammability and mechanical properties of bio-composites. *Mater. Des.* **2011**, *32*, 1990–1999. [CrossRef]
107. Artner, J.; Ciesielski, M.; Walter, O.; Döring, M.; Perez, R.M.; Sandler, J.K.W.; Altstädt, V.; Schartel, B. A Novel DOPO-Based Diamine as Hardener and Flame Retardant for Epoxy Resin Systems. *Macromol. Mater. Eng.* **2008**, *293*, 503–514. [CrossRef]
108. Martín, C.; Lligadas, G.; Ronda, J.C.; Galià, M.; Cádiz, V. Synthesis of novel boron-containing epoxy–novolac resins and properties of cured products. *J. Polym. Sci. Part A Polym. Chem.* **2006**, *44*, 6332–6344. [CrossRef]
109. Spontón, M.; Mercado, L.A.; Ronda, J.C.; Galià, M.; Cádiz, V. Preparation, thermal properties and flame retardancy of phosphorus- and silicon-containing epoxy resins. *Polym. Degrad. Stab.* **2008**, *93*, 2025–2031. [CrossRef]
110. Zhang, Z.X.; Zhang, J.; Lu, B.-X.; Xin, Z.X.; Kang, C.K.; Kim, J.K. Effect of flame retardants on mechanical properties, flammability and foamability of PP/wood–fiber composites. *Compos. Part B Eng.* **2012**, *43*, 150–158. [CrossRef]
111. Lazko, J.; Landercy, N.; Laoutid, F.; Dangreau, L.; Huguet, M.H.; Talon, O. Flame retardant treatments of insulating agro-materials from flax short fibres. *Polym. Degrad. Stab.* **2013**, *98*, 1043–1051. [CrossRef]
112. Xiong, X.; Niu, Y.; Zhou, Z.; Ren, J. Development and Application of a New Flame-Retardant Adhesive. *Polymers* **2020**, *12*, 2007. [CrossRef] [PubMed]
113. Brown, F.R.; Whitehead, T.P.; Park, J.S.; Metayer, C.; Petreas, M.X. Levels of non-polybrominated diphenyl ether brominated flame retardants in residential house dust samples and fire station dust samples in California. *Environ. Res.* **2014**, *135*, 9–14. [CrossRef] [PubMed]
114. Covaci, A.; Voorspoels, S.; de Boer, J. Determination of brominated flame retardants, with emphasis on polybrominated diphenyl ethers (PBDEs) in environmental and human samples—A review. *Environ. Int.* **2003**, *29*, 735–756. [CrossRef]
115. Hindersinn, R.R. Historical Aspects of Polymer Fire Retardance. In *Fire and Polymers*; ACS Symposium Series; American Chemical Society: Washington, WA, USA, 1990; Volume 425, pp. 87–96.
116. Abbasi, G.; Buser, A.M.; Soehl, A.; Murray, M.W.; Diamond, M.L. Stocks and flows of PBDEs in products from use to waste in the U.S. and Canada from 1970 to 2020. *Environ. Sci. Technol.* **2015**, *49*, 1521–1528. [CrossRef]

117. Covaci, A.; Harrad, S.; Abou-Elwafa Abdallah, M.; Ali, N.; Law, R.; Herzke, D.; de Wit, C. Novel Brominated Flame Retardants: A Review of Their Analysis, Environmental Fate and Behaviour. *Environ. Int.* **2011**, *37*, 532–556. [CrossRef]
118. Lewin, M.; Weil, E.D. Mechanisms and modes of action in flame retardancy of polymers. *Fire Retard. Mater.* **2001**, *1*, 31–68. [CrossRef]
119. Salamova, A.; Hermanson, M.; Hites, R. Organophosphate and Halogenated Flame Retardants in Atmospheric Particles from a European Arctic Site. *Environ. Sci. Technol.* **2014**, *48*, 6133–6140. [CrossRef]
120. Mohapatra, S.; Sarkar, P.; Bhoye, G. A low smoke zero halogen zero phosphorous fire-retardant rubber shroud. *Mater. Today Proc.* **2020**, *28*, 1423–1428. [CrossRef]
121. Bar, M.; Alagirusamy, R.; Das, A. Flame retardant polymer composites. *Fibers Polym.* **2015**, *16*, 705–717. [CrossRef]
122. Karimah, A.; Ridho, M.R.; Munawar, S.S.; Amin, Y.; Damayanti, R.; Lubis, M.A.R.; Wulandari, A.P.; Iswanto, A.H.; Fudholi, A.; Asrofi, M.; et al. A Comprehensive Review on Natural Fibers: Technological and Socio-Economical Aspects. *Polymers* **2021**, *13*, 4280. [CrossRef] [PubMed]
123. Zhu, S.; Shi, W. Thermal degradation of a new flame retardant phosphate methacrylate polymer. *Polym. Degrad. Stab.* **2003**, *80*, 217–222. [CrossRef]
124. Rothon, R.N.; Hornsby, P.R. Flame retardant effects of magnesium hydroxide. *Polym. Degrad. Stab.* **1996**, *54*, 383–385. [CrossRef]
125. Statistics, W.F. *Information Bulletin of the World*; World Fire Statistics: Ljubljana, Slovenia, 2012.
126. Kim, Y.; Lee, S.; Yoon, H. Fire-Safe Polymer Composites: Flame-Retardant Effect of Nanofillers. *Polymers* **2021**, *13*, 540. [CrossRef] [PubMed]
127. Sufianto, H.; Green, A.R. Urban Fire Situation in Indonesia. *Fire Technol.* **2012**, *48*, 367–387. [CrossRef]
128. Giebułtowicz, J.; Rużycka, M.; Wroczyński, P.; Purser, D.A.; Stec, A.A. Analysis of fire deaths in Poland and influence of smoke toxicity. *Forensic Sci. Int.* **2017**, *277*, 77–87. [CrossRef]
129. Hidayati, D.L.; Hasanah, M.; Suryani, S.I.; Dahena, N. Konseling islam untuk meningkatkan strategi coping korban bencana kebakaran di kota Samarinda. *Taujihat* **2020**, *1*, 1–21.
130. Kodur, V.; Kumar, P. Fire hazard in buildings: Review, assessment and strategies for improving fire safety. *PSU Res. Rev.* **2020**, *4*, 1–23. [CrossRef]
131. Murphy, J. Flame retardants: Trends and new developments. *Reinf. Plast.* **2001**, *45*, 42–46. [CrossRef]
132. Giancaspro, J.; Papakonstantinou, C.; Balaguru, P. Fire resistance of inorganic sawdust biocomposite. *Compos. Sci. Technol.* **2008**, *68*, 1895–1902. [CrossRef]
133. Angelini, S.; Barrio, A.; Cerruti, P.; Scarinzi, G.; Garcia-Jaca, J.; Savy, D.; Piccolo, A.; Malinconico, M. Lignosulfonates as Fire Retardants in Wood Flour-Based Particleboards. *Int. J. Polym. Sci.* **2019**, *2019*, 6178163. [CrossRef]
134. Purser, D. 3—Toxicity of fire retardants in relation to life safety and environmental hazards. In *Fire Retardant Materials*; Horrocks, A.R., Price, D., Eds.; Woodhead Publishing: Sawston, UK, 2001; pp. 69–127.
135. Reilly, T.; Beard, A. Additives used in flame retardant polymer formulations: Current practice & trends. In Proceedings of the Fire Retardants and their Potential Impact on Fire Fighter Health Workshop, NIST, Gaithersburg, MD, USA, 30 September 2009.
136. Check, L.; Marteel-Parrish, A. The fate and behavior of persistent, bioaccumulative, and toxic (PBT) chemicals: Examining lead (Pb) as a PBT metal. *Rev. Environ. Health* **2013**, *28*, 85–96. [CrossRef]
137. Kiliaris, P.; Papaspyrides, C.D. Chapter 1—Polymers on Fire. In *Polymer Green Flame Retardants*; Papaspyrides, C.D., Kiliaris, P., Eds.; Elsevier: Amsterdam, The Netherlands, 2014; pp. 1–43.
138. Morgan, A.B.; Gilman, J.W. An overview of flame retardancy of polymeric materials: Application, technology, and future directions. *Fire Mater.* **2013**, *37*, 259–279. [CrossRef]
139. Lincoln, J.D.; Shapiro, A.A.; Earthman, J.C.; Saphores, J.M.; Ogunseitan, O.A. Design and Evaluation of Bioepoxy-Flax Composites for Printed Circuit Boards. *IEEE Trans. Electron. Packag. Manuf.* **2008**, *31*, 211–220. [CrossRef]
140. Ma, C.; Qiu, S.; Yu, B.; Wang, J.; Wang, C.; Zeng, W.; Hu, Y. Economical and environment-friendly synthesis of a novel hyperbranched poly(aminomethylphosphine oxide-amine) as co-curing agent for simultaneous improvement of fire safety, glass transition temperature and toughness of epoxy resins. *Chem. Eng. J.* **2017**, *322*, 618–631. [CrossRef]
141. Salmeia, K.A.; Gooneie, A.; Simonetti, P.; Nazir, R.; Kaiser, J.-P.; Rippl, A.; Hirsch, C.; Lehner, S.; Rupper, P.; Hufenus, R.; et al. Comprehensive study on flame retardant polyesters from phosphorus additives. *Polym. Degrad. Stab.* **2018**, *155*, 22–34. [CrossRef]
142. Velencoso, M.M.; Battig, A.; Markwart, J.C.; Schartel, B.; Wurm, F.R. Molecular Firefighting—How Modern Phosphorus Chemistry Can Help Solve the Challenge of Flame Retardancy. *Angew. Chem. Int. Ed.* **2018**, *57*, 10450–10467. [CrossRef] [PubMed]
143. Howell, B.; Han, X. Effective Biobased Phosphorus Flame Retardants from Starch-Derived bis-2,5-(Hydroxymethyl)Furan. *Molecules* **2020**, *25*, 592. [CrossRef] [PubMed]
144. Arastehnejad, N.; Sulaiman, M.R.; Gupta, R.K. Nitrogen-Based Ecofriendly Flame Retardants for Polyurethane Foams. In *Polyurethane Chemistry: Renewable Polyols and Isocyanates*; ACS Symposium Series; American Chemical Society: Washington, WA, USA, 2021; Volume 1380, pp. 167–185.
145. Howell, B.A.; Daniel, Y.G. Reactive Flame Retardants from Starch-Derived Isosorbide. In *Sustainability & Green Polymer Chemistry Volume 1: Green Products and Processes*; ACS Symposium Series; American Chemical Society: Washington, WA, USA, 2020; Volume 1372, pp. 209–219.
146. Howell, B.A.; Daniel, Y.G. Isosorbide as a Platform for the Generation of New Biobased Organophosphorus Flame Retardants. *Insights Chem. Biochem.* **2020**, *1*, 1. [CrossRef]

147. Toan, M.; Park, J.-W.; Kim, H.-J.; Shin, S. Synthesis and characterization of a new phosphorus-containing furan-based epoxy curing agent as a flame retardant. *Fire Mater.* **2019**, *43*, 717–724. [CrossRef]
148. Purser, D.A. Chapter 2—Fire Safety Performance of Flame Retardants Compared with Toxic and Environmental Hazards. In *Polymer Green Flame Retardants*; Papaspyrides, C.D., Kiliaris, P., Eds.; Elsevier: Amsterdam, The Netherlands, 2014; pp. 45–86.

Review

Mycelium-Based Composites in Art, Architecture, and Interior Design: A Review

Maciej Sydor [1,*], Agata Bonenberg [2], Beata Doczekalska [3] and Grzegorz Cofta [3]

[1] Department of Woodworking and Fundamentals of Machine Design, Faculty of Forestry and Wood Technology, Poznań University of Life Sciences, 60-637 Poznań, Poland
[2] Institute of Interior Design and Industrial Design, Faculty of Architecture, Poznan University of Technology, 60-965 Poznań, Poland; agata.bonenberg@put.poznan.pl
[3] Department of Chemical Wood Technology, Faculty of Forestry and Wood Technology, Poznań University of Life Sciences, 60-637 Poznań, Poland; beata.doczekalska@up.poznan.pl (B.D.); grzegorz.cofta@up.poznan.pl (G.C.)
* Correspondence: maciej.sydor@up.poznan.pl; Tel.: +48-61-846-6144

Abstract: Mycelium-based composites (MBCs) have attracted growing attention due to their role in the development of eco-design methods. We concurrently analysed scientific publications, patent documents, and results of our own feasibility studies to identify the current design issues and technologies used. A literature inquiry in scientific and patent databases (WoS, Scopus, The Lens, Google Patents) pointed to 92 scientific publications and 212 patent documents. As a part of our own technological experiments, we have created several prototype products used in architectural interior design. Following the synthesis, these sources of knowledge can be concluded: 1. MBCs are inexpensive in production, ecological, and offer a high artistic value. Their weaknesses are insufficient load capacity, unfavourable water affinity, and unknown reliability. 2. The scientific literature shows that the material parameters of MBCs can be adjusted to certain needs, but there are almost infinite combinations: properties of the input biomaterials, characteristics of the fungi species, and possible parameters during the growth and subsequent processing of the MBCs. 3. The patent documents show the need for development: an effective method to increase the density and the search for technologies to obtain a more homogeneous internal structure of the composite material. 4. Our own experiments with the production of various everyday objects indicate that some disadvantages of MBCs can be considered advantages. Such an unexpected advantage is the interesting surface texture resulting from the natural inhomogeneity of the internal structure of MBCs, which can be controlled to some extent.

Keywords: biomaterials; bio-composites; bio design; mycelium-based composites; biopolymers; interior design; architecture; wood; mycelium; fungi; patent documents

Citation: Sydor, M.; Bonenberg, A.; Doczekalska, B.; Cofta, G. Mycelium-Based Composites in Art, Architecture, and Interior Design: A Review. *Polymers* **2022**, *14*, 145. https://doi.org/10.3390/polym14010145

Academic Editor: Petar Antov

Received: 7 December 2021
Accepted: 28 December 2021
Published: 31 December 2021

1. Introduction

Fungi can use many types of by-products as substrates for growth. When mycelium penetrates a substrate, it acts as a natural self-assembling binder, holding a loose mixture in a monolithic form, creating a solid composite of biopolymers cellulose matrix and very dense chitin reinforcement. Mycelium can fill the volume with a very dense network; one gram of soil can contain up to 600 km of hyphae [1]. The mycelium growth pattern is related to the availability of food resources, water and environmental conditions, which constantly modify the network topology. The adaptive behaviour of fungi allows them to cope with various ephemeral resources, competition, damage, and predation in a completely different manner from multicellular plants or animals [2]. In nature, the organic matter for fungal growth comes from the remains of plant and animal organisms and their metabolites. In industrial conditions, various types of biological post-consumer wastes and by-products such as wood, straws, husks, chaws, and bagasse can be used as substrates for mycelial growth [3].

Mycelium-based composites are used in construction, packaging, and in the production of various types of products. MBCs are also well suited to applied arts. Philip Ross is the author of the "Hy-Fi" tower-pavilion presented at the "MoMA's PS1" exhibition in 2014 [4], in this building structure he combined wooden beams with MBC, thus compensating for the low mechanical strength of MBC. The artist is the author of several patent applications and scientific publications in this field [5]. Pascal Leboucq designed "The Growing Pavilion" constructed by Company New Heroes in 2019, a temporary event space at Dutch Design Week constructed with panels grown from mushroom mycelium supported on a timber frame. The Redhouse Architecture Bureau (Cleveland, OH, USA) promotes the use of wood construction waste, such as panels and window frames, which can be defragmented and re-bonded with mycelium and then used to build houses [6]. In Indonesia, Mycotech, Block Research Group and the Karlsruhe Institute of Technology built a prototype spatial structure called "MycoTree" made of various biocomposites, with the addition of sugar cane and cassava root waste (2017) [7]. At Milan Design Week 2019, Carlo Ratti presented an installation entitled "Round Garden", built from a sequence of arches made of MBCs. The installation fits into the natural context and surroundings [8]. Various artists and designers have designed different mycelium-based products: e.g., Aniela Hoitink 2016 textiles, Erica Klarenbeek 2013 3D printed furniture, Jonas Edvard and Sebastian Cox 2013 lamps, Kristel Peeters and Mycofabrication 2009 shoes. Think tank Terreform ONE and non-profit organization Genspace have developed a series of seating furniture made of Mycoform (2016) [9]. Mycelium is an alternative to wood dust in 3D printing [10]. A group of British architects Blast Studio and Bio-Digital Matter Lab managed to 3D print a column of mycelium-based materials (2018) [4]. Team BioBabes printed 3D MBCs objects using polylactic acid to act as a temporary mycelium scaffold ("Hyper Articulated Myco-Morphs" 2016–2017) [11]. A number of cultural organizations research and popularize MBCs, like Futurium in series of exhibition "Mind the Fungi. Art & Design Residencies" in Berlin since 2019 [12], or Somerset House in series of cultural events "Mushrooms: The Art, Design and Future of Fungi" in London 2020 [13].

Mycelium-based composites have been reported as inventions since at least 2007 and are also the subject of scientific research. Taking into account the great potential and numerous advantages of such a material, it was considered appropriate to review the scientific literature and patent documents supplementing these sources of knowledge with our own experience in the field of manufacturing interior furnishings made of this type of interesting biocomposite. The main aim of the article is to synthesize information from the scientific literature, patent documents, and own experience to identify barriers and possibilities for an effective implementation of mycelium-based composites in industrial manufacturing, especially when applied to decorative objects used in architectural interior design of apartments.

2. Results of the Literature Review

At least 92 research papers have been published on mycelium-based composites (72 original articles [14–85], one being a hybrid of original and review articles [86], and 19 review articles [87–105]. The oldest article is from 2012 [14], the newest is from November, 2021 [81]. The analyzed articles are assigned to 19 subject areas. The two main research areas are "Materials Science" and "Engineering" (Figure 1).

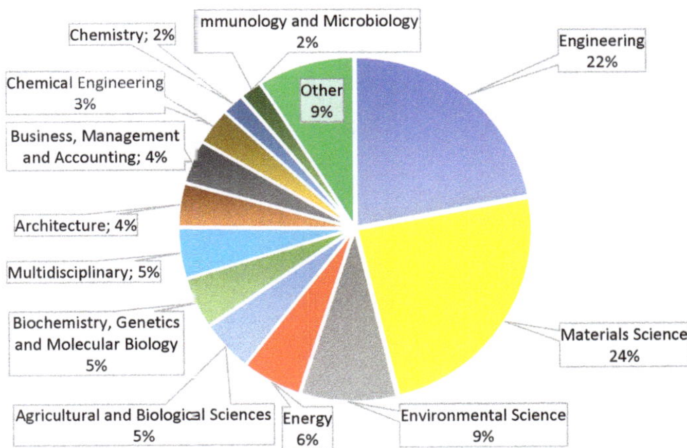

Figure 1. Subject areas of scientific articles on mycelium-based composites.

Over 130 different "author keywords" are used in the articles. Associations and frequency of co-existence for 20 most frequently used "author keywords" are shown in Figure 2.

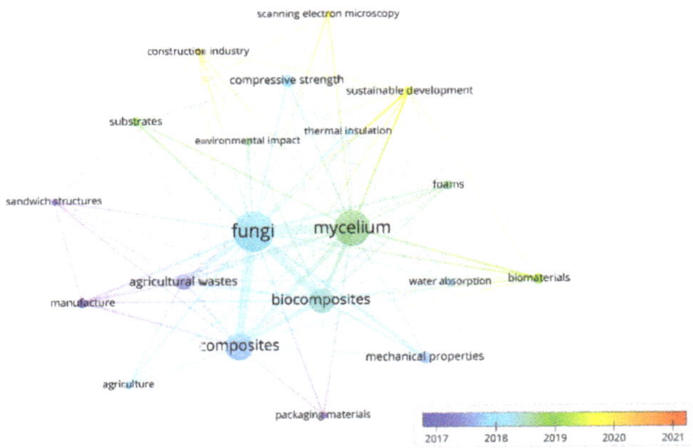

Figure 2. "Author keywords" associations in scientific articles on mycelium-based composites.

In the Figure 2, the frequency of occurrence of the keywords varies with time in color. VOSviewer was used; minor editorial changes have been made in the keywords: singular and plural forms of nouns ("fungus" = "fungi", "material" = "materials" etc.), the notation ("bio-composites" = "biocomposites" etc.), synonyms ("fungal mycelium" = "mycelium", "composite materials" = "composites", "bio-based composites" = "biocomposites" and "manufacturing process" = "manufacture"). Yellow color, turning red, indicates keywords used in the most recent articles. As can be seen, these are the words "sustainable development", "scanning electron microscopy", "construction industry" and "agricultural robots". This shows the changing research interests in this field.

The five most cited articles according to Scopus are summarized in Table 1.

There are different purposes for the research carried out. The vast majority of research focuses on finding out how to properly shape the constructional properties of the material. The objectives and results of selected research works on mycelium-based composites are collected in Table 2.

Table 1. The most frequently cited articles on mycelium-based composite materials according to Scopus.

Year	Title	Type	No. of Citations	Reference
2017	Advanced Materials from Fungal Mycelium: Fabrication and Tuning of Physical Properties	original	128	[28]
2017	Morphology and mechanics of fungal mycelium	original	80	[36]
2017	Mycelium composites: A review of engineering characteristics and growth kinetics	review	74	[90]
2012	Fungal mycelium and cotton plant materials in the manufacture of biodegradable moulded packaging material: Evaluation study of select blends of cotton by-products	original	74	[14]
2019	Fabrication factors influencing mechanical, moisture- and water-related properties of mycelium-based composites	original	66	[52]

Table 2. Parameters and aims of mycelium-based composites production in scientific research.

Fungi	Substrate	Product/Application	Main Results (MBC = Mycelium-Based Composites)	Reference
Ganoderma sp.	Cotton-based (carpel, seed hull) starch, and gypsum	Packaging material	MBC meets or exceeds the characteristics of extruded polystyrene foam	[14]
Not specified (possibly as [14])	Rice straw, hemp seed, kenaf fibre, switch grass, sorghum fibre, cotton bur fibre, flax shive	Insulation panel	Optimal performance at the noise frequency of 1000 Hz. MBC are comparable to polyurethane foam board and are better than plywood	[15]
G. lucidum, P. ostreatus	Cellulose and potato-dextrose broth (PDB)	Fibrous mycelium film	The substrate should be homogeneous. The PDB in the substrate increases the stiffness of MBC	[28]
T. versicolor	Glass fines, wheat grains, and rice hulls	Fire safe mycelium biocomposites	MBC are safer than the typical construction materials: producing much lower heat release rates, less smoke and CO_2 and longer time to flashover. Composites with glass fines had the best fire performance	[46]
T. ochracea, P. ostreatus	Beech sawdust, rapeseed straw, bran. Non-woven cotton fibre	Board	Straw-based mycelium composites are stiffer and less moisture-resistant than cotton based	[52]
T. versicolor, P. brumalis	Wheat straw, rice hulls, sugarcane bagasse, blackstrap molasses, wheat grains, malt extract	Pure mycelium	Mycelium grew slow on rice hull, sugarcane bagasse and wheat straw. Liquid blackstrap molasses accelerates growth, outperforming laboratory malt extracts.	[49]
T. versicolor	Flax dust, flax long, wheat straw dust, wheat straw, hemp fibres and pine wood shavings	Thermal insulation	The thermal conductivity and water absorption of MBCs are comparable to those of rock wool, glass wool, and extruded polystyrene. The mechanical properties depend more on the fibre arrangement than on the chemical composition of the fibres	[57]
Not specified (white-rot basidiomycete mycelium)	Mixture of spruce, pine, and fir	Particleboard	Cellulose nanofibers added to the substrate improved the mechanical properties of MBC by 5%	[53]
P. ostreatus, F. oxysporum	Sodium silicate	Pure mycelium	3% sodium silicate improve thermal stability. The *P. ostreatus* compared to the *F. oxysporum* beter improve material thermal stability (higher decomposition temperature and residual weight, lower degradation rate)	[59,84]
G. lucidum, P. ostreatus	Clay, sawdust, bleached and unbleached cellulose	Printed cylinders	The mycelium improves the 3D printing (better water resistance, material stiffness and surface hardness)	[73]

In 2016–2021, at least 20 scientific review articles were also published. The most important of these articles are listed in Table 3.

Table 3. List of scientific review publications for mycelium-based composites for art, architecture, and interior design.

Year	Reference	No. of Cited Documents	No. of Citations in Scopus	Main Findings
2016	[88]	32	22	A production cost model is described which includes labour, material and overhead costs for structured sandwich products produced from MBCs.
2017	[90]	170	74	1. MBCs are kind of biopolymer foam, but most studies admit that mechanical performance can be improved in the future. 2. Current use is limited to the packaging and chosen construction applications. New applications have been proposed (acoustic dampers, super absorbents, paper, textiles, structural and electronic parts).
2018	[91]	21	34	1. MBCs can be used for a variety of purposes with the advantage of a lower cost and the better disposal than polystyrene that is an environmental problem. 2. The biggest challenge is the negative public perception of fungus-derived products.
2019	[94]	11	26	MBCs are profitable renewable and degradable material and have the potential to replace petroleum-based materials.
2019	[92]	108	37	Improvement in know-how is expected to improve the mechanical properties and to standardize the productive process, whereas insulation and thermal properties already have shown competitive results.
2020	[86]	58	21	1. There is a correlation between raw input material composition and final material properties. 2. MBCs have implications for sustainable architecture and products. 3. The unique aesthetics of MBCs should be further explored and more clearly identified.
2020	[96]	80	44	1. Fungal biorefinery upcycles by-products into cheap and sustainable composite materials. 2. Can replace foam, timber and plastic insulation, door cores, panels, flooring, furnishings. 3. Low density and thermal conductivity, high acoustic absorption, and fire safety. 4. MBCs are suitable as thermal and acoustic insulation foams.
2021	[98]	77	6	1. MBCs are more suitable for thermal and acoustic insulation than synthetic foam and wood fibres. 2. MBCs are stiff, lightweight and biodegradable, thus are an alternative to petroleum-based packaging materials.
2021	[101]	101	0	The process of engineering affects the properties of MBCs. Bioreactor designs such as tray, packed bed and millilitre reactors, influence of mycelium growth conditions and strategies for controlling mycelium microenvironment are discussed to allow optimal process development.
2021	[102]	118	0	1. MBCs are advantageous as packaging materials with sufficient acoustic, and thermal insulation, slightly worse than expanded polystyrene. 2. The standardized process to produce an optimized material property has yet to be identified, production is less standardized than conventional engineering materials, and it is not clear how to customize the substrates for a particular species of fungi to optimize the composite mechanics.
2021	[103]	80	0	1. MBCs support a circular economy. 2. Finding the ways of enhancing their physicochemical properties will expand the application areas. 3. The properties of MBCs are competitive with those of synthetic polymers used in construction, interior architecture, and other industries.
2021	[104]	94	2	With the wide variety of fungal species and substrates available, MBCs can improve environmental sustainability of many industrial products.

3. Results of Patents Documents Analysis

A granted patent is an administrative decision: area and time limited, issued by the patent office, it provides protection for a feasible, new, non-obvious and potentially profitable solution. The basis for such a decision is a patent application, which requires an unambiguous description of the essence of the invention. The patent application also provides a priority date, i.e., the date of disclosure of the invention. The priority date can, for example, be the presentation of the invention at a trade fair or the publication of a description of the invention. Most often, however, it is the date when the invention is filed with the patent office. Some patent applications become granted patents. Inventions considered profitable by their owners are filed in many patent jurisdictions around the world. Subsequent applications may differ slightly from their prototypes in terms of content, the differences result mainly from the refinement of descriptions, as well as the rejection of some patent claims by various patent offices. The main patent documents are patent applications and granted patents from many patent offices, they form the so-called patent families. Thus, each patent family describes one invention, the date of its creation is given by its first application.

Patent documents were searched on the basis of the following keywords: mycelium; mycological; fungi; biopolymers; biomaterials; biocomposites. These words were searched for in the "TAC" sections of patent documents (TAC = title OR abstract OR claim). Searches were made in the International Patent Classification areas: C08*, C12N*, B27N* B32B* oraz B32B*, and the list of documents was reviewed, limiting it to issues related to the production of plastics such as foams, boards and blocks used in construction, furniture, the automotive industry, as packaging and as artistic products. Thus, documents dealing with the production of woven fabric, i.e., all non-structural materials used in the manufacturing technique, were omitted. Publicly available databases and analytical tools such as Google Patents, The Lens were used, and the results of queries were exported to MS Excel for further analysis.

As a result of the analysis, 212 patent documents were identified: 153 patent applications, 55 patents granted on the basis of some of these applications, and additionally 2 amended applications, 1 amended patent and 1 patent of addition. They constitute 67 extended families, and thus describe 67 different technological and product inventions related to mycelium-based composites. The oldest document was received by the United States Patent and Trademark Office on 12 December 2007 [106], while the last of the analysed documents was on 9 April 2021 [107]. The annual numbers of patent applications, according to the years of their publication, are shown in Figure 3.

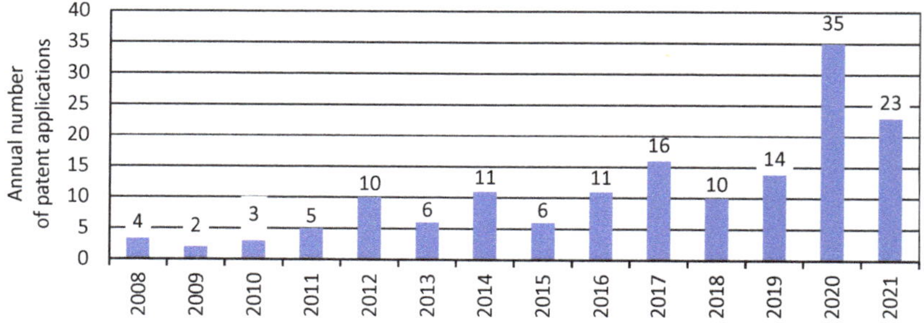

Figure 3. Annual number of patent applications according to publication year.

Data for 2021 is incomplete, not all patent documents from this year are indexed in databases. The data on the annual number of patents presented in Figure 3 show a significant increase in patent applications in the last two years.

There are significant 9 people and organizations among the owners of patent documents. This is shown in Figure 4 as shares in overall number of patent documents.

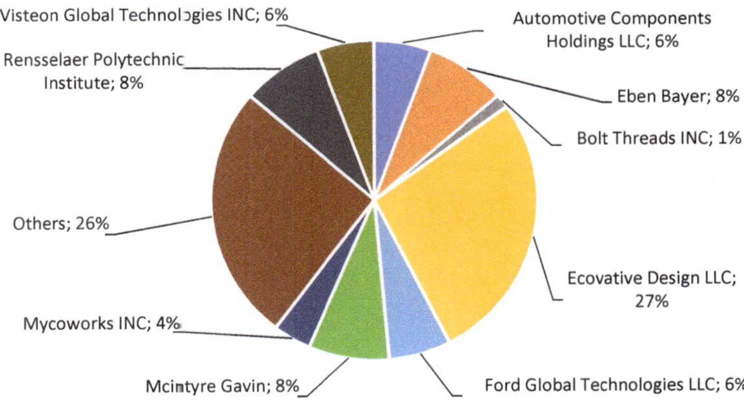

Figure 4. Shares of companies in the total number of patent applications.

The owner of the largest number of patent documents is Ecovative Design LCC (Albany, NY, USA), which has 27% of industrial property in this area (58 documents). Other persons and institutions affiliating many documents are: Eben Bayer and Mcintyre Gavin (both related to Ecovative Design LCC) and Rensselaer Polytechnic Institute (Troy, MI, USA) (17 documents each), also Ford Global Technologies LLC and Automotive Components Holdings LLC, both owned by Ford Motor Company headquartered in Dearborn, MI, USA (13 and 12 documents, respectively).

The analysis of patent documents shows 9 main countries related with the mycelium-based composites (Figure 5). The largest number of affiliated patent documents is in the USA. However, the latest documents are affiliated in Germany, Belgium and China.

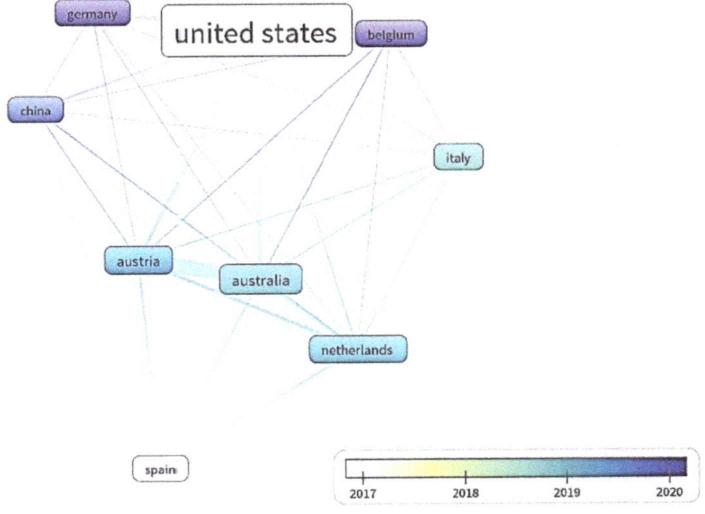

Figure 5. Links between countries in patent documents.

In 29 patent families there is at least one granted patent, these families are summarized in Table 4, presenting one selected patent from each such patents family.

Table 4. Granted patents.

Order No.	Patent No., Application Year–Granted Year, Reference	Details
1	US 9,485,917 B2, 2007–2016, [108]	ED (Ecovative Design LLC). Method for producing grown materials and products made thereby
2	US 8,001,719 B2, 2009–2011, [109]	ED. Method for producing rapidly renewable chitinous material using fungal fruiting bodies and product made thereby
3	US 8,313,939 B2, 2010–2012, [110]	FGT, ACH (Ford Global Technologies LLC, Automotive Components Holdings LLC). A method of making a moulded automotive part with a liquid fungal mixture.
4	US 8,298,810 B2, 2010–2012, [111]	
5	US 8,227,233 B2 [112]	
6	US 8,227,224 B2 [113]	FGT, ACH. Method of making moulded part comprising mycelium coupled to mechanical device
7	US 8,227,225 B2 [114]	FGT, ACH. Plasticized mycelium composite and method
8	US 8,283,153 B2 [115]	FGT, ACH. Mycelium structures containing nanocomposite materials and method
9	US 8,298,809 B2 [116]	FGT, ACH. Method of making a hardened elongate structure from mycelium
10	CN 102,329,512 B [117]	Ford Global Technologies LLC. The sheet stock mycelium of cutting and method
11	US 9,410,116 B2, 2011–2016, [118]	Mycoworks Inc. building materials
12	US 9,879,219 B2, 2012–2018, [119]	ED. A method of producing a chitinous polymer derived from fungal growth
13	CA 2,834,095 C, 2012–2018, [120]	ED. Dehydrated mycelium panels.
14	US 10,154,627 B2, 2013–2018, [121]	ED. Growing mycological biomaterials in tools that are consumed or enveloped during the growth process
15	FR 3,006,693 B1 2013–2016, [122]	Menuiseries Elva. A method of producing a composite material based on natural fibres inoculated with mycelium and parts obtained with this method
16	US 9,253,889 B2 2012–2016 [123]	ED. Sheet built-in an electrical circuit
17	US 9,085,763 B2, 2013–2015, [124]	ED. Production dehydrated mycelium elements to form tissue morphology using Pycnoporus cinnabarinus
18	AU 2013/251269 B2, 2013–2015, [125]	ED. Self-supporting composite material
19	US 10,144,149 B2, 2014–2018, [126]	ED. Stiff mycelium bound part and method of producing stiff mycelium bound parts
20	US 9,394,512 B2, 2015–2016, [127]	ED. Method for growing mycological materials
21	US 9,469,838 B2, 2015–2016, [128]	Mycoworks Inc. Set of mycelium-based materials with wood timber
22	CN 105,292,758 B 2016–2017, [129]	Shenzhen Zeqingyuan Technology Dev Service Co Ltd., Univ Sichuan Agricultural. Production method for biomass packing material
23	AU 2015/271912 B2, 2015–2020, [130]	ED. Method of manufacturing a stiff engineered composite
24	US 9,914,906 B2, 2016–2018, [131]	ED. Process for solid-state cultivation of mycelium on a lignocellulose substrate
25	CN 106,148,199 B, 2016–2019, [132]	Jiangxi University of Technology. Agricultural waste-based mycelium material with good a cushion performance and mechanical property
26	CN 106,633,989 B, 2016–2019, [133]	Shenzhen Zeqingyuan Technology Development Service Co Ltd. Using bagasse as fungi-based biomass packaging material of major ingredient and preparation method thereof
27	US 10,604,734 B2, 2017–2020, [134]	University of Alaska Anchorage. Thermal insulation material from mycelium and forestry by-products
28	KR 102,256,335 B1, 2019–2021, [135]	Lee Beom Geun. Eco-friendly packing materials comprising mushroom mycelium and the process for the preparation thereof
29	US 11,015,059 B2, 2019–2021, [107]	Bolt Threads Inc. Composite material, and methods for production thereof

More than 200 patent documents make it impossible to "intuitively" indicate the key inventions in the field of mycelium-based composites. Undoubtedly, the first patent application (US 2008/0145577 A1, "Method for producing grown materials and products made thereby" [106], filing date 12 December 2007) is important, but there are likely to be other influential inventions in this field. In the 2019 scientific article on the review of wood screw patents [136], the following criteria were proposed for the identification of important patents:

- The size of the patent family—the assumption: "only an invention with high application potential can be submitted for protection in many patent offices because the patent procedure is paid".
- Number of citations of a patent document in other, later patent documents—the assumption: "if multiple patent documents refer to a particular document, it indicates that this document describes (and perhaps at least partially solves) a significant problem.

Using these two criteria, the most influential patent documents for mycelium-based technology were listed in Table 5.

Table 5. Most influenced patent documents.

No.	Patent Document	Extended Patent Family Size	Number of Citations of the Patent Document in Other Patent Documents
1	US 2008/0145577 A1 "Method for producing grown materials and products made thereby" [106]	43	44
2	US 2012/0270302 A1 "Method for Making Dehydrated Mycelium Elements and Product Made Thereby" [137]	15	4
3	WO 2019/099474 A1 "Increased Homogeneity of Mycological Biopolymer Grown into Void Space" [138]	12	8
4	US 2012/0135504 A1 "Method for Producing Fungus Structures" [139]	11	20
5	US 2018/0282529 A1 "Solution Based Post-Processing Methods for Mycological Biopolymer Material and Mycological Product Made Thereby" [140]	9	5
6	US 2020/0024577 A1 "Method of Producing a Mycological Product and Product Made Thereby" [141]	7	4

The generalized MBCs production protocol can be compiled from research articles, patent documents, or open source manuals (e.g., [142]). Such a general protocol includes:

(1) The chosen mycelium specie is pre-grown in a Petri dish with a growth medium solidified with agar.
(2) The substrate for the culture of mycelium is homogenized (the substrate is a mix of selected biopolymers with defined granulation and proportion). The substrate is also sterilized to kill or deactivate all microorganisms in it.
(3) The pre-grown mycelium and sterile water are added to the substrate. Additional nutrients can also be added. The inoculated substrate is packed in a sterile mould (a bag or a container).
(4) The mycelium grows trough the substrate in a controlled micro-climate (temperature, air humidity, without light). The mycelium composite can be created initially in the mould to its internal reinforcement, and then outside the mould to solidify its surface.
(5) The mycelium composite is sterilized to end the growth process and then dried to the target moisture content.
(6) A pressing, machining, coating or other required product post-processing is applied.

The review of patent documents shows that biofoam composites and layered structures with mycelium-based composites can be used in building structures as structural materials (e.g., the core of sandwich panels and gap fillers), interior finishing materials (e.g., wall panels) and floors), as well as materials for portable home furnishings (furniture and other portable items) and packaging materials. They can have an insulating function due to their low heat conductivity or a sound-absorbing function. Biocomposites can therefore be an alternative to synthetic foams found in automotive bumpers, doors, roofs, engine cavities, boot linings, dashboards, and seats because the mycelium-based material has the same or better ability to absorb impacts, insulate, dampen sound and provide lightweight construction in the car from typical synthetic foams. The material also showed good fire resistance. Applications in the construction industry are mainly limited to fire-proof thermal and acoustic insulators. So far, the use of this innovative biocomposite in the construction industry has been limited only to a small scale and to exhibition installations.

Considering all the ecological advantages of mycelial and bio-substrate composites, the question arises, why such materials are not used very widely. Potential reasons for this may be problems with low mechanical properties, high water absorption, lack of Life Cycle Assessment information for this material, and lack of standard production methods and standardized methods for testing material properties.

4. Mycelium-Based Material in Elements of Interior Design—Case Study

Even though the mycelium-based composites is currently studied mostly for purposes in which visual or aesthetical aspects are insignificant, like packaging, experiments performed by the authors suggest that it can be successfully used for creating interior design elements. Mycelium-based materials embrace a new aesthetics characterized by imperfections and irregularities through natural and spontaneous growth, thus achieving a unique structure, as in wood. The physical and geometric properties of objects evolve and change slightly over time. These properties make it an unusual and challenging material. Different textures that characterize the material samples depend on how the substrate has been formed before the growth; the material's surface has visible natural fibres and dominating natural mycelium colouring: off-whites with yellow or brownish irregularities in more mature areas. The user perceives these characteristics as organic, warm, and natural, which influences the typology of products that could be created.

The first shapes obtained from mycelium-based material by Agata Bonenberg were simple panels that allowed the growth and maturation of the material to be observed (Figures 6 and 7). Then spherical objects were created to study the emergence of different textures: smooth (Figure 8), rough (Figure 9). The object shown in Figure 10 combines both; it has a smooth well-fragmented substrate at the bottom, and an uneven, rough part at the top. This opens interesting possibilities for future projects.

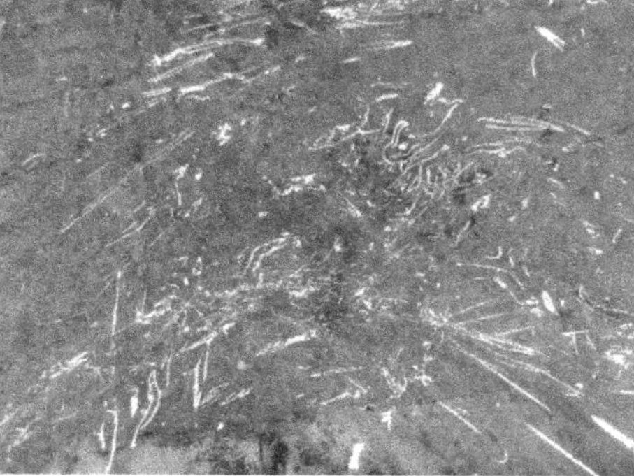

Figure 6. Mycelium-based material during growth (design and photo: A. Bonenberg).

Figure 7. Mycelium-based material after growth: smooth surface with (design and photo: A. Bonenberg).

Figure 8. Mycelium-based composite: smooth texture, with visible fibres (design and photo: A. Bonenberg).

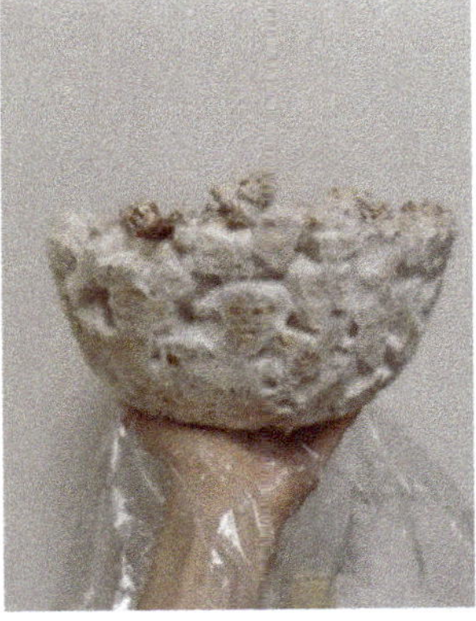

Figure 9. Mycelium-based composite: rough texture, deriving from substrate fragmentation (design and photo: A. Bonenberg).

Figure 10. Mycelium-based composite: smooth and rough textures, combined in one object (design and photo: A. Bonenberg).

Experimentation with textures and shapes of forms has led to preliminary product development and production. The designs of a table light fixture, a table bowl, and a coffee table have been executed. In each of these projects, mycelium-based elements had to be combined with other materials. The author has chosen natural components such as timber to match the design's pro-ecological spirit and give overall natural "touch". The small table lamp is a good example of this approach: a mycelium-grown, cylindrical lampshade has been fixed on a simple cubical timber base (Figures 11 and 12).

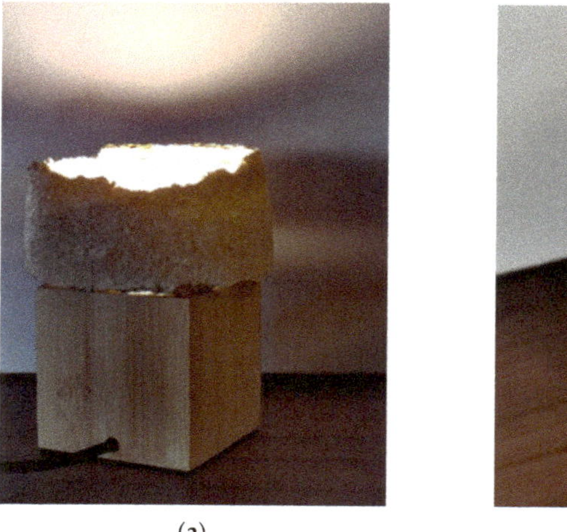

(**a**)

(**b**)

Figure 11. Mycelium-based light fixture: (**a**)—with lamp on, (**b**)—general appearance (design and photo: A. Bonenberg).

Figure 12. Mycelium-based semi-finished object (design and photo: A. Bonenberg).

Similarly, a container bowl was created, where an upper part of the object was fixed to the rough-timber torus-shaped base (Figures 12 and 13). Again, the look of the object is "organic". At the same time, the heavier wooden base gives the bowl its functional stability.

Figure 13. Finished object: mycelium-based bowl fixed to the rough-timber torus-shape base (design and photo: A. Bonenberg).

Another artifact created is a coffee table where a mycelium-based tabletop has been grown on a metal frame, ensuring its structural stability (Figure 14). The tabletop is thick but light, with well-consolidated smooth surfaces from the top and sides, but an uneven and rough texture can be perceived from the bottom. In addition, there is a clear contrast between the thin steel legs of the table and the thick-bodied top. Finally, the triangular shape gives the object expressive, characteristic looks.

Figure 14. Coffee table with mycelium-based tabletop (design and photo: A. Bonenberg).

The challenge of unconventional materials is the technique of fastening elements [143,144]. The new material requires a new approach in this field, which will be a further direction of our activities.

5. Conclusions

Regarding the current excessive dependence of the construction and production industry on hydrocarbon-containing materials occurring within Earth's crust; taking into account the abundance of waste and industrial by-products, it is necessary to make greater use of the advantages of biomaterials with a low carbon footprint [145–152]. The following conclusions can be drawn from the reviewed content presented:

1. MBCs (mycelium-based composites) offer favourable production price, ecological value, and high artistic value. Their weaknesses are insufficient design properties and not fully known reliability (quality during use), therefore both scientific research and engineering creativity, which is manifested by patents documents, are heading in this direction.
2. A review of the scientific literature shows that the material parameters of MBCs can be adjusted to the needs: by selecting the type of substrate and fungus species, by controlling the growth conditions, the method of inactivation of the mycelium after growth, and the drying method. In this way, it is possible to meet certain requirements, e.g., increase the structural load-bearing capacity to an acceptable level and reduce the affinity with water, and additionally improve the acoustic and thermal insulation. However, the problem is the almost infinite number of combinations: properties of the input biomaterials, characteristics of the mushroom species, and parameters during growth and subsequent processing of the MBC.
3. The review of patent documents shows that two current technological challenges are related to the creation of MBCs with the properties required by the final product. Especially, looking for an effective method of increasing strength, for example by increasing the density, the search for a method of obtaining a more homogeneous internal structure.
4. The described own technological experiments, consisting of the production of various everyday objects, indicate that some disadvantages of MBCs can be considered advantages. Such an unexpected advantage is the interesting and unrepeatable surface texture resulting from the natural unevenness of the internal structure of MBCs, which can be controlled to some extent.

The presented results of the analysis of a wide variety of literature and own technological experiments suggest that the share of mycelium-based composites in industrial production and construction will increase, despite certain limitations of this innovative class of materials in terms of manufacturing difficulties, insufficient strength, unknown durability and reliability, and challenges in fastening technology. These problems will be gradually solved or at least significantly minimized. This is supported by the fundamental advantages of these types of bio-composites, i.e., the ability to produce from by-products or waste, low energy requirements for production, biodegradability and artistic values.

Author Contributions: Conceptualization, A.B; methodology, M.S.; software, M.S.; validation, M.S.; formal analysis, B.D. and G.C.; investigation, A.B.; resources, A.B. and M.S.; data curation, B.D. and G.C.; writing—original draft preparation, M.S. and A.B.; writing—review and editing, M.S.; visualization, A.B.; supervision, M.S.; project administration, A.B.; funding acquisition, A.B. All authors have read and agreed to the published version of the manuscript.

Funding: This research received no external funding. The APC was funded by Poznan University of Technology, Faculty of Architecture (SDBAD 2021 "Shaping architecture and interiors in the context of contemporary cultural and social transformations–Kształtowanie architektury i wnętrz w kontekście współczesnych przeobrażeń kulturowo-społecznych" ERP 0113/SBAD/2131).

Institutional Review Board Statement: Not applicable.

Informed Consent Statement: Not applicable.

Data Availability Statement: All data, models, and code generated or used during the study appear in the submitted article.

Conflicts of Interest: The authors declare no conflict of interest.

References

1. Ekblad, A.; Wallander, H.; Godbold, D.L.; Cruz, C.; Johnson, D.; Baldrian, P.; Björk, R.; Epron, D.; Kieliszewska-Rokicka, B.; Kjøller, R. The Production and Turnover of Extramatrical Mycelium of Ectomycorrhizal Fungi in Forest Soils: Role in Carbon Cycling. *Plant Soil* **2013**, *366*, 1–27. [CrossRef]
2. Webster, J.; Weber, R. *Introduction to Fungi*, 3rd ed.; Cambridge University Press: Leiden, The Netherlands, 2007; ISBN 978-0-511-27783-2.
3. Kavanagh, K. (Ed.) *Fungi: Biology and Applications*, 3rd ed.; Wiley: Hoboken, NJ, USA, 2017; ISBN 978-1-119-37416-9.
4. Elsacker, E. Interdisciplinary Exploration of the Fabrication and Properties of Mycelium-Based Materials. Ph.D. Thesis, Vrije Universiteit Brussel, Brussel, Belgium, 2021.
5. Ţura, D.; Wasser, S.P.; Zmitrovich, I.V. Wood-Inhabiting Fungi: Applied Aspects. In *Fungi: Applications and Management Strategies*; Deshmukh, S.K., Misra, J.K., Tewari, J.P., Papp, T., Eds.; Progress in Mycological Research; CRC Press Taylor & Francis Group: Boca Raton, FL, USA, 2016; pp. 245–292. ISBN 978-1-4987-2492-0.
6. Rothschild, L.J.; Maurer, C.; Paulino Lima, I.G.; Senesky, D.; Wipat, A.; Head, J., III. *Myco-Architecture off Planet: Growing Surface Structures at Destination. NIAC 2018 Phase I Final Report*; NASA Ames Research Center: Mountain View, CA, USA, 2019; p. 58.
7. Heisel, F.; Lee, J.; Schlesier, K.; Rippmann, M.; Saeidi, N.; Javadian, A.; Nugroho, A.R.; Mele, T.V.; Block, P.; Hebel, D.E. Design, Cultivation and Application of Load-Bearing Mycelium Components: The MycoTree at the 2017 Seoul Biennale of Architecture and Urbanism. *IJSED* **2017**, *6*, 296–303. [CrossRef]
8. Ratti, C. A Living Architecture for the Digital Era. *Diségno* **2019**, *4*, 177–188.
9. Joachim, M.; Aiolova, M. *Design with Life*; ACTAR Publishers: New York, NY, USA, 2019; ISBN 978-1-948765-20-6.
10. Goidea, A.; Floudas, D.; Andréen, D. Pulp Faction: 3d Printed Material Assemblies through Microbial Biotransformation. In *Fabricate 2020: Making Resilient Architecture*; Burry, J., Sabin, J., Sheil, B., Skavara, M., Eds.; Fabricate; UCL Press: London, UK, 2020; pp. 42–49. ISBN 978-1-78735-811-9.
11. Dias, J. Hyper Articulated Myco-Morphs. Available online: https://www.biobabes.co.uk/mycomorphs (accessed on 27 December 2021).
12. Meyer, V.; Schmidt, B.; Pohl, C.; Cerimi, K.; Schubert, B.; Weber, B.; Neubauer, P.; Junne, S.; Zakeri, Z.; Rapp, R.; et al. *Mind the Fungi*; Meyer, V., Rapp, R., Eds.; Universitätsverlag der TU Berlin: Berlin, Germany, 2020; ISBN 978-3-7983-3169-3.
13. Sheldrake, M. *Entangled Life: How Fungi Make Our Worlds, Change Our Minds & Shape Our Futures*, 5th ed.; Random House: New York, NY, USA, 2020; ISBN 978-0-525-51032-1.
14. Holt, G.A.; Mcintyre, G.; Flagg, D.; Bayer, E.; Wanjura, J.D.; Pelletier, M.G. Fungal Mycelium and Cotton Plant Materials in the Manufacture of Biodegradable Molded Packaging Material: Evaluation Study of Select Blends of Cotton Byproducts. *J. Biobased Mat. Bioenergy* **2012**, *6*, 431–439. [CrossRef]
15. Pelletier, M.G.; Holt, G.A.; Wanjura, J.D.; Bayer, E.; McIntyre, G. An Evaluation Study of Mycelium Based Acoustic Absorbers Grown on Agricultural By-Product Substrates. *Ind. Crops Prod.* **2013**, *51*, 480–485. [CrossRef]
16. Arifin, Y.H.; Yusuf, Y. Mycelium Fibers as New Resource for Environmental Sustainability. *Procedia Eng.* **2013**, *53*, 504–508. [CrossRef]

17. Travaglini, S.; Noble, J.; Ross, P.G.; Dharan, C.K.H. Mycology Matrix Composites. In Proceedings of the American Society for Composites—Twenty-Eighth Technical Conference, State College, PA, USA, 9–11 September 2013; Bakis, C., Ed.; DEStech Publications, Inc.: Lancaster, PA, USA, 2013; Volume 1, pp. 9–11.
18. Velasco, P.M.; Ortiz, M.P.M.; Giro, M.A.M.; Castelló, M.C.J.; Velasco, L.M. Development of Better Insulation Bricks by Adding Mushroom Compost Wastes. *Energy Build.* **2014**, *80*, 17–22. [CrossRef]
19. Travaglini, S.; Dharan, C.K.H.; Ross, P.G. Mycology Matrix Sandwich Composites Flexural Characterization. In Proceedings of the American Society for Composites 2014—Twenty-Ninth Technical Conference on Composite Materials, La Jolla, CA, USA, 8–10 September 2014; DEStech Publications, Inc.: Lancaster, PA, USA, 2014; Volume 3, pp. 1941–1955.
20. He, J.; Cheng, C.M.; Su, D.G.; Zhong, M.F. Study on the Mechanical Properties of the Latex-Mycelium Composite. *AMM* **2014**, *507*, 415–420. [CrossRef]
21. Lelivelt, R.; Lindner, G.; Teuffel, P.; Lamers, H. The Production Process and Compressive Strength of Mycelium-Based Materials. In Proceedings of the First International Conference on Bio-based Building Materials, Clermont-Ferrand, France, 22–25 June 2015; Eindhoven University of Technology: Clermont-Ferrand, France, 2015; pp. 1–6.
22. Travaglini, S.; Dharan, C.K.H.; Ross, P.G. Thermal Properties of Mycology Materials. In Proceedings of the American Society for Composites—Thirtieth Technical Conference, East Lansing, MI, USA, 28–30 September 2015; Xiao, X., Loos, A.C., Liu, D., Eds.; DEStech Publications: Lancaster, PA, USA, 2015; Volume 3, pp. 1453–1460.
23. Jiang, L.; Walczyk, D.; McIntyre, G.; Bucinell, R. A New Approach to Manufacturing Biocomposite Sandwich Structures: Mycelium-Based Cores. In Proceedings of the ASME 2016 11th International Manufacturing Science and Engineering Conference, Blacksburg, VA, USA, 27 June–1 July 2016; American Society of Mechanical Engineers: Blacksburg, VA, USA, 2016; Volume 1, p. V001T02A025.
24. López Nava, J.A.; Méndez González, J.; Ruelas Chacón, X.; Nájera Luna, J.A. Assessment of Edible Fungi and Films Bio-Based Material Simulating Expanded Polystyrene. *Mater. Manuf. Processes* **2016**, *31*, 1085–1090. [CrossRef]
25. Mayoral González, E.; González Diez, I. Bacterial Induced Cementation Processes and Mycelium Panel Growth from Agricultural Waste. *KEM* **2016**, *663*, 42–49. [CrossRef]
26. Ziegler, A.R.; Bajwa, S.; Holt, G.; Mcintyre, G.; Bajwa, D.S. Evaluation of Physico-Mechanical Properties of Mycelium Reinforced Green Biocomposites Made from Cellulosic Fibers. *Appl. Eng. Agric.* **2016**, *32*, 931–938. [CrossRef]
27. Travaglini, S.; Dharan, C.K.H.; Ross, P.G. Manufacturing of Mycology Composites. In Proceedings of the American Society for Composites Thirty-First Technical Conference, Williamsburg, VA, USA, 19–21 September 2016; Davidson, B.D., Czabaj, M.W., Eds.; American Society for Composites (ASC): Dayton, OH, USA, 2016; Volume 4, pp. 2670–2680.
28. Haneef, M.; Ceseracciu, L.; Canale, C.; Bayer, I.S.; Heredia-Guerrero, J.A.; Athanassiou, A. Advanced Materials From Fungal Mycelium: Fabrication and Tuning of Physical Properties. *Sci. Rep.* **2017**, *7*, 41292. [CrossRef]
29. Pelletier, M.G.; Holt, G.A.; Wanjura, J.D.; Lara, A.J.; Tapia-Carillo, A.; McIntyre, G.; Bayer, E. An Evaluation Study of Pressure-Compressed Acoustic Absorbers Grown on Agricultural by-Products. *Ind. Crops Prod.* **2017**, *95*, 342–347. [CrossRef]
30. Travaglini, S.; Dharan, C.K.H.; Ross, P.G. Biomimetic Mycology Biocomposites. In Proceedings of the 32nd Technical Conference of the American Society for Composites 2017, West Lafayette, IN, USA, 23–25 October 2017; Yu, W., Pipes, R.B., Goodsell, J., Eds.; Curran Associates, Inc.: Red Hook, NY, USA, 2017; Volume 1, pp. 409–419.
31. Attias, N.; Danai, O.; Tarazi, E.; Grobman, Y. Developing Novel Applications of Mycelium Based Bio-Composite Materials for Architecture and Design. In *Books of Abstracts of the Final COST Action FP1303 International Scientific Conference, Proceedings of Building with Bio-Based Materials: Best Practice and Performance Specification, Zagreb, Croatia, 6–7 September 2017*; University of Zagreb Croatia: Zagreb, Croatia, 2017; pp. 76–78.
32. Bajwa, D.S.; Holt, G.A.; Bajwa, S.G.; Duke, S.E.; McIntyre, G. Enhancement of Termite (*Reticulitermes Flavipes* L.) Resistance in Mycelium Reinforced Biofiber-Composites. *Ind. Crops Prod.* **2017**, *107*, 420–426. [CrossRef]
33. Moser, F.J.; Wormit, A.; Reimer, J.J.; Jacobs, G.; Trautz, M.; Hillringhaus, F.; Usadel, B.; Löwer, M.; Beger, A.-L. Fungal Mycelium as a Building Material. In Proceedings of the IASS Annual Symposia, IASS 2017, Beijing, China, 10–14 September 2017; International Association for Shell and Spatial Structures (IASS): Hamburg, Germany, 2017; Volume 2017, pp. 1–7.
34. Jiang, L.; Walczyk, D.; McIntyre, G.; Bucinell, R.; Tudryn, G. Manufacturing of Biocomposite Sandwich Structures Using Mycelium-Bound Cores and Preforms. *J. Manuf. Processes* **2017**, *28*, 50–59. [CrossRef]
35. Campbell, S.; Correa, D.; Wood, D.; Menges, A. Modular Mycelia. Scaling Fungal Growth for Architectural Assembly. In Proceedings of the Computational Fabrication—eCAADe RIS 2017, Cardiff, UK, 27–28 April 2017; Spaeth, B.A., Jabi, W., Eds.; Cardiff University, Welsh School of Architecture: Cardiff, UK, 2017; pp. 125–134.
36. Islam, M.R.; Tudryn, G.; Bucinell, R.; Schadler, L.; Picu, R.C. Morphology and Mechanics of Fungal Mycelium. *Sci. Rep.* **2017**, *7*, 13070. [CrossRef]
37. Yang, Z.; Zhang, F.; Still, B.; White, M.; Amstislavski, P. Physical and Mechanical Properties of Fungal Mycelium-Based Biofoam. *J. Mater. Civ. Eng.* **2017**, *29*, 04017030. [CrossRef]
38. Travaglini, S.; Dharan, C.K.H. Advanced Manufacturing of Mycological Bio-Based Composites. In *Proceedings of the 33rd Technical Conference of the American Society for Composites 2018*; DEStech Publications Inc.: Seattle, WA, USA, 2018; Volume 5, pp. 3387–3399.
39. Läkk, H.; Krijgsheld, P.; Montalti, M.; Wösten, H. Fungal Based Biocomposite for Habitat Structures on the Moon and Mars. In Proceedings of the International Astronautical Congress (IAC 2018), Bremen, Germany, 1–5 October 2018; International Astronautical Federation: Paris, France, 2018; Volume 22, pp. 16102–16112.

40. Xing, Y.; Brewer, M.; El-Gharabawy, H.; Griffith, G.; Jones, P. Growing and Testing Mycelium Bricks as Building Insulation Materials. *IOP Conf. Ser. Earth Environ. Sci.* **2018**, *121*, 022032. [CrossRef]
41. Appels, F.V.W.; Dijksterhuis, J.; Lukasiewicz, C.E.; Jansen, K.M.B.; Wösten, H.A.B.; Krijgsheld, P. Hydrophobin Gene Deletion and Environmental Growth Conditions Impact Mechanical Properties of Mycelium by Affecting the Density of the Material. *Sci. Rep.* **2018**, *8*, 4703. [CrossRef]
42. Jones, M.; Huynh, T.; John, S. Inherent Species Characteristic Influence and Growth Performance Assessment for Mycelium Composite Applications. *Adv. Mater. Lett.* **2018**, *9*, 71–80. [CrossRef]
43. Islam, M.R.; Tudryn, G.; Bucinell, R.; Schadler, L.; Picu, R.C. Mechanical Behavior of Mycelium-Based Particulate Composites. *J. Mater. Sci.* **2018**, *53*, 16371–16382. [CrossRef]
44. Tudryn, G.J.; Smith, L.C.; Freitag, J.; Bucinell, R.; Schadler, L.S. Processing and Morphology Impacts on Mechanical Properties of Fungal Based Biopolymer Composites. *J. Polym. Environ.* **2018**, *26*, 1473–1483. [CrossRef]
45. Jones, M.; Bhat, T.; Kandare, E.; Thomas, A.; Joseph, P.; Dekiwadia, C.; Yuen, R.; John, S.; Ma, J.; Wang, C.-H. Thermal Degradation and Fire Properties of Fungal Mycelium and Mycelium—Biomass Composite Materials. *Sci. Rep.* **2018**, *8*, 17583. [CrossRef] [PubMed]
46. Jones, M.; Bhat, T.; Huynh, T.; Kandare, E.; Yuen, R.; Wang, C.H.; John, S. Waste-Derived Low-Cost Mycelium Composite Construction Materials with Improved Fire Safety. *Fire Mater.* **2018**, *42*, 816–825. [CrossRef]
47. Karana, E.; Blauwhoff, D.; Hultink, E.J.; Camere, S. When the Material Grows: A Case Study on Designing (with) Mycelium-Based Materials. *Int. J. Des.* **2018**, *12*, 119–136.
48. Pelletier, M.G.; Holt, G.A.; Wanjura, J.D.; Greetham, L.; McIntyre, G.; Bayer, E.; Kaplan-Bie, J. Acoustic Evaluation of Mycological Biopolymer, an All-Natural Closed Cell Foam Alternative. *Ind. Crops Prod.* **2019**, *139*, 111533. [CrossRef]
49. Jones, M.P.; Lawrie, A.C.; Huynh, T.T.; Morrison, P.D.; Mautner, A.; Bismarck, A.; John, S. Agricultural By-Product Suitability for the Production of Chitinous Composites and Nanofibers Utilising Trametes Versicolor and Polyporus Brumalis Mycelial Growth. *Process Biochem.* **2019**, *80*, 95–102. [CrossRef]
50. Jiang, L.; Walczyk, D.; McIntyre, G.; Bucinell, R.; Li, B. Bioresin Infused Then Cured Mycelium-Based Sandwich-Structure Biocomposites: Resin Transfer Molding (RTM) Process, Flexural Properties, and Simulation. *J. Clean. Prod.* **2019**, *207*, 123–135. [CrossRef]
51. Vidholdová, Z.; Kormúthová, D.; Ždinský, J.I.; Lagaňa, R. Compressive Resistance of the Mycelium Composite. *Ann. Wars. Univ. Life Sci.-SGGW* **2019**, *107*, 31–36. [CrossRef]
52. Appels, F.V.W.; Camere, S.; Montalti, M.; Karana, E.; Jansen, K.M.B.; Dijksterhuis, J.; Krijgsheld, P.; Wösten, H.A.B. Fabrication Factors Influencing Mechanical, Moisture- and Water-Related Properties of Mycelium-Based Composites. *Mater. Des.* **2019**, *161*, 64–71. [CrossRef]
53. Sun, W.; Tajvidi, M.; Hunt, C.G.; McIntyre, G.; Gardner, D.J. Fully Bio-Based Hybrid Composites Made of Wood, Fungal Mycelium and Cellulose Nanofibrils. *Sci. Rep.* **2019**, *9*, 3766. [CrossRef]
54. Wimmers, G.; Klick, J.; Tackaberry, L.; Zwiesigk, C.; Egger, K.; Massicotte, H. Fundamental Studies for Designing Insulation Panels from Wood Shavings and Filamentous Fungi. *BioResources* **2019**, *14*, 5506–5520. [CrossRef]
55. Bruscato, C.; Malvessi, E.; Brandalise, R.N.; Camassola, M. High Performance of Macrofungi in the Production of Mycelium-Based Biofoams Using Sawdust—Sustainable Technology for Waste Reduction. *J. Clean. Prod.* **2019**, *234*, 225–232. [CrossRef]
56. Attias, N.; Danai, O.; Tarazi, E.; Pereman, I.; Grobman, Y.J. Implementing Bio-Design Tools to Develop Mycelium-Based Products. *Des. J.* **2019**, *22*, 1647–1657. [CrossRef]
57. Elsacker, E.; Vandelook, S.; Brancart, J.; Peeters, E.; De Laet, L. Mechanical, Physical and Chemical Characterisation of Mycelium-Based Composites with Different Types of Lignocellulosic Substrates. *PLoS ONE* **2019**, *14*, e0213954. [CrossRef] [PubMed]
58. Matos, M.P.; Teixeira, J.L.; Nascimento, B.L.; Griza, S.; Holanda, F.S.R.; Marino, R.H. Production of Biocomposites from the Reuse of Coconut Powder Colonized by Shiitake Mushroom. *Ciênc. Agrotec.* **2019**, *43*, e003819. [CrossRef]
59. Zhang, X.; Han, C.; Wnek, G.; Yu, X. (Bill) Thermal Stability Improvement of Fungal Mycelium. In Proceedings of the Materials Science & Technology (MS&T) 2019, Portland, OR, USA, 29 September–3 October 2019; Association for Iron & Steel Technology (AIST): Warrendale, PA, USA, 2019; pp. 1367–1375.
60. Bhardwaj, A.; Vasselli, J.; Lucht, M.; Pei, Z.; Shaw, B.; Grasley, Z.; Wei, X.; Zou, N. 3D Printing of Biomass-Fungi Composite Material: A Preliminary Study. *Manuf. Lett.* **2020**, *24*, 96–99. [CrossRef]
61. Tacer-Caba, Z.; Varis, J.J.; Lankinen, P.; Mikkonen, K.S. Comparison of Novel Fungal Mycelia Strains and Sustainable Growth Substrates to Produce Humidity-Resistant Biocomposites. *Mater. Des.* **2020**, *192*, 108728. [CrossRef]
62. Soh, E.; Chew, Z.Y.; Saeidi, N.; Javadian, A.; Hebel, D.; Le Ferrand, H. Development of an Extrudable Paste to Build Mycelium-Bound Composites. *Mater. Des.* **2020**, *195*, 109058. [CrossRef]
63. Silverman, J.; Cao, H.; Cobb, K. Development of Mushroom Mycelium Composites for Footwear Products. *Cloth. Text. Res. J.* **2020**, *38*, 119–133. [CrossRef]
64. Antinori, M.E.; Ceseracciu, L.; Mancini, G.; Heredia-Guerrero, J.A.; Athanassiou, A. Fine-Tuning of Physicochemical Properties and Growth Dynamics of Mycelium-Based Materials. *ACS Appl. Bio Mater.* **2020**, *3*, 1044–1051. [CrossRef]
65. Zimele, Z.; Irbe, I.; Grinins, J.; Bikovens, O.; Verovkins, A.; Bajare, D. Novel Mycelium-Based Biocomposites (MBB) as Building Materials. *J. Renew. Mater.* **2020**, *8*, 1067–1076. [CrossRef]

66. Ridzqo, I.F.; Susanto, D.; Panjaitan, T.H.; Putra, N. Sustainable Material: Development Experiment of Bamboo Composite Through Biologically Binding Mechanism. *IOP Conf. Ser. Mater. Sci. Eng.* **2020**, *713*, 012010. [CrossRef]
67. Alves, R.M.E.; Alves, M.L.; Campos, M.J. Morphology and Thermal Behaviour of New Mycelium-Based Composites with Different Types of Substrates. In *Progress in Digital and Physical Manufacturing*; Almeida, H.A., Vasco, J.C., Eds.; Lecture Notes in Mechanical Engineering; Springer International Publishing: Cham, Switzerland, 2020; pp. 189–197. ISBN 978-3-030-29040-5.
68. Bhardwaj, A.; Rahman, A.M.; Wei, X.; Pei, Z.; Truong, D.; Lucht, M.; Zou, N. 3D Printing of Biomass–Fungi Composite Material: Effects of Mixture Composition on Print Quality. *JMMP* **2021**, *5*, 112. [CrossRef]
69. Saez, D.; Grizmann, D.; Trautz, M.; Werner, A. Analyzing a Fungal Mycelium and Chipped Wood Composite for Use in Construction. In Proceedings of the IASS Annual Symposium 2020/21 and the 7th International Conference on Spatial Structures Inspiring the Next Generation, Guilford, UK, 23–27 August 2021; Behnejad, S.A., Parke, G.A.R., Samavati, O.A., Eds.; IASS: Madrid, Spain, 2021; pp. 1–10.
70. Müller, C.; Klemm, S.; Fleck, C. Bracket Fungi, Natural Lightweight Construction Materials: Hierarchical Microstructure and Compressive Behavior of Fomes Fomentarius Fruit Bodies. *Appl. Phys. A* **2021**, *127*, 178. [CrossRef]
71. Irbe, I.; Filipova, I.; Skute, M.; Zajakina, A.; Spunde, K.; Juhna, T. Characterization of Novel Biopolymer Blend Mycocel from Plant Cellulose and Fungal Fibers. *Polymers* **2021**, *13*, 1086. [CrossRef] [PubMed]
72. Sivaprasad, S.; Byju, S.K.; Prajith, C.; Shaju, J.; Rejeesh, C.R. Development of a Novel Mycelium Bio-Composite Material to Substitute for Polystyrene in Packaging Applications. *Mater. Today Proc.* **2021**, *47*, 5038–5044. [CrossRef]
73. Jauk, J.; Vašatko, H.; Gosch, L.; Christian, I.; Klaus, A.; Stavric, M. Digital Fabrication of Growth-Combining Digital Manufacturing of Clay with Natural Growth of Mycelium. In Proceedings of the 26th CAADRIA Conference, Hong Kong, China, 29 March–1 April 2021; Association for Computer Aided Architectural Design Research in Asia (CAADRIA): Hong Kong, China, 2021; Volume 1, pp. 753–762.
74. Nashiruddin, N.I.; Chua, K.S.; Mansor, A.F.; Rahman, R.A.; Lai, J.C.; Wan Azelee, N.I.; El Enshasy, H. Effect of Growth Factors on the Production of Mycelium-Based Biofoam. *Clean Technol. Environ. Policy* **2021**. [CrossRef]
75. Răut, I.; Călin, M.; Vuluga, Z.; Oancea, F.; Paceagiu, J.; Radu, N.; Doni, M.; Alexandrescu, E.; Purcar, V.; Gurban, A.-M.; et al. Fungal Based Biopolymer Composites for Construction Materials. *Materials* **2021**, *14*, 2906. [CrossRef] [PubMed]
76. Elsacker, E.; Søndergaard, A.; Van Wylick, A.; Peeters, E.; De Laet, L. Growing Living and Multifunctional Mycelium Composites for Large-Scale Formwork Applications Using Robotic Abrasive Wire-Cutting. *Constr. Build. Mater.* **2021**, *283*, 122732. [CrossRef]
77. Santos, I.S.; Nascimento, B.L.; Marino, R.H.; Sussuchi, E.M.; Matos, M.P.; Griza, S. Influence of Drying Heat Treatments on the Mechanical Behavior and Physico-Chemical Properties of Mycelial Biocomposite. *Compos. Part B Eng.* **2021**, *217*, 108870. [CrossRef]
78. Jose, J.; Uvais, K.N.; Sreenadh, T.S.; Deepak, A.V.; Rejeesh, C.R. Investigations into the Development of a Mycelium Biocomposite to Substitute Polystyrene in Packaging Applications. *Arab J. Sci. Eng.* **2021**, *46*, 2975–2984. [CrossRef]
79. Stelzer, L.; Hoberg, F.; Bach, V.; Schmidt, B.; Pfeiffer, S.; Meyer, V.; Finkbeiner, M. Life Cycle Assessment of Fungal-Based Composite Bricks. *Sustainability* **2021**, *13*, 11573. [CrossRef]
80. Adamatzky, A.; Gandia, A. Living Mycelium Composites Discern Weights via Patterns of Electrical Activity. *J. Biores. Bioprod.* **2021**. [CrossRef]
81. Chan, X.Y.; Saeidi, N.; Javadian, A.; Hebel, D.E.; Gupta, M. Mechanical Properties of Dense Mycelium-Bound Composites under Accelerated Tropical Weathering Conditions. *Sci. Rep.* **2021**, *11*, 22112. [CrossRef]
82. Gou, L.; Li, S.; Yin, J.; Li, T.; Liu, X. Morphological and Physico-Mechanical Properties of Mycelium Biocomposites with Natural Reinforcement Particles. *Constr. Build. Mater.* **2021**, *304*, 124656. [CrossRef]
83. Lee, T.; Choi, J. Mycelium-Composite Panels for Atmospheric Particulate Matter Adsorption. *Results Mater.* **2021**, *11*, 100208. [CrossRef]
84. Zhang, X.; Fan, X.; Han, C.; Li, Y.; Price, E.; Wnek, G.; Liao, Y.-T.T.; Yu, X. (Bill) Novel Strategies to Grow Natural Fibers with Improved Thermal Stability and Fire Resistance. *J. Clean. Prod.* **2021**, *320*, 128729. [CrossRef]
85. Kuribayashi, T.; Lankinen, P.; Hietala, S.; Mikkonen, K.S. Dense and Continuous Networks of Aerial Hyphae Improve Flexibility and Shape Retention of Mycelium Composite in the Wet State. *Compos. Part A Appl. Sci. Manuf.* **2022**, *152*, 106688. [CrossRef]
86. Attias, N.; Danai, O.; Abitbol, T.; Tarazi, E.; Ezov, N.; Pereman, I.; Grobman, Y.J. Mycelium Bio-Composites in Industrial Design and Architecture: Comparative Review and Experimental Analysis. *J. Clean. Prod.* **2020**, *246*, 119037. [CrossRef]
87. Jiang, L.; Walczyk, D.; Mooney, L.; Putney, S. Manufacturing of Mycelium-Based Biocomposites. In Proceedings of the International SAMPE Technical Conference, Covina, CA, USA, 6–9 May 2013; Beckwith, S.W., Ed.; Society for the Advancement of Material and Process Engineering: Long Beach, CA, USA, 2013; pp. 1944–1955.
88. Jiang, L.; Walczyk, D.; McIntyre, G.; Chan, W.K. Cost Modeling and Optimization of a Manufacturing System for Mycelium-Based Biocomposite Parts. *J. Manuf. Syst.* **2016**, *41*, 8–20. [CrossRef]
89. Da Conceição Van Nieuwenhuizen, J.B.; Blauwhoff, D.R.L.M.; De Werdt, M.F.C.; Van Der Zanden, W.G.N.; Van Rhee, D.J.J.L.; Bottger, W.O.J. The Compressive Strength of Mycelium Derived from a Mushroom Production Process. *Acad. J. Civ. Eng.* **2017**, 265–271. [CrossRef]
90. Jones, M.; Huynh, T.; Dekiwadia, C.; Daver, F.; John, S. Mycelium Composites: A Review of Engineering Characteristics and Growth Kinetics. *J. Bionanosci.* **2017**, *11*, 241–257. [CrossRef]

91. Abhijith, R.; Ashok, A.; Rejeesh, C.R. Sustainable Packaging Applications from Mycelium to Substitute Polystyrene: A Review. *Mater. Today Proc.* **2018**, *5*, 2139–2145. [CrossRef]
92. Girometta, C.; Picco, A.M.; Baiguera, R.M.; Dondi, D.; Babbini, S.; Cartabia, M.; Pellegrini, M.; Savino, E. Physico-Mechanical and Thermodynamic Properties of Mycelium-Based Biocomposites: A Review. *Sustainability* **2019**, *11*, 281. [CrossRef]
93. Travaglini, S.; Parlevliet, P.P.; Dharan, C.K.H. Bio-Based Mycelium Materials for Aerospace Applications. In Proceedings of the American Society for Composites—4th Technical Conference, ASC 2019, Atlanta, GA, USA, 23–25 September 2019; DEStech Publications, Inc.: Lancaster, PA, USA, 2019.
94. Cerimi, K.; Akkaya, K.C.; Pohl, C.; Schmidt, B.; Neubauer, P. Fungi as Source for New Bio-Based Materials: A Patent Review. *Fungal Biol. Biotechnol.* **2019**, *6*, 17. [CrossRef]
95. Robertson, O.; Høgdal, F.; Mckay, L.; Lenau, T. Fungal Future: A Review of Mycelium Biocomposites as an Ecological Alternative Insulation Material. In Proceedings of the Balancing Innovation and operation, Lyngby, Denmark, 12–14 August 2020; Design Society: Glasgow, UK, 2020; Volume 101, pp. 55–68.
96. Jones, M.; Mautner, A.; Luenco, S.; Bismarck, A.; John, S. Engineered Mycelium Composite Construction Materials from Fungal Biorefineries: A Critical Review. *Mater. Des.* **2020**, *187*, 108397. [CrossRef]
97. Elsacker, E.; Vandelook, S.; Van Wylick, A.; Ruytinx, J.; De Laet, L.; Peeters, E. A Comprehensive Framework for the Production of Mycelium-Based Lignocellulosic Composites. *Sci. Total Environ.* **2020**, *725*, 138431. [CrossRef]
98. Manan, S.; Ullah, M.W.; Ul-Islam, M.; Atta, O.M.; Yang, G. Synthesis and Applications of Fungal Mycelium-Based Advanced Functional Materials. *J. Bioresour. Bioprod.* **2021**, *6*, 1–10. [CrossRef]
99. Feijóo-Vivas, K.; Bermúdez-Puga, S.A.; Rebolledo, H.; Figueroa, J.M.; Zamora, P.; Naranjo-Briceño, L. Bioproductos Desarrollados a Partir de Micelio de Hongos: Una Nueva Cultura Material y Su Impacto En La Transición Hacia Una Economía Sostenible. *RB* **2021**, *6*, 1637–1652. [CrossRef]
100. Ghazvinian, A. A Sustainable Alternative to Architectural Materials: Mycelium-Based Bio-Composites. In Proceedings of the Divergence in Architectural Research, Atlanta, GA, USA, 15 February 2021; Georgia Institute of Technology: Atlanta, GA, USA; pp. 159–167.
101. Rafiee, K.; Kaur, G.; Brar, S.K. Fungal Biocomposites: How Process Engineering Affects Composition and Properties? *Bioresour. Technol. Rep.* **2021**, *14*, 100692. [CrossRef]
102. Yang, L.; Park, D.; Qin, Z. Material Function of Mycelium-Based Bio-Composite: A Review. *Front. Mater.* **2021**, *8*, 737377. [CrossRef]
103. Angelova, G.V.; Brazkova, M.S.; Krastanov, A.I. Renewable Mycelium Based Composite–Sustainable Approach for Lignocellulose Waste Recovery and Alternative to Synthetic Materials—A Review. *Z. Nat. C* **2021**, *76*, 431–442. [CrossRef]
104. Gandia, A.; van den Brandhof, J.G.; Appels, F.V.W.; Jones, M.P. Flexible Fungal Materials: Shaping the Future. *Trends Biotechnol.* **2021**, *39*, 1321–1331. [CrossRef]
105. Javadian, A.; Ferrand, H.L.; Hebel, D.E.; Saeidi, N. Application of Mycelium-Bound Composite Materials in Construction Industry: A Short Review. *SOJ Mater. Sci. Eng.* **2020**, *7*, 1–9.
106. Bayer, E.; McIntyre, G.; Swersey, B.L. Method for Producing Grown Materials and Products Made Thereby. Patent Application US 2008/0145577, 30 March 2016.
107. Smith, M.J.; Goldman, J.; Boulet-Audet, M.; Tom, S.J.; Li, H.; Hurlburt, T.J. Composite Material, and Methods for Production Thereof. U.S. Patent 11015059 B2, 25 May 2021.
108. Bayer, E.; McIntyre, G. Method for Producing Grown Materials and Products Made Thereby. U.S. Patent 9485917 B2, 8 November 2016.
109. Bayer, E.; McIntyre, G. Method for Producing Rapidly Renewable Chitinous Material Using Fungal Fruiting Bodies and Product Made Thereby. U.S. Patent 8001719 B2, 23 August 2011.
110. Kalisz, R.E.; Rocco, C.A.; Tengler, E.C.J.; Petrella-Lovasik, R.L. Injection Molded Mycelium and Method. U.S. Patent 8313939 B2, 20 November 2012.
111. Rocco, C.A.; Kalisz, R.E. Mycelium Structure with Self-Attaching Coverstock and Method. U.S. Patent 8298810 B2, 30 October 2012.
112. Kalisz, R.E.; Rocco, C.A. Method of Making Foamed Mycelium Structure. U.S. Patent 8227233 B2, 24 July 2012.
113. Kalisz, R.E.; Rocco, C.A. Method of Making Molded Part Comprising Mycelium Coupled to Mechanical Device. U.S. Patent 8227224 B2, 24 July 2012.
114. Rocco, C.A.; Kalisz, R.E. Plasticized Mycelium Composite and Method. U.S. Patent 8227225 B2, 24 July 2012.
115. Rocco, C.A.; Kalisz, R.E. Mycelium Structures Containing Nanocomposite Materials and Method. U.S. Patent 8283153 B2, 9 October 2012.
116. Kalisz, R.E.; Rocco, C.A. Method of Making a Hardened Elongate Structure from Mycelium. U.S. Patent 8298809 B2, 30 October 2012.
117. Kalisz, R.E.; Rocco, C.A. The Sheet Stock Mycelium of Cutting and Method. Patent CN 102329512 B, 1 June 2016.
118. Ross, P. Method for Producing Fungus Structures. U.S. Patent 9410116 B2, 9 August 2016.
119. McIntyre, G.; Bayer, E.; Flagg, D. Method of Producing A Chitinous Polymer Derived from Fungal Mycelium. U.S. Patent 9879219 B2, 30 January 2018.

120. Bayer, E.; Mcintyre, G. Method for Making Dehydrated Mycelium Elements and Product Made Thereby. Patent CA 2834095 C, 11 December 2018.
121. McIntyre, G.; Bayer, E.; Palazzolo, A. Method of Growing Mycological Biomaterials. U.S. Patent 10154627 B2, 18 December 2018
122. Laboutiere, L.; Lemiere, E.; Mechineau, A. Process for Manufacturing a Composite Material Based On Natural Fibres Seeded with Mycelium and Part Obtained with Such a Process. Patent FR 3006693 B1, 1 April 2016.
123. Bayer, E.; McIntyre, G. Method of Growing Electrically Conductive Tissue. U.S. Patent 9253889 B2, 2 February 2016.
124. Winiski, J.M.; Hook, S.S.V. Tissue Morphology Produced with the Fungus Pycnoporus Cinnabarinus. U.S. Patent 9085763 B2, 21 July 2015.
125. Bayer, E.; Mcintyre, G.; Swersey, B. Self Supporting Composite Material. Patent AU 2013/251269 B2, 1 October 2015.
126. McIntyre, G.R.; Tudryn, G.; Betts, J.; Winiski, J. Stiff Mycelium Bound Part and Method of Producing Stiff Mycelium Bound Parts U.S. Patent 10144149 B2, 4 December 2018.
127. Bayer, E.; McIntyre, G.R. Method for Growing Mycological Materials. U.S. Patent 9394512 B2, 19 July 2016.
128. Schaak, D.D.; Lucht, M.J. Biofilm Treatment of Composite Materials Containing Mycelium. U.S. Patent 9469838 B2, 18 October 2016.
129. Chen, Q.; Peng, Z.; Chen, Q.; Liu, H.; Chen, C.; Zhang, Y.; Liu, Y. Production Method for Biomass Packing Material. Patent CN 105292758 B, 8 December 2017.
130. Betts, J.D.; Mcintyre, G.R.; Mooney, L.; Tudryn, G.J. Method of Manufacturing a Stiff Engineered Composite. AU 2015/271912 B2, 20 February 2020.
131. Winiski, J.; Hook, S.V.; Lucht, M.; McIntyre, G. Process for Solid-State Cultivation of Mycelium on a Lignocellulose Substrate. U.S. Patent 9914906 B2, 13 March 2018.
132. Deng, Y.; Xu, Y.; Gao, C.; Yu, S.; Dong, W. Method for Preparing Degradable Buffer Material by Using Macro Fungi Mycelia. Patent CN 106148199 B, 18 October 2019.
133. Chen, Q.; Peng, Z.; Chen, Q.; Liu, H.; Chen, C.; Zhang, Y.; Liu, Y. Using Bagasse as Fungi Based Biomass Packaging Material of Major Ingredient and Preparation Method Thereof. Patent CN 106633989 B, 24 May 2019.
134. Amstislavski, P.; Yang, Z.; White, M.D. Thermal Insulation Material from Mycelium and Forestry Byproducts. U.S. Patent 10604734 B2, 31 March 2020.
135. Lee, B.G. An Eco-Friendly Packing Materials Comprising Mushroom Mycelium and the Process for the Preparation Thereof. Patent KR 102256335 B1, 26 May 2021.
136. Sydor, M. Geometry of Wood Screws: A Patent Review. *Eur. J. Wood Prod.* **2019**, *77*, 93–103. [CrossRef]
137. Bayer, E.; McIntyre, G. Method for Making Dehydrated Mycelium Elements and Product Made Thereby. Patent Application US 20120270302, 25 October 2012.
138. Kaplan-Bie, J.H.; Bonesteel, I.T.; Greetham, L.; McIntyre, G.R. Increased Homogeneity of Mycological Biopolymer Grown into Void Space. Patent Application WO 2019/099474 A1, 23 May 2019.
139. Ross, P. Method for Producing Fungus Structures. Patent Application US 2012/0135504 A1, 9 August 2016.
140. Kaplan-Bie, J.H. Solution Based Post-Processing Methods for Mycological Biopolymer Material and Mycological Product Made Thereby. Patent Application US 2018/0282529 A1, 4 October 2018.
141. Carlton, A.; Bayer, E.; McIntyre, G.; Kaplan-Bie, J. Method of Producing a Mycological Product and Product Made Thereby. Patent Application US 2020/0024577 A1, 23 January 2020.
142. Elsacker, E.; Peters, K.; Poncelet, W. *Hard Mycelium Materials Manual*; Manuals & guides; BioFabForum: Sint-Denijs-Westrem, Belgium, 2018.
143. Branowski, B.; Zabłocki, M.; Sydor, M. Experimental Analysis of New Furniture Joints. *Bioresources* **2018**, *13*, 370–382. [CrossRef]
144. Branowski, B.; Starczewski, K.; Zabłocki, M.; Sydor, M. Design Issues of Innovative Furniture Fasteners for Wood-Based Boards. *Bioresources* **2020**, *15*, 8472–8495. [CrossRef]
145. Sydor, M.; Wieloch, G. Construction Properties of Wood Taken into Consideration in Engineering Practice. *Drewno* **2009**, *52*, 63–73.
146. Antov, P.; Savov, V. Possibilities for Manufacturing Eco-Friendly Medium Density Fibreboards from Recycled Fibres—A Review. In Proceedings of the 30th International Conference on Wood Science and Technology, ICWST 2019 and 70th Anniversary of Drvna industrija Journal: Implementation of Wood Science in Woodworking Sector, Zagreb, Croatia, 12–13 December 2019; Lucic, R.B., Zivkovic, V., Barcic, A.P., Vlaovic, Z., Eds.; University of Zagreb, Faculty of Forestry: Zagreb, Croatia, 2019; pp. 18–24.
147. Branowski, B.; Zabłocki, M.; Sydor, M. The Material Indices Method in the Sustainable Engineering Design Process: A Review. *Sustainability* **2019**, *11*, 5465. [CrossRef]
148. Vukoje, M.; Itrić Ivanda, K.; Kulčar, R.; Marošević Dolovski, A. Spectroscopic Stability Studies of Pressure Sensitive Labels Facestock Made from Recycled Post-Consumer Waste and Agro-Industrial By-Products. *Forests* **2021**, *12*, 1703. [CrossRef]
149. Jivkov, V.; Simeonova, R.; Antov, P.; Marinova, A.; Petrova, B.; Kristak, L. Structural Application of Lightweight Panels Made of Waste Cardboard and Beech Veneer. *Materials* **2021**, *14*, 5064. [CrossRef] [PubMed]
150. Bekhta, P.; Noshchenko, G.; Réh, R.; Kristak, L.; Sedliačik, J.; Antov, P.; Mirski, R.; Savov, V. Properties of Eco-Friendly Particleboards Bonded with Lignosulfonate-Urea-Formaldehyde Adhesives and PMDI as a Crosslinker. *Materials* **2021**, *14*, 4875. [CrossRef] [PubMed]

151. Antov, P.; Savov, V.; Trichkov, N.; Krišťák, Ľ.; Réh, R.; Papadopoulos, A.N.; Taghiyari, H.R.; Pizzi, A.; Kunecová, D.; Pachikova, M. Properties of High-Density Fiberboard Bonded with Urea–Formaldehyde Resin and Ammonium Lignosulfonate as a Bio-Based Additive. *Polymers* **2021**, *13*, 2775. [CrossRef]
152. Pędzik, M.; Janiszewska, D.; Rogoziński, T. Alternative Lignocellulosic Raw Materials in Particleboard Production: A Review. *Ind. Crops Prod.* **2021**, *174*, 114162. [CrossRef]

 polymers

Article

Application of Failure Criteria on Plywood under Bending

Miran Merhar

Department of Wood Science and Technology, Biotechnical Faculty, University of Ljubljana, Jamnikarjeva 101, 1000 Ljubljana, Slovenia; Miran.Merhar@bf.uni-lj.si; Tel.: +386-1-320-36-29

Abstract: In composite materials, the use of failure criteria is necessary to determine the failure forces. Various failure criteria are known, from the simplest ones that compare individual stresses with the corresponding strength, to more complex ones that take into account the sign and direction of the stress, as well as mutual interactions of the acting stresses. This study investigates the application of the maximum stress, Tsai-Hill, Tsai-Wu, Puck, Hoffman and Hashin criteria to beech plywood made from a series of plies of differently oriented beech veneers. Specimens were cut from the manufactured boards at various angles and loaded by bending to failure. The mechanical properties of the beech veneer were also determined. The specimens were modelled using the finite element method with a composite modulus and considering the different failure criteria where the failure forces were calculated and compared with the measured values. It was found that the calculated forces based on all failure criteria were lower than those measured experimentally. The forces determined using the maximum stress criterion showed the best agreement between the calculated and measured forces.

Keywords: max stress; Tsai-Wu; Tsai-Hill; Puck; Hoffman; Hashin; failure criteria; beech; finite element modelling; composites

Citation: Merhar, M. Application of Failure Criteria on Plywood under Bending. *Polymers* **2021**, *13*, 4449. https://doi.org/10.3390/polym13244449

Academic Editors: Pavlo Bekhta, Petar Antov, Yonghui Zhou and Viktor Savov

Received: 29 November 2021
Accepted: 15 December 2021
Published: 18 December 2021

1. Introduction

The use of wood as a sustainable material is increasing. Wood has anisotropic mechanical properties [1], but when the principal axes coincide with the orientation of the tissue, wood can be considered orthotropic and, in certain cases, transversely isotropic. One such example is wood composite, including plywood. The ratios of mechanical properties between the longitudinal and transverse directions of wood are between 10 and 1 [1], which are greatly reduced in plywood where the veneer sheets are glued together at different angles [2,3].

Plywood is a very common and long-used building material. It can be made from veneers of different tree species, which affects the physical and mechanical properties of plywood [4–8]. Important factors influencing the properties of plywood include design features such as the number, thickness and orientation of individual layers and the technological process of panel production [9–15].

Since different combinations of the listed factors can be used to obtain different properties, many authors have already dealt with the determination of the properties. Some have determined the modulus of elasticity [4,16,17], others the shear modulus [18,19], while still others have measured the strength of already manufactured plywood.

However, since plywood is composed of individual layers of veneer, it is important to know the mechanical properties in all directions of the wood, as the mechanical properties can vary greatly between longitudinal and transverse direction. In addition, the strength of wood also varies when it is subjected to tension or compression. Therefore, even for uniaxial stress conditions, a distinction must be made between compressive and tensile loads [1]. For a multi-axial stress state, in addition to the distinction between compression and tension, the strength in different directions must also be considered, for both normal and shear strength.

The relationships between the actual stresses and the corresponding strengths, and the consideration of whether or not the specimen will fail under specific load, are governed by various failure criteria. One of the most commonly used criteria are max stress [20], Tsai-Hill [21], Hoffman [22], Hashin [22], Tsai-Wu [23] and Puck [24,25], shown in Table 1. While the Tsai-Hill, Hoffman and Tsai-Wu criteria can only be used to determine the failure load or to determine whether the specimen will fail at a particular combination of stresses, the max stress, Puck and Hashin criteria can also be used to determine the failure mode, i.e., fibre or inter-fibre failure.

Table 1. Failure criteria. σ_x and σ_y—normal stresses in x and y directions; τ_{xy}—shear stress; X and Y—normal strengths (compressive or tensile) in the x and y directions; S—shear strength; X_t and X_c—normal tensile and compressive strength in x direction; Y_t and Y_c—normal tensile and compressive strength in y direction; a_{xy}—interaction coefficient; . $R_{\perp}^{(+)}$ and $R_{\perp}^{(-)}$—normal tensile and compressive strength perpendicular to the fibres; $R_{\perp\parallel}$—shear strength; $p_{\parallel\perp}^{(-)}$, $p_{\parallel\perp}^{(+)}$, $p_{\perp\perp}^{(-)}$ and $p_{\perp\perp}^{(+)}$—slopes of the failure curves.

Criterion	Equation	No.
max stress [20]	$\left(\left\lvert\frac{\sigma_x}{X}\right\rvert, \left\lvert\frac{\sigma_y}{Y}\right\rvert, \left\lvert\frac{\tau_{xy}}{S}\right\rvert\right) = 1$	(1)
Tsai-Hill [21]	$\left(\frac{\sigma_x}{X}\right)^2 - \frac{\sigma_x\sigma_y}{X^2} + \left(\frac{\sigma_y}{Y}\right)^2 + \left(\frac{\tau_{xy}}{S}\right)^2 = 1$	(2)
Hoffman [22]	$\frac{\sigma_x{}^2}{X_tX_c} + \frac{\sigma_y{}^2}{Y_tY_c} + \frac{\tau_{xy}{}^2}{S^2} - \frac{\sigma_x\sigma_y}{X_tX_c} + \frac{\sigma_x}{X_tX_c} + \frac{\sigma_y}{Y_tY_c} = 1$	(3)
Hashin [22]	$\begin{gathered} \left(\frac{\sigma_x}{X_t}\right)^2 + \left(\frac{\tau_{xy}}{S}\right)^2 = 1, \sigma_x > 0; -\frac{\sigma_x}{X_c} = 1, \sigma_x < 0 \\ \left(\frac{\sigma_y}{Y_t}\right)^2 + \left(\frac{\tau_{xy}}{S}\right)^2 = 1, \sigma_y > 0 \\ \left(\frac{\sigma_y}{2S}\right)^2 + \left(\frac{\tau_{xy}}{S}\right)^2 + \left[\left(\frac{Y_c}{2S}\right)^2 - 1\right]\frac{\sigma_y}{Y_c} = 1, \sigma_y < 0 \end{gathered}$	(4)
Tsai-Wu [23]	$\left(\frac{1}{X_t} - \frac{1}{X_c}\right)\sigma_x + \left(\frac{1}{Y_t} - \frac{1}{Y_c}\right)\sigma_y + \frac{1}{X_tX_c}\sigma_x^2 + \frac{1}{Y_tY_c}\sigma_y^2 + 2a_{xy}\sqrt{\left(\frac{1}{X_tX_c}\frac{1}{Y_tY_c}\sigma_x\sigma_y\right)} + \frac{1}{S^2}\tau_{xy}^2 = 1$	(5)
Puck [24,25]	$\begin{gathered} \left\lvert\frac{\sigma_x}{X_t}\right\rvert = 1, \sigma_x > 0; \left\lvert\frac{\sigma_x}{X_c}\right\rvert = 1, \sigma_x < 0 \\ \sqrt{\left(\frac{\tau_{xy}}{R_{\perp\parallel}}\right)^2 + \left(1 - p_{\perp\parallel}^{(+)}\frac{R_{\perp}^{(+)}}{R_{\perp\parallel}}\right)^2\left(\frac{\sigma_y}{R_{\perp}^{(+)}}\right)^2} + p_{\perp\parallel}^{(+)}\frac{\sigma_y}{R_{\perp\parallel}} = 1, \sigma_y \geq 0 \\ \frac{1}{R_{\perp\parallel}}\left(\sqrt{\tau_{xy}^2 + \left(p_{\parallel\perp}^{(-)}\sigma_y\right)^2} + p_{\parallel\perp}^{(-)}\sigma_y\right) = 1, \sigma_y < 0, \left\lvert\frac{\sigma_y}{\tau_{xy}}\right\rvert \leq \frac{R_{\perp\perp}^A}{\lvert\tau_{xyc}\rvert} \\ \left[\left(\frac{\tau_{xy}}{2\left(1+p_{\perp\perp}^{(-)}\right)R_{\perp\parallel}}\right)^2 + \left(\frac{\sigma_y}{R_{\perp}^{(-)}}\right)^2\right]\frac{R_{\perp}^{(-)}}{-\sigma_y} = 1, \sigma_y < 0, \left\lvert\frac{\sigma_y}{\tau_{xy}}\right\rvert \geq \frac{R_{\perp\perp}^A}{\lvert\tau_{xyc}\rvert} \end{gathered}$	(6)

Most researchers have studied the application of failure criteria to unidirectional (UD) composites [26–34]. In these cases, the criteria predict the failure loads more or less satisfactorily. In a UD composite, the failure of the matrix usually implies the failure of the entire specimen. In a composite material, such as a plywood, where the individual layers are arranged at different angles, the failure of a particular layer in the transverse direction (matrix) does not necessarily imply the failure of the entire specimen. Thus, an adjacent layer whose fibres are oriented at a certain angle to the preceding layer may arrest the progression of the failure of the matrix, which in some cases can also carry the load of the preceding layer broken by the matrix.

Furthermore, the described criteria have already been used in the calculation of failure forces in oriented wood specimens loaded with different combinations of normal and shear stresses [35–37]. In doing so, the researchers analytically determined the normal and shear stresses as the function of grain direction and loading force and then inserted them into various criteria. However, as far as the author is aware, the application of these criteria in determining the failure forces of wood composites such as plywood, where the individual veneer layers are oriented differently, cannot be found anywhere in the literature. Thus, to determine the failure force of the entire plywood specimen, one must know the magnitude and direction of the stress in individual veneer layers and then determine the failure load for each individual layer. This is only possible by using the finite element method to get the result with the required accuracy.

The aim of this study is therefore to investigate the applicability or accuracy of various failure criteria in determining the failure forces of plywood loaded in bending. Due to the structure of the panel with differently oriented veneers, each ply exhibits a different biaxial stress state, which makes the application of classical bending mechanics very difficult. From this point of view, the use of failure criteria in combination with the finite element method proves to be the most suitable when it comes to determining the failure force in each layer of the panel. These forces are compared with the experimentally determined values, and verified the application of failure criteria's in wood composites such as plywood.

2. Materials and Methods

2.1. Plywood Processing

Peeled beech (*Fagus sylvatica*) veneer of 600 mm × 600 mm with tangential texture and nominal thickness of 1.5 mm was used to produce the plywood. The veneer was free from visual defects with homogeneous texture and was obtained from a single log with uniform growth of annual rings. The veneer was conditioned in the laboratory at a constant temperature of 22 °C and relative air humidity of 45%. After conditioning, the veneer had an average moisture content of 6.7%.

The veneers were used to produce 11-, 7- and 3- ply plywood panels with the veneer orientations shown in Table 2 and Figure 1. The panels marked 11E were manufactured with the same orientations of all veneer plies. They were used to fabricate test specimens to determine the mechanical properties in the principle directions of the wood tissue.

Table 2. Tissue orientations (θ) of individual layers for 3-, 7- and 11-layer beech veneer plywood.

Ply no.	11 Layers			7 Layers		3 Layers
	11E (°)	**11A (°)**	**11P (°)**	**7A (°)**	**7P (°)**	**3P (°)**
1	0	0	0	0	0	0
2	0	30	90	45	90	90
3	0	−30	0	−45	0	0
4	0	60	90	90	90	-
5	0	−60	0	−45	0	-
6	0	90	90	45	90	-
7	0	−60	0	0	0	-
8	0	60	90	-	-	-
9	0	−30	0	-	-	-
10	0	30	90	-	-	-
11	0	0	0	-	-	-

Meldur H97 melamine-urea-formaldehyde (MUF) adhesive was used, provided by Melamin d.d. (Kočevje, Slovenia). According to the manufacturer, MUF adhesive consisted of 62% ± 2% dry content, viscosity (as per SIST EN ISO 2431 (2019) ϕ4, 20 °C) was 80 s to 200 s and consisted of maximum 0.5% free formaldehyde. To the adhesive 1% NH_4Cl catalyst and 5% filler (rye flour) were added to increase viscosity. The mixture was then stirred for 15 min until a homogeneous mixture was obtained.

The adhesive application to the individual veneer layers was 180 g/m^2. The pressing of the board was 1.6 MPa, the temperature 130 °C and the pressing time was 13 min, 10 min and 7 min for 11-, 7- and 3- layer plywood, respectively. After pressing, the boards were stacked, weighed and conditioned for 1 week.

After conditioning, the boards were cut to the size of 550 mm × 550 mm, and their thickness was measured. The seven-layer board was 9.9 mm thick, while the 11-layer board was 15.6 mm thick, corresponding to 1.42 mm per layer.

Mechanical properties of plywood building material, i.e., veneer sheets in longitudinal, radial and tangential directions were determined. Tensile strength was determined on specimens of single veneers and compressive strength, shear strength, modulus of elasticity and shear modulus on specimens cut from 11E plywood.

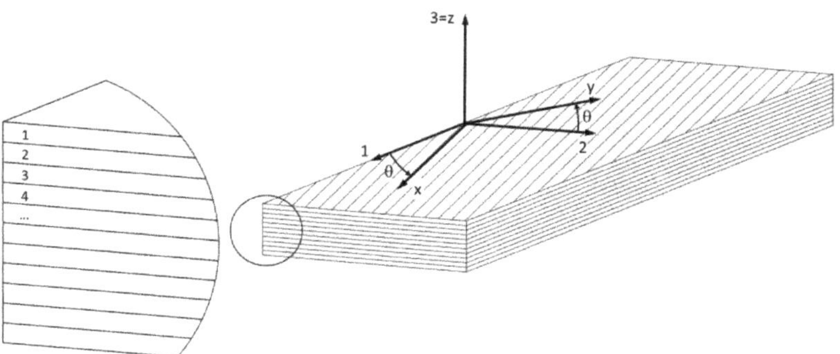

Figure 1. Plywood with specimen coordinate system (1-2-3) and principal material coordinate system (x-y-z).

2.2. Determination of Principal Mechanical Properties of Plywood Building Material (i.e., Veneer Sheets)

2.2.1. Tensile Strength

Tensile strength in longitudinal (L) direction was determined in accordance with the SIST EN 408:2010 standard [38]. The test specimens were made of single layers of veneer with dimensions 200 mm × 30 mm × 1.5 mm (L × T × R). The tensile strength in tangential direction was determined according to the ASTM D143:2000 [39] standard, where the test specimens with dimensions 60 mm × 60 mm × 15.6 mm and a width of 25 mm in the narrower part of the cross section were cut from 11E panel. The specimens were clamped in the jaws of the Zwick Z005 universal testing machine (Zwick Roell Group, Ulm, Germany) and tensioned at a rate of 0.3 mm/min. The tensile strength was then calculated using the following equation

$$\sigma_t = \frac{F_{t\,max}}{b\,h} \tag{7}$$

where F_{tmax} is the failure force and b and h are the width and thickness of the specimens, respectively. The tensile strength in radial direction was not determined experimentally because there are no fractures in the radial direction in the fabricated flexural specimens as plane stress state in the TL plane is expected. Therefore, the same value was used for the radial tensile strength as for the tangential one.

2.2.2. Compressive Strength

The compressive strength in the longitudinal (L), tangential (T) and radial (R) directions was determined on specimens measuring 15.6 mm × 15.6 mm × 15.6 mm which were cut from 11-ply boards (11E). Three specimens were loaded in the longitudinal direction, 3 in the tangential direction and 3 in the radial direction where the loading rate was 0.3 mm/min. The EN408:2010 [38] standard was used to determine compressive strength with the following equation

$$\sigma_c = \frac{F_{c,prop}}{b\,l} \tag{8}$$

where $F_{c,prop}$, is the failure force determined as the intersection between the loading curve and the parallel of the trend line from the initial linear part between 10% and 40% of the failure force with an offset of 0.01 × h, and b and l are the width and length of the specimen, respectively.

2.2.3. Shear Strength

The shear strength τ_{TL} was determined by the asymmetric four-point bending (AFPB) test [40]. Specimens measuring 150 mm × 50 mm × 15.6 mm (T × L × R) were made from 11E plywood. The specimens were cut in the middle so that the cross-sectional height was 15.7 mm and the loading rate was 2 mm/min. The shear strength was calculated using equation [40]

$$\tau_{TL} = \frac{P}{2bh} \tag{9}$$

where P is the force at which shear failure occurs, b is the cross-sectional height, which was 15.7 mm, and h is the cross-sectional thickness with the value of 15.6 mm.

The shear strengths in the RL and TR directions were calculated using Equation (10) [41], as well as the shear strength in the TL direction and compared with the experimentally determined values

$$\begin{gathered}\tau_{RL} = \tfrac{1}{3}\left[3\,\sigma_{uR}\sigma_{uL}\right]^{0.5},\ \ \tau_{TL} = \tfrac{1}{3}\left[3\,\sigma_{uT}\sigma_{uL}\right]^{0.5} \\ \tau_{RT} = \left[\frac{16}{(K-1)^2\,(\sigma_{uR}+\sigma_{uT})^2} + \frac{1}{\sigma_{uR}\sigma_{uT}} - \frac{1}{\sigma_{uR}{}^2} - \frac{1}{\sigma_{uT}{}^2}\right]^{-0.5}\end{gathered} \tag{10}$$

where σ_{uL}, σ_{uR} and σ_{uT} are the strengths in longitudinal, radial and tangential directions, respectively, and K is a constant which equals to 0.2 for hardwood.

2.2.4. Modulus of Elasticity

The modulus of elasticity was determined on 11-layer specimens cut from 11E plates. Three specimens with dimensions 410 mm × 40 mm × 15.6 mm in L × T × R and three specimens in T × L × R direction were cut from the plate to determine the elastic modulus in longitudinal and tangential directions, respectively. The specimens were loaded to four-point bending according to the EN408:2010 [38] standard, where the loading rate was 2.7 mm/min. The specimens were loaded to failure and then the modulus of elasticity was determined in the linear range between 0.2 and 0.3 F_{max} using the following equation:

$$E_{L,T} = \frac{3al^2 - 4a^3}{4bh^3\left(\frac{w_2-w_1}{F_2-F_1}\right)} \tag{11}$$

where F_1 and F_2 are the forces at displacement w_1 and w_2, respectively, b and h are the width and thickness of the specimen, l is the distance between supports, which was 276 mm, and a is the distance between the support and the location of loading, which was 90 mm.

The modulus of elasticity in the radial direction was taken from literature [1], as exact value was not necessary because the plane stress state in TL plane was predicted.

2.2.5. Shear Modulus

The shear modulus in the TL direction was determined using the plate twist method [19]. Three plates measuring 145 mm × 145 mm were made from plywood 11E. The plates were then diagonally supported as well as loaded at distance of 195 mm. The shear modulus was then calculated using the following equation:

$$G_{TL} = \frac{3\,B^2\,s}{4h^3}\cdot\frac{P}{\delta} \tag{12}$$

where P and δ are the shear force and deflection, respectively, and s is the correction factor related to the position of the clamp or load with respect to the diagonal of the specimen, calculated using

$$s = 3\cdot\left(\frac{S}{D}\right)^2 - 2\cdot\left(\frac{S}{D}\right) - 2\cdot\left(1-\frac{S}{D}\right)^2\cdot ln\left(1-\frac{S}{D}\right) \tag{13}$$

where S is the distance between supports or between loads and D is the diagonal of the specimen.

The shear moduli in RL and TR directions were calculated according to Bachtiar et. al. [41] as well as the TL shear modulus, and compared with the measured values:

$$G_{RL} = \left(\frac{\nu_{RL}+1}{E_L} + \frac{\nu_{LR}+1}{E_R}\right)^{-1}, \quad G_{TL} = \left(\frac{\nu_{TL}+1}{E_L} + \frac{\nu_{LT}+1}{E_T}\right)^{-1}$$
$$G_{Rt} = \left(\frac{\nu_{TR}-1}{E_R} + \frac{\nu_{RT}-1}{E_T} + \frac{8}{(1-K)\,(E_R+E_T)}\right)^{-1} \quad (14)$$

2.3. *Plywood Failure Bending Forces Determination*

Specimens, 40 mm wide and 410 mm, 270 mm and 110 mm long, were prepared from 11-, 7- and 3-ply plates, respectively. 11A and 7A specimens were cut with an angular distribution of 22.5°: 0°, 22.5°, 45°, 67.5°, 90°, −22.5°, −45° and −67.5° (Figure 2), while the 11P, 7P and 3P specimens were cut only at the following angles due to symmetry: 0°, 22.5°, 45°, 67.5° and 90°. Since the boards were made from veneer with uniform mechanical properties, obtained from a log with uniform annual growth, only four specimens were made for each combination of tissue orientations. If there were major differences in mechanical properties between samples with the same combinations of tissue orientations, additional samples would be made.

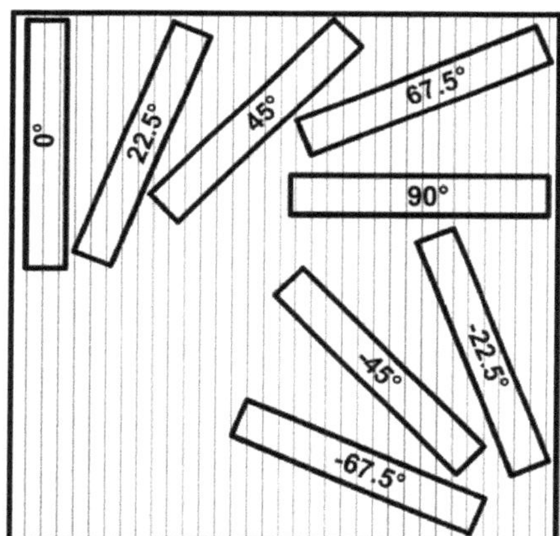

Figure 2. Scheme of cutting specimens from the boards. Numbers designate the first layer tissue orientation according to the individual specimen coordinate system as shown in Figure 1.

The specimens were tested with four-point bending to the specimen failure. The distance between supports was 276 mm, 180 mm and 80 mm for the 11-, 7- and 3-ply plates, respectively, while the distance between loads and supports was 90 mm, 58 mm and 26 mm, respectively, and the loading rate was 2.7 mm/min.

The maximum force at which the specimen failed and the force at which the linear part of the curve changed to a nonlinear one was then determined from the measurements. The force at which the transition from the linear to the nonlinear part occurred was used as the failure force for the two reasons.

The first reason was that when wood specimens are subjected to bending loads, compression failure may occur first at the top of the specimen, but the specimen does not break in two. It is characteristic of wood that the compressive strengths are lower than the tensile one [1]. This means that when the wood is bent, the ultimate strength on the compressed top side is reached earlier than on the tensile bottom side. Although compressive failure occurs on the top side of the specimen, the specimen withstands the loads further where densification of wood tissue occurs. The compressive failure of the tissue can be identified from the loading curve as a transition from the linear to the nonlinear part, which is identical to the situation in the compression test. As the load increases, the tensile stresses at the bottom of the specimen further increase, where they eventually break when they reach the tensile strength or appropriate combination of normal and shear stresses.

The second reason for considering the transition force from the linear to the nonlinear part is that the tensile failure of the veneer in tangential direction may occur first on the bottom side of the specimen and only then compressive failure on the top side of the specimen. As shown by the results of the experimental work, veneer, unlike solid wood, has a lower tensile strength than compressive strength in tangential direction which is due to the way it is manufactured. Since the veneer in plywood is glued at different angles, local tensile failures in tangential direction are usually inhibited by the adjacent layer with more longitudinally oriented fibres, which also bear the load from the broken layer, which is not a major problem as the longitudinal strength is up to 30-times greater than the tangential transverse strength. Although the specimen has not yet collapsed on the lower tensile side, the undamaged cross-section of the specimen is smaller, which can be seen from the increase in the compliance of the specimen as it transitions from the linear to the nonlinear part of the loading curve.

The transition from the linear to the nonlinear part of the curve, and thus the determination of the failure force, was determined in a similar way to the compression test. From the linear part of the loading curve, a trend line was obtained from the measurements between 0.2 and 0.3 F_{max}, and then a parallel was drawn representing the strain of 0.0002 mm/mm. The force of tissue failure was determined from the intersection between the offset line and the measurement curve.

2.4. Finite Element Modelling

All specimens tested experimentally for four-point bending with different fabric orientations and numbers of plies were also modelled using the finite element method with the ANSYS v17.2 (ANSYS, Inc., Canonsburg, PA, USA), Composite Module software (Figure 3). The size of the specimens and the spacing between supports and loads were the same as those in the experimental work. The specimens were modelled as planar elements with a thickness of 1.418 mm per layer and the size of elements of 3 mm.

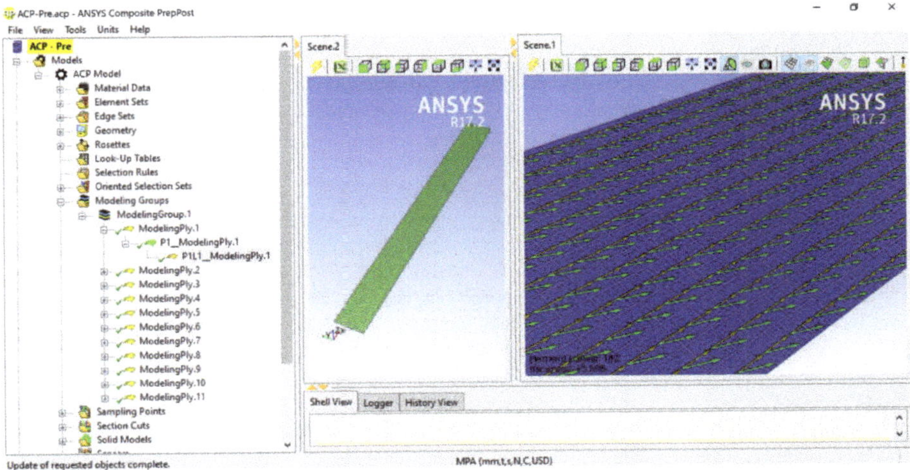

Figure 3. Composite finite element model.

Since a linear model was used, only the determination of the failure of the first most critical layer is credible. No further sequence of failure of other layers is possible with a linear model, as it does not include nonlinear tissue failure. The latter would only be possible with a non-linear model, which is beyond the scope of this study.

In the model, a force of 1000 N was applied to the specimen. Factors of safety (FOS) were calculated, defined as the ratio between the actual force and failure force where a safety factor of 1 means the failure of the layer. The safety factors were calculated for each layer using the max stress, Tsai-Wu, Puck, Tsai-Hill, Hoffmann and Hashin failure criteria (Table 1).

The failure criterion of Tsai-Wu (Table 1) includes a constant of a_{xy} that must be between -1 and 1, but despite numerous studies by different authors [32,42,43], there is still no universal equation for its determination. The following values were used in the study: -1, -0.7, -0.3, 0, 0.3, 0.6 and 1. For each value, the factors of safety and the corresponding failure force of the critical layer were calculated, and then compared with the measured force, where the coefficient of determination R^2 between the calculated and the average of measured values was determined.

For the Puck criterion (Table 1), it is also necessary to determine the slopes $p_{\parallel\perp}^{(-)}$, $p_{\parallel\perp}^{(+)}$, $p_{\perp\perp}^{(-)}$ and $p_{\perp\perp}^{(+)}$ (in this paper designated as p) of the fracture curves. In the literature, there are only constants given for glass fibre/epoxy and carbon fibre/epoxy [25] between 0.2 and 0.3. Since the data for wood have not been found yet in the literature, the values 0.01, 0.15, 0.3 and 0.5 were considered in the study. For each individual value, the factor of safety and the failure force were calculated, and then compared with the measured values.

For the elastic modulus, shear moduli and strengths in the different directions, the values obtained from experimental work were used. The Poisson's ratios were taken from the literature [1] where the values for ν_{LT}, ν_{LR} and ν_{TR} were 0.518, 0.448 and 0.359, respectively.

Minimum safety factors for the weakest layer of the plate were determined using all the failure criteria, which were then multiplied by the load force. The calculated failure forces were then compared with the forces obtained from the experiment, so that for each criterion the coefficient of determination R^2 between the calculated and the average of measured values was calculated.

2.5. Statistical Evaluation of Data

The average, standard deviation and coefficient of variation (COV) were calculated for the measured values of each group. The correlation between different groups of data (experimentally determined and theoretically calculated ones) was determined using the coefficient of determination (R^2), defined as the proportion of variance in the first group of variables that is correlated with the second group of variables.

3. Results

3.1. Principle Mechanical Properties of Plywood Building Material (Veneer Sheets)

Figure 4 shows the compressive strength measurements in longitudinal, tangential and radial directions along with the trend lines used to determine the failure force for the compressive strength calculation, while Figure 5 shows compressed specimens. In the initial phase of the measurement, the forces increase nonlinearly due to the adaptation of the specimen to the table of the testing machine. The nonlinear part is followed by a linear part and then again, a nonlinear part due to the local tissue failure.

Table 3 shows the test results. The average compressive strengths in the longitudinal, tangential and radial directions are 65.4 MPa, 11.4 MPa and 11.1 MPa, respectively, while the average tensile strengths in the longitudinal and tangential directions are 96.8 MPa and 3.7 MPa, respectively. The tensile strength in the longitudinal direction is greater than the compressive strength and comparable to that in the literature [44,45] while the tensile strength in the tangential direction is much lower than the compressive one and also lower than stated in the literature which gives 9 MPa for beech in the tangential direction [1]. The reason for the lower tensile strength are microcracks that form when the veneer is peeled, and due to the microcracks, stress concentrations occur at tensile load. For the tensile strength in radial direction, the same value as in tangential direction was used as the plane stress state in TL plane is considered.

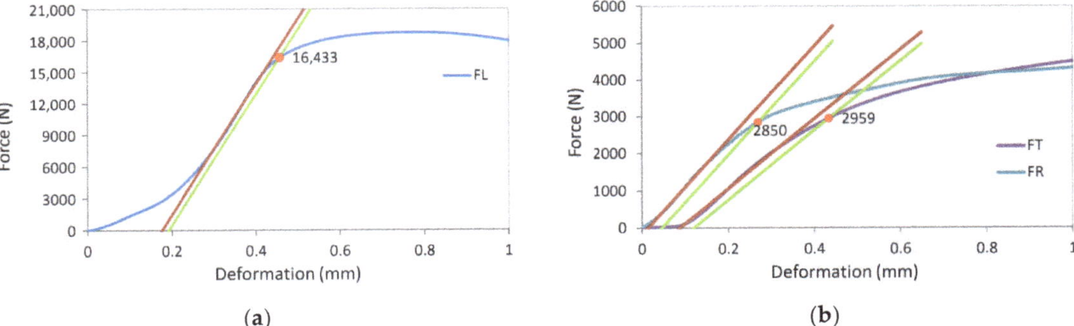

Figure 4. Measured forces and deformations in compression strength determination experiment. Red line represents linear trend between 10% and 40% of the failure force. Green line represents 1% strain offset of the red line: (**a**) Longitudinal direction; (**b**) Tangential and radial direction.

Figure 5. Samples for determining the pressure strength in longitudinal, tangential and radial direction.

Table 3. Experimentally determined normal and shear strength in longitudinal (L), tangential (T) and radial (R) direction.

Specimen	Compression (MPa)			Tension (MPa)			Shear (MPa)			
	L	T	R	L	T	R	TL	TL [41]	RL	TR
1	66.9	11.5	11.5	99.9	3.9	-	9.2	-	-	-
2	63.6	11.3	11.0	115.1	3.8	-	9.9	-	-	-
3	65.6	11.4	10.9	95.3	3.4	-	8.7	-	-	-
4	-	-	-	94.2	3.7	-	-	-	-	-
5	-	-	-	99.2	-	-	-	-	-	-
6	-	-	-	93.1	-	-	-	-	-	-
7	-	-	-	100.7	-	-	-	-	-	-
Avg	65.4	11.4	11.1	96.8	3.7	3.7	9.3	10.9	9.3	1.6
Std. dev.	1.7	0.1	0.3	10.0	0.2	-	0.6	-	-	-
COV (%)	2.6	0.9	2.6	10.3	5.7	-	6.7	-	-	-

The average shear strength in the TL direction determined from the asymmetric four-point bending test is 9.3 MPa, which is higher than the literature data for LVL beech specimens [46,47]. The shear strength in the TL direction was also calculated using Equation (10), where the results was 10.9 MPa. Since the difference was not great a reasonable applicability of Equation (10) can be confirmed and further used to calculate shear strength in the TR direction where the result was 1.6 MPa, while the shear strength in RL direction was taken the same as in TL direction.

The results for the modulus of elasticity and shear modulus are shown in Table 4 The average value of modulus of elasticity in longitudinal and tangential directions is 14,854 MPa and 984 MPa, respectively, and is comparable with the literature [44–46], while the modulus of elasticity in radial direction was taken from the literature [1]. The average value of the shear modulus in TL direction is 593 MPa, while the theoretically calculated value according to Equation (14) is 619 MPa, which corresponds to a difference of 4.2%. Since the difference was negligible, the applicability of Equation (14) was confirmed, and it was used to calculate the moduli in RL and RT directions, which were 1464 MPa and 388 MPa, respectively. The standard deviations of the measurements as well as the coefficient of variation were very small, which can be attributed to the relatively homogeneous specimens.

Table 4. Modulus of elasticity and shear modulus (in MPa).

Specimen	Experiment		Kollman [1]	Experiment	Equation 14		
	E_L	E_T	E_R	G_{TL}	G_{TL}	G_{RL}	G_{RT}
1	14,578	991	-	586	-	-	-
2	14,597	992	-	590	-	-	-
3	14,578	968	-	604	-	-	-
Avg	14,584	984	2380	593	619	1464	388
St. dev.	11	14	-	9	-	-	-
COV (%)	0.08	1.38	-	1.59	-	-	-

3.2. Failure of Four-Point Plywood Bending Specimens

Figure 6 shows the forces and deformations of the four-point bending loaded 11-layer specimen cut from plate 11A and with the direction of the first layer of 0° and −45°, while Figure 7 shows the corresponding specimens. The transition from the linear to the nonlinear part occurs for specimens 11A0° and 11A−45° at failure forces of 1688 N and 1209 N, while the final breaking force is 3003 and 2007 N, respectively.

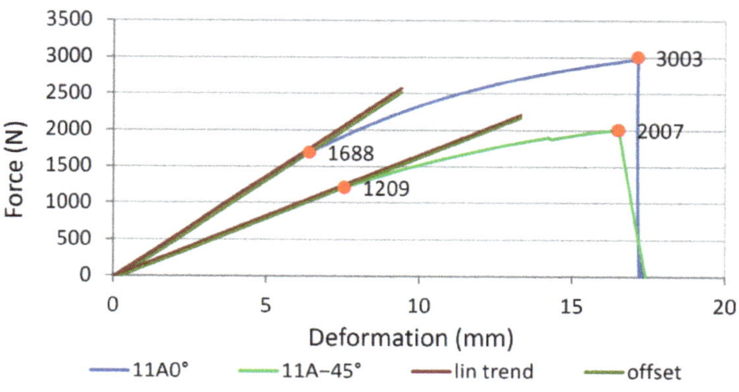

Figure 6. Forces and deformations of the four-point bending specimen 11A0° and 11A−45°.

Figure 7. 4-point bending specimens 11A0° and 11A−45°: (**a**) Whole specimens; (**b**) detail of the specimen underside with layer numbering.

Tables 5 and 6 show the min-max mid-sample FOS values for each layer for samples 11A0° and 11A−45°, respectively, for different failure criteria along with the failure mode according to the Puck, max stress and Hashin criteria, where the Tsai-Wu criterion has a constant a_{xy} equal to 1 and the Puck criterion has a constant p equal to 0.01. In Table 5, the sample 11A0° has the weakest upper layer No. 1, as it has the lowest min FOS value for all criteria and ranges from 1.33 to 1.54 for different criteria. The distribution of the FOS values together with failure mode is shown in Figure 8. For the first layer, all three criteria determined the same failure mode of compression failure in the fibre direction, which was expected since the angle of the fibres was 0°. The next weakest layer was the lowest layer No. 11, with the same fibre angle of 0°, but with tensile stresses. Again, all three criteria predict failure in the fibre direction, which is confirmed by Figure 7b, which nicely shows the failure of the lowest layer No. 11 in the fibre direction. This is to be expected, as compressive strength in longitudinal direction is lower than the tensile one. According to the measurement, the compressive failure of the top layer No. 1 occurred at the force of 1688 N, where the force of 1540 N was calculated by the Puck criteria, but the specimen did not break into two pieces as the top layer can still bear the compressive load despite the local compressive failure. By increasing the force and deformation, the failure of the bottom layer No. 11 followed at a measured force of 3003 N. According to the Puck criterion, the failure force of the lower layer is expected to be 2300 N, and according to the Tsai-Wu failure criterion, it is expected to be 2910 N. The theoretical forces are valid only under the assumption that the stress increases linearly with the force, which is not true in our case because the compressive stress in the upper layer does not increase with increasing force due to the failure of the tissue, and as such the calculated failure force for the lower layer No. 11 cannot be considered as completely valid. In the event that we wish to determine the actual failure force of the specimen at which the bottom layer would also fail, a nonlinear finite element model would be required, which is beyond the scope of the current research.

Table 5. Calculated factor of safety (FOS) at loading force of 1000 N for various failure criteria for sample 11A0°. (Failure modes: Puck: f—fibre failure, mA—inter-fibre failure in mode A, mB—inter-fibre failure in mode B; max stress: 1c and 1t—compression and tension failure in direction 1; 2c and 2t—compression and tension failure in direction 2; 12—shear failure; Hashin: f—fibre failure, m—matrix failure).

Ply no.	Ply Orientation (°)	Tsai-Wu		Tsai-Hill		Hoffman		Puck			Max Stress			Hashin		
		FOS		FOS		FOS		FOS		Failure	FOS		Failure	FOS		Failure
		Min	Max	Min	Max	Min	Max	Min	Max		Min	Max		Min	Max	
1	0	1.45	1.46	1.46	1.47	1.33	1.35	1.54	1.55	f	1.54	1.55	1c	1.54	1.55	f
2	30	2.61	2.69	2.71	2.79	2.58	2.69	3.14	3.24	mA	3.56	3.77	1c	3.56	3.77	f
3	−30	3.09	3.17	3.21	3.30	3.02	3.11	3.66	3.79	mA	3.88	4.10	1c	3.88	4.10	f
4	60	10.1	10.4	6.56	6.60	8.30	8.36	7.93	8.14	-	7.93	8.22	-	6.91	7.04	-
5	−60	15.9	16.2	11.0	11.2	13.9	14.1	12.4	12.7	-	12.4	12.6	-	11.0	11.2	-
6	90	16.3	16.7	15.8	16.1	13.9	14.1	18.7	19.1	-	22.3	22.9	-	22.3	22.9	-
7	−60	7.55	7.74	8.04	8.30	7.02	7.26	8.88	8.97	-	12.4	12.6	-	8.89	8.98	-
8	60	4.25	4.35	4.43	4.57	3.87	3.98	5.10	5.22	-	6.60	6.97	-	5.13	5.24	-
9	−30	4.59	4.63	4.10	4.15	4.44	4.50	4.80	4.87	-	5.66	5.81	-	4.11	4.16	-
10	30	3.52	3.58	3.29	3.34	3.52	3.53	3.86	3.91	mB	4.14	4.20	-	3.30	3.34	f
11	0	2.91	2.94	2.26	2.26	2.49	2.50	2.30	2.31	f	2.30	2.31	1t	2.30	2.31	f

Table 6. Calculated factor of safety (FOS) at loading force of 1000 N for various failure criteria for sample 11A−45°. (Failure modes: Puck: f—fibre failure, mA—inter-fibre failure in mode A, mB—inter-fibre failure in mode B; max stress: 1c and 1t—compression and tension failure in direction 1, 2c and 2t—compression and tension failure in direction 2, 12—shear failure; Hashin: f—fibre failure, m—matrix failure).

Ply No.	Ply Orientation (°)	Tsai-Wu		Tsai-Hill		Hoffman		Puck			Max Stress			Hashin		
		FOS		FOS		FOS		FOS		Failure	FOS		Failure	FOS		Failure
		Min	Max	Min	Max	Min	Max	Min	Max		Min	Max		Min	Max	
1	−45	1.93	2.01	1.84	1.89	2.11	2.18	2.12	2.25	mC	2.33	2.65	12	1.73	1.78	m
2	−15	1.22	1.27	1.21	1.27	1.37	1.51	1.27	1.41	f	1.27	1.41	1c	1.27	1.41	f
3	−75	4.44	5.28	3.21	3.36	3.44	3.74	3.54	3.97	mC	3.54	4.05	2c	3.53	3.88	m
4	15	1.71	1.87	1.68	1.85	1.89	2.02	1.68	1.85	f	1.68	1.85	1c	1.68	1.85	f
5	−105	6.79	9.64	6.56	7.58	7.51	8.83	7.71	8.70	-	9.01	10.7	-	6.21	7.15	-
6	45	9.44	11.1	10.6	12.8	10.9	12.7	11.5	14.6	-	11.5	16.7	-	11.5	14.9	-
7	−105	2.95	3.25	3.08	3.29	3.08	3.18	3.08	3.31	mA	3.28	3.47	2t	3.08	3.31	m
8	15	1.91	2.24	2.26	2.58	2.14	2.45	2.50	2.75	f	2.50	2.75	1t	2.50	2.73	f
9	−75	1.15	1.29	1.12	1.25	1.07	1.18	1.15	1.31	mA	1.15	1.32	2t	1.15	1.31	m
10	−15	1.06	1.25	1.24	1.46	1.23	1.41	1.44	1.68	mA	1.77	2.02	1t	1.53	1.72	f
11	−45	0.80	0.95	0.82	1.01	0.82	1.01	0.82	1.02	mA	0.87	1.13	2t	0.82	1.02	m

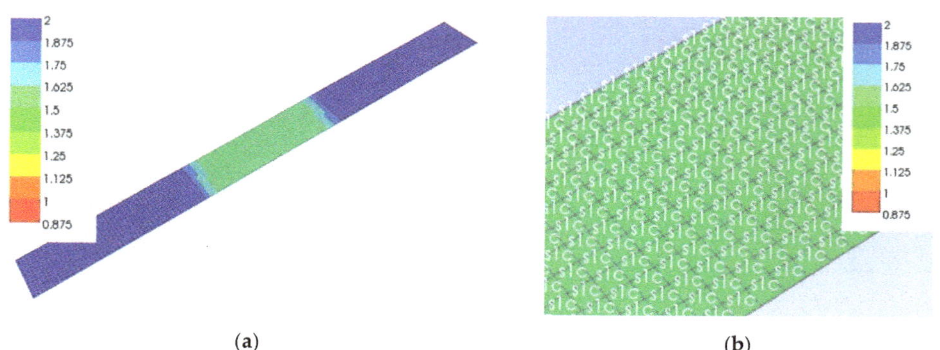

Figure 8. Distribution of safety factor (FOS) of layer No. 1 determined by the max stress criterion for specimen 11A0° loaded with a force of 1000 N: (**a**) Whole specimen; (**b**) detail of the middle part.

The situation is opposite for sample 11A−45°, whose min-max mid-sample FOS are listed in Table 6. Here, layer No. 11 is the weakest, having the lowest min FOS for various criteria ranging from 0.8 to 0.87. Unlike sample 11A0°, where layer No. 1 had a uniform FOS distribution across the width, the FOS distribution in layer No. 11 of sample 11A−45° varied considerably from 0.87 to 1.13 for the max stress criterion as shown in Figure 9. The transverse tensile failure of the lower layer No. 11 occurs first, followed by layer No. 9, and only then occurs compressive failure at the top part of the specimen, as the tensile and compressive strengths in tangential direction were 3.7 MPa and 11.4 MPa, respectively.

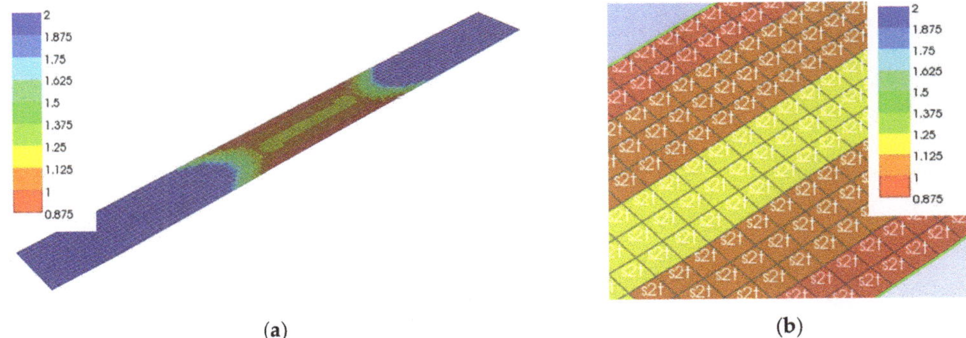

Figure 9. Distribution of the safety factor (FOS) of the layer No. 11 determined by the max stress criterion for specimen 11A−45° loaded with a force of 1000 N: (**a**) Whole specimen; (**b**) detail of the middle part.

All the criteria predicted the failure of layers No. 11 and 9 in the transverse or shear direction, which can also be seen in Figure 7b. Puck's criterion predicted failure under the combination of normal and shear stress (mode A), while the max stress and Hashin's criterion predicted matrix failure due to transverse tension. According to Puck's criterion, the lowest layer No. 11 would fail at a force of 820 N, whereas a force of 1209 N was determined in the experiment. The reason for the difference could be that despite the failure of layer No. 11 in the transverse direction, the stiffness drop is not affected because the load carrying capacity of layer No. 11 in the transverse direction is much lower than the load carrying capacity of layer No. 10, which transmits the load in the longitudinal direction of the failing layer No. 11.

The cross-section distribution of longitudinal and tangential stresses for sample 11A0° at a load of 1688 N is shown in Figure 10a. For all layers, the longitudinal stress predominates, while the maximum tangential stress is 1.5 MPa, much lower than the tensile and compressive strengths, which are 3.7 MPa and 11.4 MPa, respectively. In addition, the maximum shear stresses do not exceed 4 MPa, which is again much lower than the shear strength of 9.3 MPa. Due to the predominant longitudinal stresses, all the outer layers break in the direction of the fibres, which is also predicted by all three failure criteria.

The situation is different for the sample of 11A−45 ° shown in Figure 10b for the loading force of 1209 N. The stresses in the outer layers are larger in tangential direction and are −4.6 and 4.6 MPa on the compression and tension sides of the specimen, respectively, which is more than the tensile strength of −3.7 MPa and implies to the failure of the corresponding layer.

Figure 11 shows the experimentally determined and theoretically calculated forces of the four-point bending loaded 11-ply specimens. The measured forces vary a little and show a clear trend related to the different orientations of layers. The maximum load capacity for 11A specimen is at 0° and −22° for first ply orientation angle and then the forces decrease. Figure 11a,b shows the theoretically calculated failure forces according to the criteria of Tsai-Wu and Puck, respectively, based on a minimum factor of safety over the entire specimen cross-section for different values of the constants a_{xy} and p. The differences between individual values calculated with Tsai-Wu are not large, but they agree best with the criterion with constant $a_{xy} = 1$. The latter can also be seen in Table 7, where the coefficient of determination R^2 between the calculated values and the average of the measured forces is 0.551. In addition, for the Puck calculations, the differences are minimal for different constant values of p, where R^2 for $p = 0.01$ is 0.636.

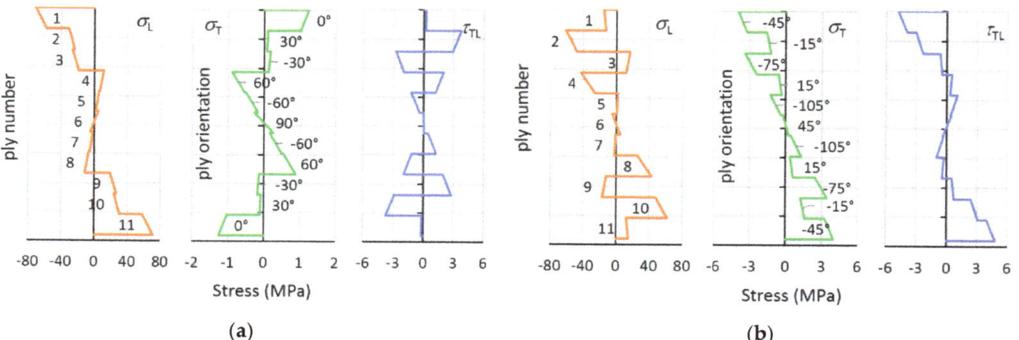

Figure 10. Longitudinal, tangential and shear stress distribution: (**a**) Specimen 11A0°, loading force 1688 N; (**b**) specimen 11A−45°, loading force 1209 N.

The results for 11P samples are shown in Figure 11d-f. The Tsai-Wu criterion (Figure 11d) predicts the failure values well at smaller orientations, while at higher angles the deviation between the predicted and measured values is larger. The best prediction is at $a_{xy} = 0$, where R^2 is 0.367. Even using the Puck criterion (Figure 11e), the forces differ only slightly at different p values, and the R^2 is highest at 0.421 for a value of $p = 0.01$.

Figure 11c,f shows the results of all criteria for 11A and 11P samples, respectively. The max stress failure criterion has the best correlation with the highest R^2, which is 0.674 for 11A samples and 0.538 for 11P samples. Max stress is followed by Hashin and Puck criteria in both groups. For the 11A samples, the difference is not very large, while it is larger for the 11P samples. For the 11A samples, then the criteria of Tsai Hill, Tsai Wu and Hoffman are followed, while for the 11P samples, the criteria of Hoffman, Tsai Wu and Tsai Hill are followed.

The results for seven ply samples are shown in Figure 12. For 7A specimens the forces calculated by using the Tsai-Wu criterion (Figure 12a) differ only slightly for different values of a_{xy}, having the best R^2 (Table 7) of 0.133 for an $a_{xy} = 0.3$, while the forces calculated with the Puck criterion (Figure 12b) practically do not differ from each other and the highest coefficient of determination of 0.338 has a criterion with $p = 0.01$, as was the case for the 11-layer samples. The calculations for 7P samples are shown in Figure 12d,e. As with the 11P samples, Tsai-Wu (Figure 12d) predicts the forces well at smaller angles, while the deviation is larger at larger angles. The largest R^2 has a criterion with $a_{xy} = -1$, while Puck's criterion (Figure 12e) again has the largest R^2 at a value of $p = 0.01$.

Figure 12c,f shows the results of all criteria for 7A and 7P samples, respectively. As with the 11-layer plates, the max stress criterion has the best R^2, followed by Hashin, Puck, Tsai-Hill, Tsai-Wu and Hoffman for the 7A specimens, and Tsai-Wu, Hoffmann, Hashin and Tsai-Hill for the 7P specimens.

Figure 13 shows the results for the 3-layer samples. Tsai-Wu criterion (Figure 13a) has slightly larger deviations even at smaller angles and the best R^2 at $a = -1$, similar to the 7P samples, while the Puck criterion (Figure 13b) has the largest R^2 at $p = 0.01$. Comparison of all criteria is shown in Figure 13c. Tsai-Wu and Hoffmann criteria have a slightly better R^2 than the max stress criterion, followed by the Hashin, Puck and Tsai-Hill failure criteria.

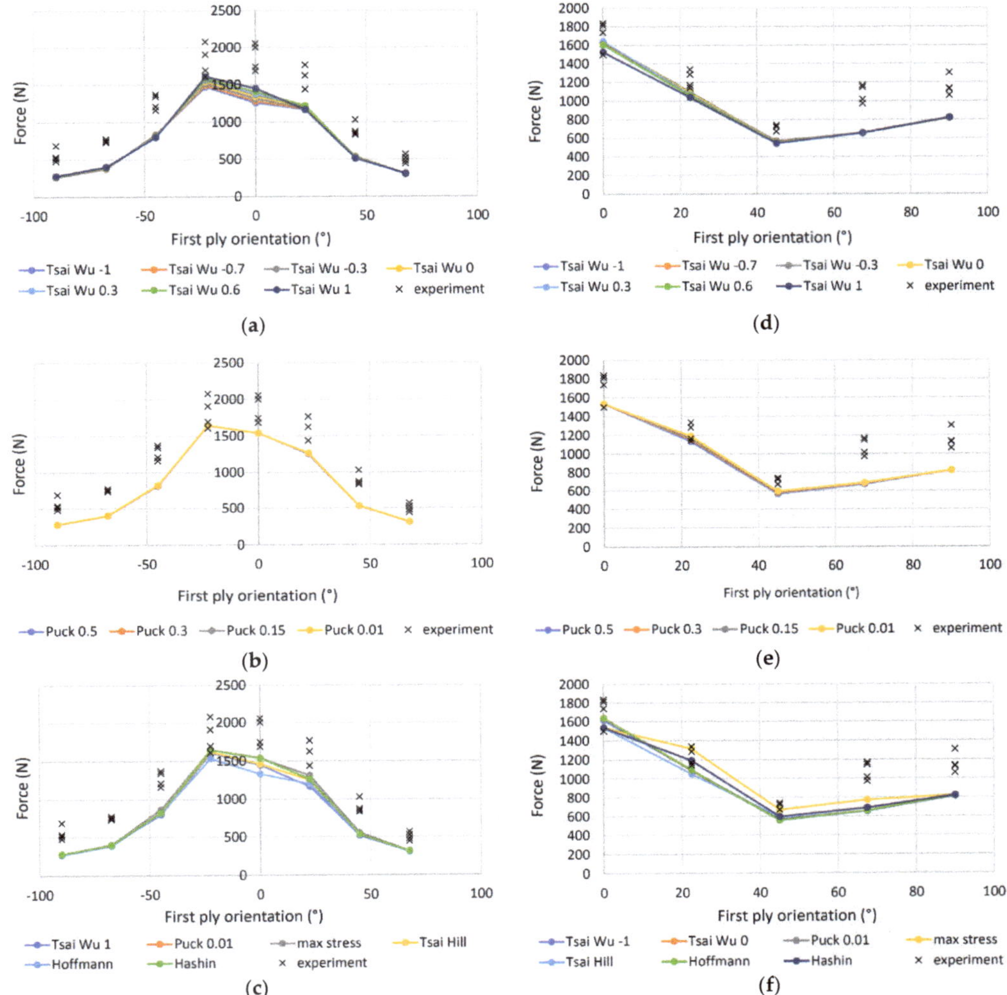

Figure 11. Measured and calculated forces of the four-point bending loaded 11-ply specimens: (**a**) Specimens 11A, Tsai-Wu criterion; (**b**) specimens 11A, Puck criterion; (**c**) specimens 11A, various criteria; (**d**) specimens 11P, Tsai-Wu criterion; (**e**) specimens 11P, Puck criterion; (**f**) specimens 11P, various criteria.

Table 7. The coefficient of determination (R^2) for different failure criteria between the calculated and the average of measured values of the failure forces. The green colour indicates the criterion with the best correlation of all criteria, and the bold values the best correlation within each criterion.

	a_{xy} **(Tsai-Wu)/*p* (Puck)**	**11A**	**11P**	**7A**	**7P**	**3P**
Tsai Wu	−1	0.431	0.358	0.060	***0.024***	***0.696***
	−0.7	0.456	0.364	0.081	0.016	0.680
	−0.3	0.487	0.371	0.109	0.005	0.659
	0	0.508	***0.367***	0.132	−0.008	0.643
	0.3	0.530	0.362	***0.133***	−0.039	0.626
	0.6	0.547	0.339	0.128	−0.086	0.609
	1	***0.551***	0.282	0.122	−0.165	0.587

Table 7. *Cont.*

	a_{xy} (Tsai-Wu)/p (Puck)	11A	11P	7A	7P	3P
Puck	0.5	0.627	0.371	0.318	−0.087	0.625
	0.3	0.631	0.394	0.329	−0.067	0.627
	0.15	0.634	0.409	0.334	−0.057	0.627
	0.01	***0.636***	***0.421***	***0.338***	***−0.053***	***0.628***
max stress		0.674	0.538	0.405	0.052	0.644
Tsai Hill		0.600	0.361	0.242	−0.133	0.585
Hoffmann		0.503	0.369	0.126	−0.002	0.648
Hashin		0.636	0.422	0.347	−0.042	0.628

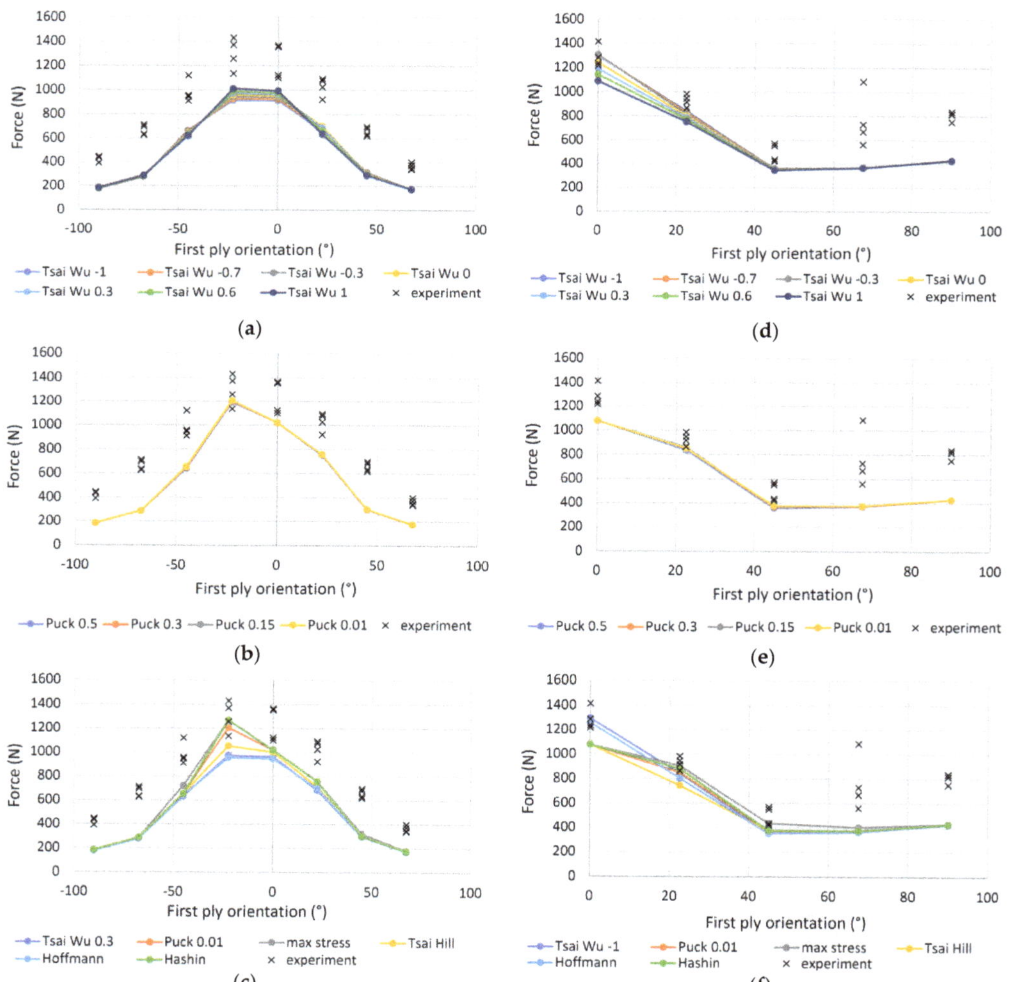

Figure 12. Measured and calculated forces of the 4-point bending loaded 7-ply specimens: (**a**) Specimens 7A, Tsai-Wu criterion; (**b**) specimens 7A, Puck criterion; (**c**) specimens 7A, various criteria; (**d**) specimens 7P, Tsai-Wu criterion; (**e**) specimens 7P, Puck criterion; (**f**) specimens 7P, various criteria.

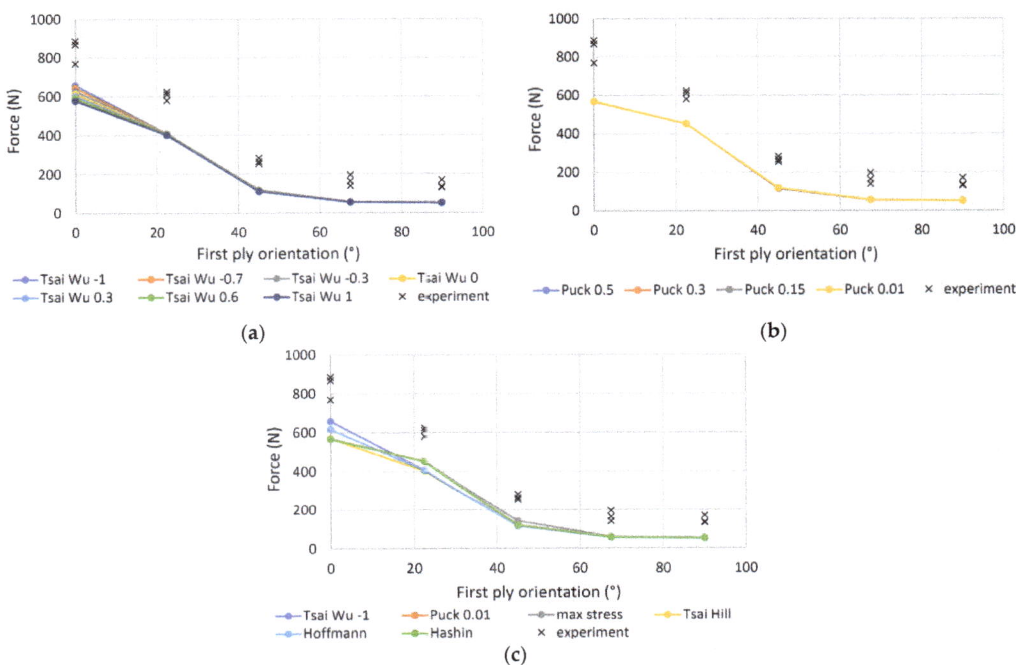

Figure 13. Measured and calculated forces of the four-point bending loaded 3P specimens: (**a**) Tsai-Wu criterion; (**b**) Puck criterion; (**c**) various failure criteria.

4. Discussion

Comparison of the coefficients of determination shows that all criteria agree better for the three- and 11-layer samples and worse for the seven-layer samples. For the 11- and seven-layer samples, the maximum stress criterion with the highest coefficient of determination has the best agreement between the calculated and measured values. In contrast, Tsai-Wu for the three-layer samples has a slightly higher R^2 than the maximum stress criterion, while for the 11-layer and seven-layer samples, the R^2 for Tsai-Wu is lower than most of the other criteria. It can be concluded from the study, that the maximum stress criterion is the most accurate criterion for predicting the failure criteria of wood-based materials such as plywood. The latter result is somewhat surprising since the maximum stress criterion is the simplest of all those studied and also does not take into account mutual interactions of normal stresses in different directions and the interaction of normal and shear stresses, which are essential for other unidirectional composites [25]. Hashin and Puck's criteria also have high R^2 values which, together with the maximum stress criterion, have similar conditions for failure of the tissue in the fibre direction, i.e., they compare only longitudinal stresses with corresponding strengths.

Similar results were obtained by others [22,48], who also investigated the application of the Tsai-Wu, Tsai-Hill, Hoffmann and maximum stress criteria to solid wood specimens. They found that the maximum stress and the Tsai-Hill criterion had the best correlation, followed by the Hoffman criterion, while the Tsai-Wu criterion had the worst correlation.

The study also found that the calculated values of Tsai-Wu criterion did not differ significantly at different values of a_{xy} in the entire range from −1 to 1. Slightly different results were obtained by others [21,49] for solid wood, where the agreements differed significantly at various a_{xy}. Little effect on the results was also obtained for the p values of the Puck criterion, where there were practically no differences between the calculated values at different p.

Regardless of the type of criterion used, the theoretically calculated forces are smaller than the measured ones. One of the reasons could be that the failure criteria take into account the relationship between the actual stress and the strengths, with the material failing when stresses reach the strength. This is true for unidirectional composites and solid wood. The situation could be different for composites consisting of layers with various tissue orientations. Here, the adjacent layers prevent the layers from failing, especially when the fabric fails in the transverse direction, because the adjacent tissue, which has a more prolonged longitudinal tissue orientation, prevents the failure progression of the adjacent layer.

However, in such cases, the studies performed may not provide an accurate prediction of tissue failure, especially if a linear model is used that predicts a linear increase in stresses with increasing force. In such cases it would be necessary to use a nonlinear model that takes into account the constraining effect of adjacent layers in the event of failure, as well as the limit stresses that certain layers should not exceed in the event of local failure.

The use of failure criteria, like any model, depends on the input data. However, if the correct input data is not precisely known, one may end up with inaccurate calculations, which is a particular problem with plywood. These are made up of individual layers of veneer, usually produced by peeling, and when veneer is produced there are large bending deformations in the tangential direction, resulting in local cracks. Because of the small cracks, the tensile strength in the tangential direction is much lower than the tensile strength of solid wood. For example, the tensile strength of solid wood in the tangential direction can be up to three-times higher than the tensile strength of peeled veneer. Therefore, it is important to consider real data in the modelling, which can be a problem if it is not available, and that the values of solid wood from the literature are taken into account.

5. Conclusions

The use of failure criteria is an important tool for combined loading that can be used to determine the force at which failure of the specimen occurs. These are important for composites as well as for solid wood, which is a distinctly orthotropic material with different strengths in different directions as well as different strengths depending on the sign of the load. In most cases, it is important to consider mutual interactions, which are taken into account by more demanding criteria that are essential for UD composites made of man-made materials. However, upon investigation on plywood specimens, the max stress criterion was found to have the best correlation between the calculated and measured values, followed by the Puck and Hashin criteria. A similar finding was made by some other authors for wood.

Regardless of the criterion, the calculated values for plywood are smaller than those determined experimentally, which ultimately means that we are on the safe side.

The method presented for applying failure criteria to plywood proved to be a suitable novelty, but it should be emphasized that knowledge of the exact mechanical properties of the material from which the plywood is made is essential for the accuracy of the failure force determination.

By using the failure criteria in combination with the finite element method, both of which are based on a linear model, a failure force is obtained at which the weakest layer fails. For plywood made of wood with a lower compressive strength than tensile one, this usually means that in the flexural test the compressive failure occurs first on the upper compression side of the specimen and only later on the lower tensile side of the specimen. Despite the initial compression failure, the specimen does not break in two but continues to resist the increase in force. If it is necessary to obtain a force that breaks the specimen in two, a nonlinear model must be used.

Funding: The research was supported by the P2-0182 Programs, co-financed by the Slovenian Research Agency.

Institutional Review Board Statement: Not applicable.

Informed Consent Statement: Not applicable.

Data Availability Statement: Not applicable.

Acknowledgments: The author wishes to thank Dominik Koderman for his help with sample preparation.

Conflicts of Interest: The author declares no conflict of interest.

References

1. Kollmann, F.F.P.; Côte, W.A. *Principles of Wood Science and Technology. Solid Wood*; Springer: Berlin, Germany, 1975; p. 592.
2. Merhar, M. Determination of Elastic Properties of Beech Plywood by Analytical, Experimental and Numerical Methods. *Forests* **2020**, *11*, 1221. [CrossRef]
3. Kallakas, H.; Rohumaa, A.; Vahermets, H.; Kers, J. Effect of different hardwood species and lay-up schemes on the mechanical properties of plywood. *Forests* **2020**, *11*, 649. [CrossRef]
4. Bal, B.C. Some physical and mechanical properties of reinforced laminated veneer lumber. *Constr. Build. Mater.* **2014**, *68*, 120–126. [CrossRef]
5. Zhou, H.Y.; Wei, X.; Smith, L.M.; Wang, G.; Chen, F.M. Evaluation of Uniformity of Bamboo Bundle Veneer and Bamboo Bundle Laminated Veneer Lumber (BLVL). *Forests* **2019**, *10*, 921. [CrossRef]
6. Mohammadabadi, M.; Jarvis, J.; Yadama, V.; Cofer, W. Predictive models for elastic bending behavior of a wood composite sandwich panel. *Forests* **2020**, *11*, 624. [CrossRef]
7. Yu, X.; Xu, D.; Sun, Y.; Geng, Y.; Fan, J.; Dai, X.; He, Z.; Dong, X.; Dong, Y.; Li, Y. Preparation of wood-based panel composites with poplar veneer as the surface layer modified by in-situ polymerization of active monomers. *Forests* **2020**, *11*, 893. [CrossRef]
8. Jakob, M.; Stemmer, G.; Czabany, I.; Müller, U.; Gindl-Altmutter, W. Preparation of High Strength Plywood from Partially Delignified Densified Wood. *Polymers* **2020**, *12*, 1796. [CrossRef] [PubMed]
9. Král, P.; Klímek, P.; Mishra, P.K.; Rademacher, P.; Wimmer, R. Preparation and characterization of cork layered composite plywood boards. *BioResources* **2014**, *9*, 1977–1985. [CrossRef]
10. Salca, E.A.; Bekhta, P.; Seblii, Y. The effect of veneer densification temperature and wood species on the plywood properties made from alternate layers of densified and non-densified veneers. *Forests* **2020**, *11*, 700. [CrossRef]
11. Merhar, M. Determination of dynamic and static modulus of elasticity of beech plywood. *Les/Wood* **2020**, *69*. [CrossRef]
12. Bekhta, P.; Sedliačik, J.; Bekhta, N. Effect of Veneer-Drying Temperature on Selected Properties and Formaldehyde Emission of Birch Plywood. *Polymers* **2020**, *12*, 593. [CrossRef]
13. Jorda, J.; Kain, G.; Barbu, M.-C.; Petutschnigg, A.; Král, P. Influence of Adhesive Systems on the Mechanical and Physical Properties of Flax Fiber Reinforced Beech Plywood. *Polymers* **2021**, *13*, 3086. [CrossRef] [PubMed]
14. Réh, R.; Krišťák, Ľ.; Sedliačik, J.; Bekhta, P.; Božiková, M.; Kunecová, D.; Vozárová, V.; Tudor, E.M.; Antov, P.; Savov, V. Utilization of Birch Bark as an Eco-Friendly Filler in Urea-Formaldehyde Adhesives for Plywood Manufacturing. *Polymers* **2021**, *13*, 511. [CrossRef] [PubMed]
15. Bekhta, P.; Sedliačik, J.; Bekhta, N. Effects of Selected Parameters on the Bonding Quality and Temperature Evolution Inside Plywood During Pressing. *Polymers* **2020**, *12*, 1035. [CrossRef] [PubMed]
16. Yoshihara, H. Influence of the specimen depth to length ratio and lamination construction on Young's modulus and in-plane shear modulus of plywood measured by flexural vibration. *BioResources* **2012**, *7*, 1337–1351.
17. Wilczyński, M.; Warmbier, K. Elastic moduli of veneers in pine and beech plywood. *Drewno* **2012**, *188*, 47–56.
18. Avilés, F.; Couoh-Solis, F.; Carlsson, L.A.; Hernández-Pérez, A.; May-Pat, A. Experimental determination of torsion and shear properties of sandwich panels and laminated composites by the plate twist test. *Compos. Struct.* **2011**, *93*, 1923–1928. [CrossRef]
19. Yoshihara, H. Edgewise shear modulus of plywood measured by square-plate twist and beam flexure methods. *Constr. Build. Mater.* **2009**, *23*, 3537–3545. [CrossRef]
20. Aicher, S.; Klöck, W. Linear versus quadratic failure criteria for inplane loaded wood based panels. *Otto-Graf Journal* **2001**, *12*, 187–200.
21. Cabrero, J.M.; Gebremedhin, K.G. Evaluation of failure criteria in wood members. In Proceedings of the 11th World Conference on Timber Engineering 2010, Trentino, Italy, 20–24 June 2010; pp. 1274–1280.
22. Mascia, N.T.; Simoni, R.A. Analysis of failure criteria applied to wood. *Engineering Failure Analysis* **2013**, *35*, 703–712. [CrossRef]
23. van der Put, T.A.C.M. The tensorpolynomial failure criterion for wood. *Delft Wood Sci. Found. Publ. Ser.* **2005**, *2*, 31.
24. Puck, A.; Schürmann, H. Failure analysis of FRP laminates by means of physically based phenomenological models. In *Failure Criteria in Fibre-Reinforced-Polymer Composites*; Elsevier: Amsterdam, The Netherlands, 2004; pp. 264–297. [CrossRef]
25. Puck, A.; Kopp, J.; Knops, M. Guidelines for the determination of the parameters in Puck's action plane strength criterion. *Compos. Sci. Technol.* **2002**, *62*, 371–378. [CrossRef]

26. Chybiński, M.; Polus, Ł. Experimental and numerical investigations of laminated veneer lumber panels. *Arch. Civ. Eng.* **2021**, *67*, 351–372. [CrossRef]
27. Gong, Y.; Huang, T.; Zhang, X.; Jia, P.; Suo, Y.; Zhao, S. A reliable fracture angle determination algorithm for extended puck's 3d inter-fiber failure criterion for unidirectional composites. *Materials* **2021**, *14*, 6325. [CrossRef] [PubMed]
28. Gong, Y.; Huang, T.; Zhang, X.; Suo, Y.; Jia, P.; Zhao, S. Multiscale analysis of mechanical properties of 3d orthogonal woven composites with randomly distributed voids. *Materials* **2021**, *14*, 5247. [CrossRef]
29. Song, C.; Jin, X. Fracture angle prediction for matrix failure of carbon-fiber-reinforced polymer using energy method. *Compos. Sci. Technol.* **2021**, *211*. [CrossRef]
30. Wei, L.; Zhu, W.; Yu, Z.; Liu, J.; Wei, X. A new three-dimensional progressive damage model for fiber-reinforced polymer laminates and its applications to large open-hole panels. *Compos. Sci. Technol.* **2019**, *182*. [CrossRef]
31. Sun, Q.; Zhou, G.; Meng, Z.; Guo, H.; Chen, Z.; Liu, H.; Kang, H.; Keten, S.; Su, X. Failure criteria of unidirectional carbon fiber reinforced polymer composites informed by a computational micromechanics model. *Compos. Sci. Technol.* **2019**, *172*, 81–95. [CrossRef]
32. Imtiaz, H.; Liu, B. An efficient and accurate framework to determine the failure surface/envelop in composite lamina. *Compos. Sci. Technol.* **2021**, *201*, 108475. [CrossRef]
33. Li, N.; Ju, C. Mode-Independent and Mode-Interactive Failure Criteria for Unidirectional Composites Based on Strain Energy Density. *Polymers* **2020**, *12*, 2813. [CrossRef]
34. Khan, M.S.; Abdul-Latif, A.; Koloor, S.S.R.; Petrů, M.; Tamin, M.N. Representative Cell Analysis for Damage-Based Failure Model of Polymer Hexagonal Honeycomb Structure under the Out-of-Plane Loadings. *Polymers* **2021**, *13*, 52. [CrossRef] [PubMed]
35. Pramreiter, M.; Bodner, S.C.; Keckes, J.; Stadlmann, A.; Feist, F.; Baumann, G.; Maawad, E.; Müller, U. Predicting strength of Finnish birch veneers based on three different failure criteria. *Holzforschung* **2021**, *75*, 847–856. [CrossRef]
36. Yoshihara, H. Failure conditions of solid wood on off-axis compression testing. *Holzforschung* **2019**, *73*, 251–258. [CrossRef]
37. Akter, S.T.; Bader, T.K. Experimental assessment of failure criteria for the interaction of normal stress perpendicular to the grain with rolling shear stress in Norway spruce clear wood. *Eur. J. of Wood Wood Prod.* **2020**, *78*, 1105–1123. [CrossRef]
38. EN 408*Timber structures—Structural timber and glued laminated timber—Determination of some physical and mechanical properties*, BSI: Brussels, Belgium, 2010.
39. ASTM D143*Standard Test Methods for Small Clear Specimens of Timber*, Intertek: Philadelphia, PA, USA, 2000.
40. Yoshihara, H. Shear properties of wood measured by the asymmetric four-point bending test of notched specimen. *Holzforschung* **2009**, *63*, 211–216. [CrossRef]
41. Bachtiar, E.V.; Rüggeberg, M.; Hering, S.; Kaliske, M.; Niemz, P. Estimating shear properties of walnut wood: A combined experimental and theoretical approach. *Mater. Struct./Mater. et Constr.* **2017**, *50*. [CrossRef]
42. Li, S.G.; Sitnikova, E.; Liang, Y.N.; Kaddour, A.S. The Tsai-Wu failure criterion rationalised in the context of UD composites. *Compos. Part a-Appl. Sci. Manuf.* **2017**, *102*, 207–217. [CrossRef]
43. Chen, X.; Sun, X.; Chen, P.; Wang, B.; Gu, J.; Wang, W.; Chai, Y.; Zhao, Y. Rationalized improvement of Tsai–Wu failure criterion considering different failure modes of composite materials. *Compos. Struct.* **2021**, *256*. [CrossRef]
44. Ozyhar, T.; Hering, S.; Niemz, P. Moisture-dependent elastic and strength anisotropy of European beech wood in tension. *J. Mater. Sci.* **2012**, *47*, 6141–6150. [CrossRef]
45. Hering, S.; Keunecke, D.; Niemz, P. Moisture-dependent orthotropic elasticity of beech wood. *Wood Sci. Technol.* **2012**, *46*, 927–938. [CrossRef]
46. Aicher, S.; Ohnesorge, D. Shear strength of glued laminated timber made from European beech timber. *Eur. Journal Wood Wood Prod.* **2011**, *69*, 143–154. [CrossRef]
47. Aicher, S.; Christian, Z.; Hirsch, M. Rolling shear modulus and strength of beech wood laminations. *Holzforschung* **2016**, *70*, 773–781. [CrossRef]
48. Mascia, N.T.; Nicolas, E.A. Evaluation of Tsai-Wu criterion and Hankinson's formula for a Brazilian wood species by comparison with experimental off-axis strength tests. *Wood Mater. Sci. Eng.* **2012**, *7*, 49–58. [CrossRef]
49. Mascia, N.T.; Nicolas, E.A.; Todeschini, R. Comparison between tsai-wu failure criterion and hankinson's formula for tension in wood. *Wood Res.* **2011**, *56*, 499–510.

MDPI
St. Alban-Anlage 66
4052 Basel
Switzerland
Tel. +41 61 683 77 34
Fax +41 61 302 89 18
www.mdpi.com

Polymers Editorial Office
E-mail: polymers@mdpi.com
www.mdpi.com/journal/polymers

www.ingramcontent.com/pod-product-compliance
Lightning Source LLC
LaVergne TN
LVHW071005170726
843515LV00006B/1080